ION IMPLANTATION
Science and Technology
Second Edition

ION IMPLANTATION
Science and Technology
Second Edition

Edited by

J. F. Ziegler

IBM Thomas J. Watson Research Center
Yorktown Heights, New York

ACADEMIC PRESS, INC.
Harcourt Brace Jovanovich, Publishers
Boston   San Diego   New York
Berkeley   London   Sydney
Tokyo   Toronto

PHYSICS

ACADEMIC PRESS, INC.
1250 Sixth Avenue, San Diego, CA 92101

United Kingdom Edition published by
ACADEMIC PRESS INC. (LONDON) LTD.
24-28 Oval Road, London NW1 7DX

Library of Congress Cataloging-in-Publication Data

Ion implantation: science and technology / edited by J.F. Ziegler. —
    2nd ed.
        p.   cm.
    Bibliography : p.
    Includes index.
    ISBN 0-12-780621-0
    1. Ion implantation.   I. Ziegler, J. F. (James F.)
QC702.7.I55I59 1988   539'.6 – dc19   88-6382
                                                           CIP

88 89 90 91   9 8 7 6 5 4 3 2 1
Printed in the United States of America

Contents

v

Contributors

Numbers in parentheses indicate the pages on which the authors' contributions begin.

M. I. Current (377), *Applied Materials, Implant Division, 3050 Bowers Avenue MS 0107, Santa Clara, California 95054*

William R. Ghen (439), *Varian/Extrion Division, Blackburn Industrial Park, PO Box 1266, Gloucester, Massachusetts 01930*

Jozsef Gyulai (93), *Central Research Institute for Physics, PO Box 49, H-1525 Budapest, Hungary*

P. Hamers (415), *IBM, Boblingen, Federal Republic of Germany*

P. L. F. Hemment (165), *Department of Electronic and Electrical Engineering, University of Surrey, Guildford, Surrey GU2 5XH, England*

W. A. Keenan (377), *Prometrix Corporation, 3255 Scott Boulevard, Building 6, Santa Clara, California 95054*

M. E. Mack (313), *Eaton Semiconductor Equipment Division, 108 Cherry Hill Drive, Beverly, Massachusetts 01915*

Siegfried Mader (63), *IBM Thomas J. Watson Research Center, PO Box 218, Yorktown Heights, New York 10598*

Constantine J. Maletskos (439), *Varian/Extrion Division, Blackburn Industrial Park, PO Box 1266, Gloucester, Massachusetts 01930*

Kenneth H. Purser (291), *General Ionex Corporation, 19 Graf Road, Newburyport, Massachusetts 01950*

H. Ryssel (415), *Fraunhofer Arbeitsgruppe fur Integrierte Schaltungen, Artilleriestrasse 12, D-8520 Erlangen, Federal Republic of Germany*

T. C. Smith (345), *Motorola Semiconductor Group, 2200 W. Broadway, Mesa, Arizona 85202*

Ken G. Stephens (221), *Department of Electronic and Electrical Engineering, University of Surrey, Guildford, Surrey GU2 5XH, England*

Nicholas R. White (291), *General Ionex Corporation, 19 Graf Road, Newburyport, Massachusetts 01950*

C. B. Yarling (377), *Applied Materials, Implant Division, 3050 Bowers Avenue MS 0107, Santa Clara, California 95054*

J. F. Ziegler (3), *IBM Thomas J. Watson Research Center, PO Box 218, Yorktown Heights, New York 10598*

Preface

Ion implantation is the major technology used to introduce impurities into solids in a uniform and reliable way. Its primary application is in the semiconductor industry where it is usually the technique of choice for the electrical doping of semiconductors.

This book is a tutorial presentation of the science, techniques, and machines of ion implantation. Its purpose is both to introduce this complex field in a simple way, and to act as a reference work which can lead to the thousands of scientific papers of the field. For this purpose there is an extensive index, and each chapter lists excellent review papers on specialized subjects.

The first section of this book concerns the science of ion implantation. It covers the historical development of the field, and the basic theory of energetic ion penetration of solids. The major concentration of this section is to explain the nature of the creation of damage in crystalline silicon during ion implantation, and the methods which can be used to recover the original crystallinity. Especially helpful are the TEM photographs scattered throughout this section which show the many phases of the morphology of ion implantation damage. Methods are described which allow the quantitative evaluation of the success of the implantation and the recovery of the semiconductor.

The last half of this book describes the ion accelerators (implanters) used in ion implantation, with a detailed presentation of the major components which require maintenance. A large part of this section concerns the methods of quantitatively evaluating the performance of ion implanters. A chapter is devoted to the extensive safety hazards of implanters and methods to maintain safe operation.

J. F. ZIEGLER

Ion Implantation Science

THE STOPPING AND RANGE OF IONS IN SOLIDS

J. F. Ziegler

IBM-Research
Yorktown, New York
10598   USA

This chapter reviews the physics associated with the penetration of energetic ions into solids. It describes the quantitative evaluation of how the ions lose energy to the solid and the final distribution of these ions after they stop within the solid. Also considered are the first order effects on the atoms of the solid, particularly the electronic excitation of the atoms, the displacement of lattice atoms by energetic collisions (lattice damage) and the production of plasmons and phonons in the solid by the passing ions. No evaluation is made of thermal effects in the solid, especially redistribution of lattice atoms or implanted ions by thermal or vacancy induced diffusion.

This broad field is presented in an historical context so it can be seen how various aspects became important because of specific scientific interests in atomic theory, quantum mechanics, radioactive atoms, nuclear physics and ion implantation.

Finally, the theory is illustrated by describing the computer program TRIM which allows calculations to be performed on small PC type computers.

CHAPTER SUMMARY

1 -   INTRODUCTION AND HISTORICAL REVIEW

2 -   NUCLEAR STOPPING CROSS-SECTIONS

3 -   ELECTRONIC STOPPING CROSS-SECTIONS

4 -   TRIM CALCULATIONS OF RANGE DISTRIBUTIONS

ION IMPLANTATION:
SCIENCE AND TECHNOLOGY

3

1 -   INTRODUCTION AND HISTORICAL REVIEW

HISTORICAL SUMMARY

For seventy-five years the stopping of energetic ions in matter has been a subject which has received great theoretical and experimental interest. The theoretical treatment of the stopping of ions in matter is due greatly to the work of Bohr (13a,15a,48a), Bethe (30a,32a,34a), Bloch (33a,33b), Firsov (57a,57b,58a,58b) and Lindhard (53b,54a,63a,64a,68a,68b). It has been reviewed by Bohr (48a), Whaling (58c), Fano (63b), Jackson (62a,75a), Bichsel (70e), Sigmund (75b), Ahlen (80e) and Ziegler, et al. (78a,80a,80b,82b). Soon after the discovery of energetic particle emission from radioactive materials, there was interest in how these corpuscles were slowed down in traversing matter. In 1900, Marie Curie stated (00a) the hypothesis that "les rayons alpha sont des projectiles materiels susceptibles de perdre de leur vitesse en traversant la matiere." Many scientists immediately realized that since these particles could penetrate thin films, such experiments might finally unravel the secrets of the atom. Early attempts to create a particle energy loss theory were inconclusive for there was not yet an accurate proposed model of the atom.

The theoretical treatment for the scattering of two point charges was derived by J. J. Thomson in his classic book on electricity (03a). Much of the traditional particle energy-loss symbolism can be traced to this book which introduced a comprehensive treatment for classical Coulombic scattering between energetic charged particles. This work, however, did not attempt to calculate actual stopping powers.

Enough experimental evidence of radioactive particle interactions with matter was collected in the next decade to make stopping power theory one of the central concerns of those attempting to develop an atomic model. In 1909 Geiger and Marsden were studying the penetration of alpha-particles through thin foils, and the spread of the trajectories after emerging from the back side. They hoped to determine the distribution of charges within the foil by the angular spread of the transmitted beam. There are conflicting histories as to who made the suggestion that they look for backscattered particles - but the subsequent startling data reversed the current thought on atomic structure. They reported that about .01% of the heavy alpha-particles were scattered back from the target, and from an analysis of the data statistics such backscattered events had to be from isolated single collisions. Two years later Rutherford was able to demonstrate theoretically (11a) that the backscattering was indeed due to a single event, and by analyzing this and electron scattering data he was able to first calculate that the *nucleus* of Al atoms must have a charge of about 22 and about 138 for platinum!

J. J. Thomson, director of the prestigious Cavendish Laboratory, and Niels Bohr, a fresh post-doctoral scientist at Rutherford's Manchester Laboratory, both published almost simultaneously (12a,13a) an analysis of the stopping of charged particles by matter, and they illustrated much of their divergent ideas on the model of an atom. Thomson incredibly ignored in his paper the Rutherford alpha-particle scattering theory (11a) of a year before. But the nuclear atom with a heavy positively-charged core was the basis of Bohr's ideas. (13a,15a).

Bohr's early work is instructive because for the first time a unified theory of stopping was attempted, and we can see this and similar works the essential problems of stopping theory:

(a)   How does an energetic charged particle (a point charge) lose energy to the quantized electron plasma of a solid (inelastic energy loss)?

(b)   How do you incorporate into this interaction simultaneous distortion of the electron plasma caused by the particle (target polarization)?

(c)   How can you extend the point charge-plasma interaction to that for a finite moving atom in a plasma?

(d)   How do you estimate the degree of ionization of the moving atom and describe its electrons when it is both ionized and within an electron plasma?

(e)   How do you calculate the screened Coulomb scattering of the moving atom with each heavy target nucleus it passes?

(f)   How do you include relativistic corrections to all of the above?

This is a brief list of the major problems encountered, and scientific interest shifts back and forth between them over the decades because of external scientific tidal forces. Examples might be (a) the development of quantal scattering in the nineteen twenties, (b) the study of nuclear fission in the thirties and forties, (c) the study of nuclear physics in the fifties, (d) the technological applications of ion implantation for material modification in the sixties, and the use of ion beams in material analysis in the seventies. This ebb and flow of interest continues because of the recurrent importance of the problem, and the difficulty of calculating the penetration of energetic atoms in solids from first principles. We briefly review some of the historical milestones in this field below.

One of Bohr's original conclusions was that the energy loss of ions passing through matter could be divided into two components: nuclear stopping (energy loss to the medium's atomic positive cores) and electronic stopping (energy loss to the medium's light electrons). Bohr, in his first papers, correctly deduced that the electronic stopping would be far greater than the nuclear stopping for energetic light ions such as are emitted by radioactive sources. This conclusion was based on recoil kinematics considering only the relative masses and abundances of the target electrons and nuclei.

Bohr further introduced atomic structure into stopping theory by giving target electrons the orbit frequencies obtained from optical spectra and calculating the energy transferred to such harmonic oscillators. He noted that the experimentally reported stopping powers for heavy atom targets indicated that many electrons in these targets must be more tightly bound than the optical data suggested. He also realized that his accounting of the energy loss process was seriously limited by a lack of knowledge of the charge state of the ion inside the matter, i.e., its effective charge in its interaction with the target medium.

A major advance in understanding stopping powers came 20 years later when Bethe (30a,32a,34a) and Bloch (33a,33b) restated the problems from the perspective of quantum mechanics, and derived in the Born approximation the fundamental equations for the stopping of very fast particles in a quantized electron plasma. This theoretical approach remains the basic method for evaluating the energy loss of light particles with velocities of 10 MeV/amu - 2 GeV/amu. This restriction in velocity is because below these velocities the ion projectile may not be fully stripped of its electrons (which is assumed by the theory), and above this velocity there are additional relativistic corrections.

In the late 1930's a renewed interest was taken in energy loss with the discovery of nuclear fission and the energetic heavy particles which resulted from nuclear disintegration. Various theoretical studies were published by Bohr, Lamb, Knipp, Teller and Fermi.

The problem presented by the fission fragment data was how to treat the interaction of a *partially stripped* heavy ion. This is called the 'effective-charge' problem, for it was hoped that if a degree of ionization for the projectile could be estimated, then the traditional stopping power theories could be used. Bohr suggested (40d,41a) that the ion be considered to be stripped of all electrons with velocities lower than the ion velocity, and using the Thomas Fermi atom he could show that

$$Z_1^* = Z_1^{1/3} \, v/v_o \qquad (1-1)$$

where Z_1 is the atomic number of the ion, and Z_1^* is its effective charge in energy loss to the target electrons. v is the ion velocity and v_o is the Bohr velocity ($\sim 2.2 \times 10^8$ cm/sec). Further, he estimated a screening distance, a_{12}, between two colliding atoms which limits the energy transfer between nuclei

$$a_{12} = a_o/(Z_1^{2/3} + Z_2^{2/3})^{1/2} \qquad (1-2)$$

where a_o is the Bohr radius, .59 Å. These two expressions form the basis of much of the stopping theory of the next 30 years. Lamb (40b) considered the same problem as Bohr, and suggested a similar effective charge approximation, but based on the energy rather than the velocity of the ion's electrons. Lamb also got a similar, but less detailed, expression for stopping power assuming Thomas-Fermi atoms. He suggested that the target electron velocity distribution would significantly alter the stopping of the fission fragment.

Fermi (40c) considered the same points as Lamb and Bohr, but concentrated upon evaluating the interaction of a charged particle with the dielectric plasma of a solid, and the electric polarization of the medium by the particle. This polarization of the target medium had been suggested first by Swann (38a), and Fermi was able to reduce this difficult problem to a form which could be calculated. He showed that stopping powers were universally proportional to target density, on an equal - mass - traversed basis.

A detailed suggestion for scaling stopping powers was shown by Knipp and Teller (41b), who successfully used the effective charge concept of Bohr and Lamb to scale H stopping values to equivalent He ion stopping powers.

These theoretical studies had only limited success with the fission fragment problem, and the primary practical result was to provide scaling relationships for the heavy ion stopping and ranges. That is, it allowed for interpolation and modest extrapolation from existing data into systems with different ions/targets/energies. Basically, the dominant effects were the target material and the ion's velocity. If you knew the stopping of one ion, say a proton, at two velocities in a material, and you knew the stopping of a heavy ion in that material at one of these velocities, then its stopping at the second velocity would be a direct proportionality. This law was applicable over a wide variety of velocities and ions. In fact it became so widely used that it stimulated several papers of objections to its theoretical shortcomings (63b,72b).

During the 1950's there were fundamental papers on evaluating both the energy transfer from slow particles to quantized electron plasmas, and in the energy loss to target nuclei. The study of particle stopping in a free electron gas is the first step in calculating the energy loss of an ion to a target's electrons. This problem was evaluated in Bohr's earliest papers (13a,15a) where he considered the electrons to be charged harmonic oscillators, with orbit frequencies established by the analysis of optical data. This interaction of a particle with an electron plasma was extended to quantized plasmas and then to Thomas-Fermi atoms by Bethe (30a,32a) and Bloch (33a,33b). An excellent review of relativistic particle stopping powers has recently been made by Ahlen (80e) and this chapter will not cover in detail this subject. For energies above 10 MeV/amu you should consult ref. (80a). Fermi then considered how the fast charged particle would polarize a classical electron medium of the target and hence modify the particle/plasma interaction (40c). This work was extended by Fermi and Teller (47b) to a degenerate free electron gas and they found that for slow particles the energy loss would be directly proportional to the particle's velocity. Bohr pointed out (48a) that behind the particle there would be an oscillating wake of charge, and this was evaluated more rigorously by Bohr (48b), and by Neufeld and Ritchie (55a). A full treatment of a charged particle penetrating a quantized electron plasma was presented at about the same time by Lindhard (54a), Neufeld and Ritchie

(55a) and Fano (56c). The Lindhard approach concentrates on non-relativistic particle interactions with a free-electron gas and provides a full general treatment with the following assumptions:

- The free electron gas consists of electrons at zero temperature (single electrons are described by plane waves) on a fixed uniform positive background with overall charge neutrality.

- The initial electron gas is of constant density.

- The interaction of the charged particle is a perturbation on the electron gas.

- All particles are non-relativistic.

The Lindhard approach is widely cited in the literature as it formed part of the first unified theory of ion penetration of solids (63a), and it has been widely used as the basis for calculating the electronic stopping of ions in matter (see, for example, 67a, 70b, 72f, 74a, 75k, 77h, 78a, 79d).

The energy loss to target nuclei is basically the study of screened Coulomb collisions between two colliding atoms. In the 1950's, major advances were made in the elastic energy loss of the ion to target nuclei. Bohr summarized much of the earlier work in 1948 (48a) which used the Thomas Fermi model to estimate the screened Coulomb potential, $V(r)$ between atoms:

$$V(r) = \frac{Z_1 Z_2}{r} \exp(-r/a) \qquad (1-3)$$

where Z_1 and Z_2 are the atomic numbers, r is their separation, and a is a "screening parameter". This screening parameter is an important concept in much of the theory which follows. It essentially increases the size of an atom by moderating the effect of the nuclear positive charge on the outer electrons because the inner electrons shield some of the nuclear charge. This screening parameter then leads to a "screening function" which is the reduction of potential at a point due to the inner electron screening. Once the screening parameter is specified, then the classical scattering and energy transfer can be calculated. Bohr argued that a reasonable approximation might be:

$$a = a_o/(Z_1^{2/3} + Z_2^{2/3})^{1/2}. \qquad (1-4)$$

but this approximation was not derived.

Firsov took a more practical approach and used numerical techniques to derive the interatomic potentials of two colliding Thomas-Fermi atoms (58a,b). After finding

the numeric values of the potentials as a function of the atomic separation he then
fitted these potentials with eq. (1-3) and found that the best fit was obtained with:

$$a = a_o/(Z_1^{1/2} + Z_2^{1/2})^{2/3} \tag{1-5}$$

Another problem which received wide attention in the 1950's was the degree
of ionization of the ion as it goes through materials. As we noted before, Bohr and
others suggested that one simple criterion would be to assume that ions lose electrons
whose orbital velocities would be less than the ion velocity. He suggested that the ion
charge *fraction*, Z^*/Z, would be

$$Z^*/Z = v/(v_o Z_1^{2/3}) \tag{1-6}$$

This relation comes from the Thomas-Fermi atom which assumes the electronic charge
densities of atoms are similar with a common unit of length being proportional to
$Z^{-1/3}$. The charge density is proportional to Z^2, and the total binding energy scales as
$Z^{7/3}$. Therefore the binding per electron scales as $Z^{4/3}$ and the electron velocities are
proportional to $Z^{2/3}$. Lamb had proposed (40b) the electron binding energy was the
important stripping criterion, while Bohr suggested it was the electron velocity. A
definitive clarification was made by Northcliffe (60c) who reduced a wide variety of
experimental data by dividing each ion/target/energy experimental stopping power by
the stopping power of protons in the same target and at the same *velocity*. In perturbation theory this ratio should scale as $(Z^*)^2$ where Z^* is the number of electrons left
on the ion. He found a large amount of data could be accurately described using the
relation:

$$(Z^*/Z) = 1 - a \exp\left[\frac{b}{Z^{2/3}} \cdot \frac{v}{v_o}\right] \tag{1-7}$$

where a and b are fitting constants. The expression expands to be the Bohr relationship.

By the end of the 1950's the status may be summarized as:

(a)   A good treatment of the energy loss of a charged particle to a quantized
        electron plasma. The theory includes both polarization of the medium about
        the charge, and discussions of extensions of particle interactions with electron
        plasmas to electrons in atomic matter.
(b)   A good calculation of interatomic potentials and the energy transferred during a
        scattering collision between two atoms.
(c)   A good evaluation of the effective charge of heavy ions in solids for the
        intermediate velocity range $3(v_o < E < 30 v_o)$

Problems left to be solved included:

(d)   How to extend the electron plasma point-charge interaction theory to the interaction with a finite sized ion?

(e)   How to derive fundamentally the effective charge of a moving ion (where effective charge is defined as a combination of ion charge state plus target polarization)?

(f)   And finally, how to you modify all of the above to use more realistic Hartree-Fock atoms rather than statistical atoms?

In 1963 the first unified approach to stopping and range theory was made by Lindhard, Scharff and Schiott (63a) and their approach is commonly called the LSS-theory. This work brought together all the pieces, and bridging approximations were made so that calculations of stopping and range distributions could, for the first time, be made within a single model. This remarkable achievement was the result of over a decade of study by Lindhard and collaborators (53b, 54a, 63a, 64a, 68a, 68b), with the later publications deriving in detail some of the major equations of LSS theory. LSS theory was the peak of stopping and range theory based on statistical atoms. With this theory it was possible to predict the range of ions in solids within a factor of two - a remarkable achievement considering it was applicable over the entire range of atomic species and energies up to the stopping power maximum (70a, 70f, 75e, 75f, 75g, 77a). Since it was based on Thomas-Fermi atoms it was most accurate for atoms with many electrons in the intermediate range where they are neither fully stripped nor almost neutral. The theory naturally shows no shell effects.

During the 1960's and 70's the primary advances came by applying numerical methods to traditional theoretical approaches. The use of computers permitted the incorporation of more realistic Hartree-Fock atoms into the theory and gave significant improvements. These important steps were initiated by Rousseau, Chu and Powers (70b) in electronic stopping, and Wilson, Haggmark and Biersack (77c) in nuclear stopping.

One way to evaluate these theoretical steps is to estimate the accuracy with which one can calculate stopping powers. After the work by Bethe-Bloch in the 1930's, the stopping of high velocity protons could be calculated to about 20%. By the late 1950's, the excellent review article by Whaling (58c) pointed out that little could be calculated for anything heavier than a proton. This changed abruptly with the LSS theory in 1963 (63a), which created a unified approach to the stopping of low energy heavy ions. With this approach, most stopping powers could be estimated within a factor of 2 or 3, and the ranges of these ions in single-element targets could be calculated within a factor of 2.

The LSS theory was the last of the comprehensive theories based on statistical models of atom-atom collisions. Improvements in calculating stopping and ranges over the next twenty years were made by using numerical techniques and removing some of the approximations used by Bohr, Firsov and Lindhard. One new theoretical insight which has had profound implications was made by Brandt and Kitagawa (82a) where they revised the Bohr suggestion of the degree of ionization of ions traveling within solids.   Bohr had suggested that the ion's electrons which had orbital velocities less than the instantaneous velocity of the ion would be stripped off, leaving the ion only with its inner high-velocity electrons.   Brandt and Kitagawa suggested that this stripping criteria should be modified to consider the ions's electron velocity only relative to the Fermi velocity of the solid.   They then proceeded to develop the formalism to allow the full evaluation of this new concept which has proved to be quite accurate.

Stopping powers in 1988 (the date of this chapter) can now be calculated with an average accuracy of better than 10% for low energy heavy ions, and to better than 2% for high velocity light ions.   Range distributions for amorphous elemental targets have about the same accuracy.

2 - NUCLEAR STOPPING CROSS-SECTIONS

In this section we shall review the mathematics of the collision of two charged particles and then the collisional scattering of two atoms with emphasis on the elastic energy transferred to the stationary atom.   The collision kinematics are calculated from the atom-atom interatomic potentials.   We discuss various models of atoms and show their potentials and their interatomic potentials.   It is shown how these can be reduced to a single analytic function which is called a universal interatomic potential.   This function is applied to generate new universal nuclear stopping cross-sections and scattering functions which can be used to calculate the physics of ion penetration of solids.

Introduction To Two Atom Scattering

The classical transfer of energy between a moving and a stationary charged particle depends only the mass and charge of the two particles, and the moving particle's initial speed and direction. While the moving charge passes, the stationary particle recoils and absorbs energy. The moving particle is deflected. The final velocities and trajectories can be simply found from the conservation of momentum and energy of the system.   Mathematically this problem is called the asymptotic orbit problem - it has analytical solutions for simple screened potentials between the particles.   Numerical techniques have been developed to evaluate the more complex collision of quantal atoms with shell effects, and the absorption of energy into the Pauli promotion of the electrons as will be discussed later.

Classical Two Particle Scattering

We will first summarize the equations relating the initial and final states of the elastic scattering of two particles.   For convenience we shall call the incident particle an 'ion' and the stationary particle an 'atom'.   This will allow a smooth transition from this section to the later physics of ion/target atom collisions. In classical non-relativistic elastic collisions, the following relations hold for laboratory coordinates (see figure 2-1).

Conservation of Energy:

$$E_o = \frac{1}{2}M_1V_o^2 = \frac{1}{2}M_1V_1^2 + \frac{1}{2}M_2V_2^2 \qquad (2-1)$$

where E_o is the initial ion kinetic energy, V_o is the incident velocity of the ion with mass M_1, and V_1 is the ion's final velocity after striking the target atom of mass M_2, which recoils with velocity V_2.

Conservation of Momentum:

Longitudinal: $\quad M_1V_o = M_1V_1 \cos \vartheta + M_2V_2 \cos \phi.$ \qquad (2-2)

Lateral: $\quad 0 = M_1V_1 \sin \vartheta + M_2V_2 \sin \phi.$ \qquad (2-3)

where ϑ is the final angle of deflection of the ion and ϕ is the final recoil angle of the target ion.

The solution of these three equations can be made in various forms depending on which variables are cancelled by substitution. For example, eliminating the target recoil angle ϕ we have:

$$\left(\frac{V_1}{V_2}\right)^2 - 2\left(\frac{V_1}{V_o}\right)\frac{M_1}{M_1+M_2} \cos \vartheta - \frac{M_2-M_1}{M_2+M_1} = 0 \qquad (2-4)$$

which is independent of the way the forces between the particles vary with separation.

We will now restate the above in center-of-mass coordinates. The basic reason for this transformation is to show that no matter how complex the force is between the two particles, so long as it acts only along the line joining them (no transverse forces), the relative motion of the two particles can be reduced to that of a single particle moving in a central potential (later called the interatomic potential) centered at the origin of the center-of-mass coordinates. This point is essential to the rest of this section.

For center-of-mass (CM) coordinates we define the system velocity, V_c, such that in this coordinate system there is zero net momentum (see figure 2-1):

$$M_1V_o = (M_1 + M_2)V_c. \qquad (2-5)$$

For convenience, we also define in CM coordinates a reduced mass, M_c, by the relation:

$$\frac{1}{M_c} = \frac{1}{M_1} + \frac{1}{M_2} \qquad (2-6)$$

or

LABORATORY COORDINATES

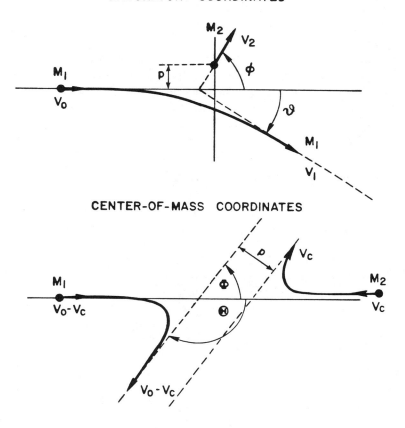

CENTER-OF-MASS COORDINATES

Figure (2-1)   The upper figure defines the scattering variables in a two-body collision. The projectile has mass M_1 and an initial velocity V_o, and an impact parameter, p, with the target particle. The projectile's final angle of deflection is ϑ and its final velocity is V_1. The target particle with mass, M_2, recoils at an angle ϕ with velocity V_2.

The lower figure is the same scattering in center-of-mass (CM) coordinates in which the *total momentum of the system is zero*. The coordinate systems moves with velocity V_c relative to the laboratory coordinates, and the new angles of scatter and recoil are Θ and Φ

$$M_c = \frac{M_1 M_2}{M_1 + M_2}. \tag{2-7}$$

We can now solve for the CM velocity as

$$V_c = V_o M_c / M_2. \tag{2-8}$$

This relation shows an advantage of using CM coordinates - the system velocity remains constant and is independent of the final angle of scatter between the two particles. Hence, the total linear momentum of the system is always zero, and the particle velocities are inversely proportional to their masses (see figure 2-1):

$$\frac{V_o - V_c}{V_c} = \frac{M_2}{M_1} \tag{2-9}$$

and the CM initial kinetic energy, E_c, is simply

$$E_c = \frac{1}{2} M_c V_o^2. \tag{2-10}$$

Using these equations, we can find many relations between the initial parameters and the final scattering angles.

The conversion of scattering angles from the CM system to the laboratory system is shown in figures (2-2) and (2-3). In figure (2-2) the conversion for the target-particle recoil is shown to be very simple because its initial laboratory velocity is zero. Hence its laboratory final velocity vector, V_2, is related to its CM velocity vector, V_c, by the translation vector between the two systems, V_c, as defined in equation (2-5) to make the total momentum of the system to be zero.

Using this relation, we can now obtain the energy transferred, T, in the collision from the incident projectile to the target particle:

$$T \equiv \frac{1}{2} M_2 V_2^2$$

$$T = \frac{M_2}{2} \left(\frac{2 V_o M_c \cos \phi}{M_2} \right)^2.$$

$$T = \frac{2}{M_2} (V_o M_c \cos \phi)^2. \tag{2-15}$$

This can be related to the angle of scatter of the projectile by using equation (2-13),

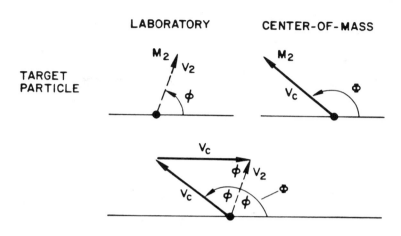

Figure (2-2)   Angular conversion of center-of-mass coordinates to laboratory coordinates for the target particle. The upper two figures shows the final particle vectors in both systems. The two systems are related since the CM system moves laterally with velocity, V_c, as defined by eq. (2-5) which defines V_c to make the total momentum of the CM system to be zero.

The lower figure then relates the final target velocity, V_2 in the Lab and V_c in CM, by the lateral vector V_c which is their difference. Since the triangle is isosceles, $\Phi = 2\phi$, which is independent of the particle masses and velocities.

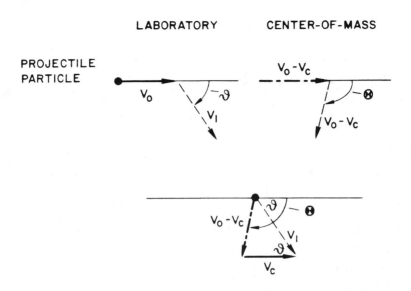

Figure (2-3)   Angular conversion of center-of-mass coordinates to laboratory coordinates for the projectile particle.   The upper two figures show the initial and final vectors for both systems.   The lower figure relates the two systems since the CM system is defined to move laterally with velocity V_c relative to the laboratory system.

$2\phi = (\pi - \Theta)$, giving the relations

$$T = \frac{2}{M_2}\left(V_0 M_c \sin \frac{\Theta}{2}\right)^2 = \frac{4E_c M_c}{M_2}\sin^2\frac{\Theta}{2}$$
$$= \frac{4E_0 M_1 M_2}{(M_1+M_2)^2}\sin^2\frac{\Theta}{2} \qquad (2-16)$$

These relations contain the energy lost by the projectile and hence will lead to the stopping power of the projectile when we evaluate the energy loss cross-section.

Similarly, the projectile parameter conversion is shown in figure (2-3), but it is more complicated because of the initial velocity, V_0, of the projectile. The relation of the scattering angles can be derived from figure (2-3) to be

$$\tan \vartheta = \frac{(V_0 - V_c)\sin \Theta}{V_c + (V_0 - V_c)\cos \Theta}$$

but we may use equation (2-9) which relates $(V_0 - V_c)/V_c = M_2/M_1$, to simplify the angular relationship to

$$\tan \vartheta = \frac{(M_2/M_1)\sin \Theta}{1+(M_2/M_1)\cos \Theta}$$

or: $$\tan \vartheta = \frac{M_2 \sin \Theta}{M_1+M_2 \cos \Theta}. \qquad (2-17)$$

Another way to analyze this relationship is to evaluate figure (2-3) using the law of sines, and we can obtain the equivalent relation:

$$\frac{\sin (\Theta - \vartheta)}{\sin \vartheta} = \frac{V_c}{V_0 - V_c}$$

which can be rearranged using equation (2-9) to be

$$\sin (\Theta - \vartheta) = \frac{M_1}{M_2} \sin \vartheta. \qquad (2-18)$$

Thus the CM to laboratory angular conversion is shown by equation (2-17) or (2-18) for the projectile. The relation between each of these scattering angle pairs can be made with equation, $\Phi = \pi - \Theta$.

The discussion above concerns asymptotic values, and is valid for all symmetric elastic forces between particles. If the force is of the electrostatic type, $1/r^2$, then the total path of the two particles can be determined to be hyperbolas. The only part of the path which will interest us later is the closest point of approach of the two particles.

Calling this minimum radius of separation, r_{min} , we can show for repulsive inverse square fields that this point is where the potential energy between the two particles exactly equals the initial CM kinetic energy:

$$\frac{1}{2}M_c V_o^2 = \frac{2Z_1 Z_2 e^2}{d} \qquad (2-19)$$

where d is called the *collision diameter* and relates the Coulomb energy to the kinetic energy of the system as shown above and where Z_1 and Z_2 are the number of charges on the two particles. The collision diameter, d, is defined by equation (2-19). For the minimum separation during the scattering we can find r_{min} from just the collision diameter, d, and the final angle of scatter:

$$r_{min} = \frac{d}{2}\left(csc \frac{\Theta}{2} \pm 1 \right) \qquad (2-20)$$

where the $\pm$ sign is for repulsive (+) or attractive (-) fields. For a head-on repulsive collision we have the condition for closest approach, r_{min} , that the kinetic energy has all been converted to Coulomb energy and

$$r_{min} = d. \qquad (2-21)$$

Two-Body Central Force Scattering

The above discussion reviewed the elastic scattering of two particles and the final-state formulae are valid for all such collisions which have conservation of energy and momentum. The equations above which are most applicable to stopping powers are eq. (2-15) and (2-16) which show the energy loss from the projectile particle to the target particle during a scattering event. But to get the *cross-section* for the energy transferred we need to know the probability for each final scattering angle, and this is obtained only by evaluating the details of the scattering trajectory and hence the probability of scatter into each scattering angle.

The most important simplification of this problem is to assume that the force between the two particles acts only along the line joining them, and there are no transverse forces. Then, as discussed above, the use of center-of-mass (CM) coordinates reduces any two-body problem to a one-body problem, namely the interaction of a particle with mass M_c and velocity V_c with a static potential field centered at the origin of the CM coordinates. This is because in the CM system *the total linear momentum of the particles is always zero*, the paths of the two particles are symmetric as shown in figure 2-1, and the evaluation of one particle's path directly gives the path of the other particle. The conversion from CM scattering angles to laboratory angles is then done with relations such as equations (2-12) and (2-17, 2-18).

The derivation of the complete particle scattering path is usually done using Lagrangian mechanics in polar coordinates. Since we are dealing with only two particles and no transverse forces, the problem is two-dimensional with the plane defined by the projectile's initial velocity vector and the initial target particle position. We call the azimuthal polar coordinate Θ and the radial coordinate r in the following discussion..

Only two new equations are needed to determine the trajectory of the particle in the CM coordinate system. We use the notation of $\dot{r} \equiv dr/dt$ and $\dot{\Theta} \equiv d\Theta/dt$ for the time differentials of motion in polar coordinates. From conservation of energy we have for the system

$$E_c = \frac{1}{2}M_c(\dot{r}^2 + r^2\dot{\Theta}^2) + V(r) \qquad (2-24)$$

where E_c is the CM energy [defined by equation (2-10) to be $E_c = \frac{1}{2} M_c V_o^2$], and this is merely restated in equation (2-24) with CM polar coordinates.

The second equation states the conservation of angular momentum for the system, and in CM coordinates this is:

$$J_c = M_c r^2 \dot{\Theta} \qquad (2-25)$$

where J_c is the constant of angular momentum. This is the polar coordinate relation which we in cartesian coordinates is

$$J_c = M_c V_c p \qquad (2-11)$$

where p is the impact parameter defined in figure (2-1).

With these three equations we can solve, for example, for $d\Theta$ as a function of dr. This relation is most important for it will then allow us to calculate directly the energy transferred in any collision for *any* central force potential, $V(r)$.

We substitute equation (2-11) into equation (2-25) to remove the variable J_c, then we insert this into equation (2-24) and solve for $\dot{r}$. The result is the radial equation of motion and it can take the form

$$\dot{r} = \frac{dr}{dt} = V_c \left[1 - \frac{V(r)}{E_c} - (\frac{p}{r})^2 \right]^{1/2}. \qquad (2-26)$$

From equations (2-25)-(2-26) the angular coordinate will be determined by

$$\dot{\Theta} = \frac{d\Theta}{dt} = \frac{V_c p}{r^2} \qquad (2-27)$$

Combining (2-26) and (2-27) we have the result:

$$\frac{d\Theta}{dr} = \frac{d\Theta}{dt}\frac{dt}{dr} = \frac{p}{r^2\left[1 - \frac{V(r)}{E_c} - \frac{p^2}{r^2}\right]^{1/2}} \qquad (2-28)$$

Integrating this we have the final relation

$$\Theta = \int_{-\infty}^{\infty} \frac{p\,dr}{r^2\left[1 - \frac{V(r)}{E_c} - \frac{p^2}{r^2}\right]^{1/2}}$$

$$= 2\int_{r_{min}}^{\infty} \frac{p\,dr}{r^2\left[1 - \frac{V(r)}{E_c} - \frac{p^2}{r^2}\right]^{1/2}} \qquad (2-29)$$

This equation allows us to evaluate the final angle of scatter, Θ, in terms of the initial CM energy, E_c, the potential, $V(r)$, and the impact parameter, p.

Later, we will use this equation to calculate the energy transferred to the target atom using equations (2-16), and then evaluate the cross-section for such scattering by integrating over all impact parameters using equation (2-29). Equation (2-29) is called the general orbit equation for two-body central force scattering. The only conditions on its applicability are, essentially, that the central force potentials of each particle must not vary with time, nor depend on the particle's motion. The potentials must be spherically symmetric. Finally, the laws of conservation of energy and momentum must hold for the system as a whole.

Interatomic Potentials

Because of their universal applicability, statistical models of interatomic interaction have been widely used in calculating nuclear stopping. The most widely used is the Sommerfeld approximation to the Thomas-Fermi potential (32b), the Moliere approximation (47a), the Lenz-Jensen (32c), and the Bohr potential (48a). Each of these potentials may be considered as a Coulombic term $(1/r)$ multiplied by a "screening" function. The Coulombic term arises from the positive point nucleus, and the electronic screening reduces its value for all radii. The screening function may be defined as the ratio of the actual atomic potential at some radius to the potential caused by an unscreened nucleus:

$$\Phi = \frac{V(r)}{(Ze/r)} \qquad (2-30)$$

where Φ is the screening function, $V(r)$ is the potential at radius r, Z is the atomic number and e is the electronic charge. The implications of the spherical symmetry of this definition are discussed in the section on Solid State Charged Distributions.

There are three general approaches which have been used to calculate screening functions (and hence interatomic potentials). One way is to use experimental data and to try to work back to the potential. A second is to incorporate full quantal treatments for all collisional interactions. A third way is to isolate the few interactions of importance to energetic collisions and to estimate the potential from limited calculations. These techniques are described below.

First, the interatomic potential can be derived from crystal data such as phonon dispersion curves, elastic constants, compressibility and the x-ray lattice constants. Although one can obtain excellent potentials for atoms at normal solid-state separations, this method can not be used in obtaining potentials for the small nuclear separations which one finds in atomic collisions. Further, the only potentials which can be obtained are those for atoms of materials which can be fabricated in crystalline form, i.e. for like-atom collisions, e.g. Si-Si, and for the few binary crystals such as GaAs. Finally, the technique can only be applied to stable crystal elements, and has not been used for H or He atoms. The various methods which have been developed to obtain potentials from crystal data has been reviewed by Johnson (73c).

The second technique is at the opposite extreme of computational difficulty. One begins with two Hartree-Fock atoms and as they merge one recalculates the electronic orbits based, for example, on multi-configurational self-consistent-field methods. This type of calculation is the most accurate known and can treat all atomic combinations. It further allows for the treatment of excited and ionized states in either of the atoms. This approach is overly-sophisticated for collisional studies, but it can be used to obtain bench-marks to establish the accuracy of simpler approaches. That is, the complete calculation can be made for a few systems, and we can require that any simpler approach should produce potentials which agree accurately to the best calculation which can be done.

A third technique is to use the simplified quantum-mechanical approach which was suggested by Gombas (49a). One of the earliest calculations using this approach was by Wedepohl (67c), and a wide study was made by Wilson, Haggmark and Biersack (77c). This method begins with two atomic charge distributions, and for the calculation of the interatomic potential the local density approximation is assumed without any reconfiguration of the atomic structures as a whole. For any volume element of either atom, the number of its electrons in that volume does not change as the two atoms merge. The actual number of electrons in a specific volume element may change if it is part of the overlap volume of the two atoms. For this overlap volume there will be electrons from both atoms, and these are treated as a free electron gas within the volume element. All overlap volumes absorb energy (decrease the attractive potential) because the Pauli principle demands that for an increased electron density there must be promotion of electrons into higher energy levels.

This approximate interatomic potential calculation therefore has two parts: one is the Coulombic interaction between all the electrons and the two nuclei, and the second is the increased quantal energy which goes into excitation and exchange effects for the electrons in the volume of atomic overlap. The full treatment of this subject is given in reference (84a).

The *interatomic* screening function definition is:

$$\Phi_1 \equiv \frac{V(r)}{(Z_1 Z_2 e^2/r)} \qquad (2-31)$$

which relates the interatomic screening function, Φ_1, with the potential, $V(r)$. This is a natural extension of the screening function of a single atom, equation (2-30).

We have found a type of reduced radial coordinate which allows the development of a single analytic function to accurately calculate any interatomic potential (84a):

$$a_U = .8854 \ a_o/(Z_1^{.23} + Z_2^{.23}) \qquad (2-32)$$

Note the similarity fo equations (1-4) and (1-5), and this screening function can be fit by:

$$\Phi_U = .1818e^{-3.2x} + .5099e^{-.9423x} + .2802e^{-.4028x}$$
$$+ .02817e^{-.2016x} \qquad (2-33)$$

This curve is shown in figure (2-4) with the statistical atom screening functions. This universal screening function is an improvement to one found by Wilson et al. (77f) which they identify as a C-Kr potential.

Energy Transfer from Projectile Atom to Target Atom

We shall first briefly review the formulae we will need which were derived at the beginning of this section. The energy transferred during the screened Coulomb collision of two atoms will be described as a function of two variables, the projectile atom's initial energy, E, and its impact parameter, p. These are identified in figure (2-1), with p being defined as the projected offset of the original path of Z_1 from Z_2. If these two variables are known, then the energy transfer, T, to the target atom was determined simply from conservation of energy and momentum, as was shown in the derivation of equation (2-16):

$$T = \frac{4M_1 M_2}{(M_1 + M_2)^2} \ E_o \sin^2 \frac{\Theta}{2} = \frac{4E_c M_c}{M_2} \sin^2 \frac{\Theta}{2} \qquad (2-34)$$

where M_1 and M_2 are the masses of atoms, Z_1 and Z_2, and where Θ is the projectile's scattering angle in center-of-mass coordinates, which is related to the lab frame by:

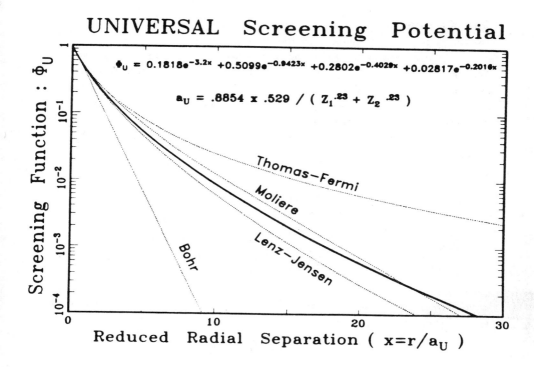

Figure (2-4)   The reduced screening function have been fitted to the analytic expression shown above with four exponential terms. This screening function is identified as Φ_u, a universal screening function with its argument, x, being defined as $x \equiv r/a_u$, where a_u is the universal screening length shown above.

$$\vartheta = \tan^{-1}\{\sin \Theta /[\cos \Theta + M_1/M_2]\}. \qquad (2-35)$$

where ϑ is the laboratory final deflection angle of the projectile, see equation (2-17). The energy transfer is therefore proportional to $\sin^2 \Theta/2$.

As discussed at the beginning of this section, the problem of a two body collision may be reduced to that of a particle in a single central-force field if the following conditions are met: (a) The two potentials are each spherically symmetric and do not vary with time or with either particle's velocity, and (b) The laws of the conservation of energy and momentum are conserved for the system as a whole. With these conditions, the scattering angle of deflection, Θ, was derived in equation (2-29) to be:

$$\Theta = \int_{-\infty}^{\infty} \frac{p dr}{r^2\left[1 - \dfrac{V(r)}{E_c} - \dfrac{p^2}{r^2}\right]^{1/2}}$$
$$= 2\int_{r_{min}}^{\infty} \frac{p dr}{r^2\left[1 - \dfrac{V(r)}{E_c} - \dfrac{p^2}{r^2}\right]^{1/2}} \qquad (2-36)$$

where $V(r)$ is the interatomic potential of the two atoms and E_c is the center-of-mass energy defined as

$$E_c = E_o M_2/(M_1 + M_2),$$

and r_{min} is the distance of closest approach during the collision. Equation (2-36) is the general solution for a particle in a spherically symmetric central force field and the difficult problem of two atom scattering has been reduced to this simple form. The energy transferred to the target atom is now a function p and E, and is found by inserting the solution of Eq. (2-36) into Eq. (2-34).

In order to solve the integral (2-36) in an universal way, especially in a way independent of the ion-target combination, we use the substitutions $x = r/a$, $b = p/a$, and $\varepsilon = E_c/(Z_1 Z_2 e^2/a)$ and obtain:

$$\theta = \pi - 2\int_{ro/a}^{\infty} \frac{b dx}{x^2\left[1 - \dfrac{\Phi(x)}{x\varepsilon} - \left(\dfrac{b}{x}\right)^2\right]} \qquad (2-37)$$

where we have also replaced the interatomic potential with the previously discussed screening function $V = (Z_1 Z_2 e^2/r)\Phi(r/a)$, see equation (2-31). Now equation (2-37) allows the calculation of the final scattering angle with the new parameters ε and b, and the individual atomic variables, Z_1, Z_2, M_1 and M_2 have been eliminated (see further comments on this derivation in references (57a,b, 63a, and 64c).

Universal Nuclear Stopping Powers

The energy lost by the ion per unit path length $\equiv dE/dR$. This is related to the nuclear stopping cross-section, $S_n(E)$, by the relation $dE/dR = NS_n(E)$, where N = the atomic density of the target.

The nuclear stopping power, $S_n(E)$, is the average energy transferred when summed over all impact parameters, so from eq. (2-34) we have:

$$S_n(E) = \int_0^\infty T(E,p)2\pi p dp = 2\pi\gamma E_o \int_0^{p_{max}} \sin^2\frac{\Theta}{2} p dp \qquad (2-38)$$

with the integration's upper limit being the sum of the two atomic radii, p_{max} beyond which the interatomic potential, and T, is zero. We use the center-of-mass transformation unit:

$$\gamma \equiv 4M_1 M_2/(M_1 + M_2)^2. \qquad (2-39)$$

In order to show clearly the results of using classical charge distributions and also those using solid state distributions, it is easiest to again plot the nuclear stopping in reduced units in which a single curve describes all combinations of classical atom-atom collision. Lindhard et al. have discussed at length the calculation of nuclear stopping using Thomas-Fermi atoms (63a,68a). They suggested a reduced coordinate system for nuclear stopping which we will extend to our new calculations. Using the formalism of Lindhard we define:

$$S_n(\varepsilon) = \frac{\varepsilon}{\pi a_U^2 \gamma E_o} S_n(E) \qquad (2-40)$$

where ε is a reduced energy introduced by Lindhard et al. (63a, 68a) defined as

$$\varepsilon \equiv a_U M_2 E_o/Z_1 Z_2 e^2 (M_1 + M_2) \qquad (2-41)$$

and a_U is the universal screening length of equation (2-32).

Equation (2-38) defines nuclear stopping in physical units, and eq. (2-41) converts it to LSS reduced units.

We show in figure (2-6) the reduced nuclear stopping for the four classical atom screening functions and for the universal screening function shown in figure (2-5). The universal nuclear stopping is calculated by restating eq. (2-38) in reduced units using the reduced impact parameter $b = p/a_1$:

$$S_n(\varepsilon) = \varepsilon b^2 \int_0^\infty \sin^2\frac{\Theta}{2} d(b^2) \qquad (2-42)$$

(see, for example, ref. 68a or 77f for expansions of this equation).

The dotted lines in figure (2-5) reproduce within a few percent the similar calculations of ref. 77f for the classical interatomic potentials and confirm the accuracy of our computer program. The small circles on the plot are solutions of eq. (2-42)

UNIVERSAL Reduced Nuclear Stopping

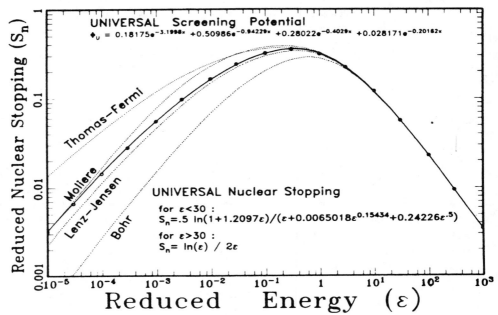

Figure (2-5)   The universal screening function, figure (2-4) can be used to calculate the nuclear stopping power using equation (2-38). The result is shown above in reduced coordinates in the spirit of LSS theory (ref. 63a) but with the screening length of LSS theory, a_l, being replaced with that of figure (2-4), a_u, see equation (2-40). The reduced nuclear stopping power calculations are shown as small circles over the 8 decades of energy. Through these points has been fitted the analytic formula identified in the figure as: Universal Nuclear Stopping. This function agrees with the calculated nuclear stopping within a few percent. Also shown are the nuclear stopping calculations based on the four classical atomic models.

using the universal screening function, equation (2-33). The solid line is an analytic expression fitted to the points with the form:

$$S_n(\varepsilon) = \frac{\ln(1 + a\varepsilon)}{2(\varepsilon + b\varepsilon^c + d\varepsilon^{1/2})} \qquad (2-43)$$

where a, b, c and d are fitting coefficients as shown on the figure. This equation does not have the proper high energy properties where nuclear stopping must become like Rutherford scattering, so for reduced energies above 30 it is better to shift to unscreened nuclear stopping:

$$S_n(\varepsilon) = \ln \varepsilon / 2\varepsilon \qquad (2-44)$$

The figure shows for all potentials the nuclear stopping is identical for $\varepsilon > 10$. For lower values of epsilon the universal stopping curve falls between the Moliere and Lenz-Jensen curves.

For practical calculations, the universal nuclear stopping is

$$S_n(E_o) = \frac{8.462 \times 10^{-15} Z_1 Z_2 M_1 S_n(\varepsilon)}{(M_1 + M_2)(Z_1^{.23} + Z_2^{.23})} \quad eV/(atom/cm^2) \qquad (2-45)$$

with the reduced energy, ε, being calculated as

$$\varepsilon = \frac{32.53 \, M_2 E_o}{Z_1 Z_2 (M_1 + M_2)(Z_1^{.23} + Z_2^{.23})} \qquad (2-46)$$

and the reduced nuclear stopping being calculated as:

For $\varepsilon \leq 30$: $S_n(\varepsilon)$

$$= \frac{\ln(1 + 1.1383\varepsilon)}{2[\varepsilon + .01321 \, \varepsilon^{.21226} + .19593 \, \varepsilon^{.5}]} \qquad (2-47)$$

For $\varepsilon > 30$: $S_n(\varepsilon) = \frac{\ln(\varepsilon)}{2\varepsilon} \qquad (2-48)$

3 - ELECTRONIC STOPPING CROSS-SECTIONS

Introduction

The total stopping cross-section of ions in solids is divided into two parts: the energy transferred by the ion to the target electrons (called electronic stopping) and to the target nuclei (called nuclear stopping). The nuclear stopping component is usually considered separately because the heavy recoiling target nucleus can be assumed to be unconnected to its lattice during the passage of the ion, and the elastic recoil energy which is transferred to it can be treated simply as the kinetic scattering of two heavy screened particles (see section 2 on Nuclear Stopping). Separation of the energy loss of the ion into two separate components ignores the possible *correlation* between hard nuclear collisions and large inelastic losses to electronic excitation. It is felt that this correlation probably is not significant when many collisions are averaged over, as when an ion penetrates a solid, but is of importance for single scattering studies (79g), and for very thin targets (83a, 83b).

Interaction of a Particle with a Free Electron Gas

Both Thomson (12a) and Darwin (12c) treated energy loss as the energy transferred by a moving charged particle to a free electron. Since such a collision has an infinite energy-loss cross-section without shielding, they were both led into the problem of specifying maximum impact parameters based on atomic densities. Bohr based his study of electronic stopping cross-sections on a model which considered the target a collection of harmonic oscillators whose frequency was determined by optical absorption data (13a). He was able to show, in contrast to the earlier work, that a natural energy-loss cross-section cut-off can occur and if one limits the collision time, p/v, to less than the atomic orbital time, $1/\nu$, where p is the impact parameter of a particle of velocity v on a harmonic oscillator of frequency ν. For longer collision times the interaction becomes adiabatic and no energy is transferred from the ion to the target atom.

During the next decade several attempts were made to bring quantum-mechanics into this problem, and it was done finally by Bethe (30a). This work was extended to relativistic particles by Bethe (32a) and Moller (32d), and then further expanded by Bethe and Bloch. What is sometimes called the Bethe-Bloch theory (30a,31a,32abcd,33ab) considers a particle interacting with an isolated atom of harmonic oscillators. This approach solved the charged-particle/atom energy-loss problem quantum mechanically in the first Born approximation. The atomic nature of various target elements is concentrated into a single number representing the mean excitation of the atomic electrons of that element. The results of the earlier Bohr work differed significantly from the results of Bethe and this led several authors to comment on the discrepancy (31a,33a,33b). Bloch found that the Bohr distant collision theory was quantum-mechanically correct as the mean energy loss averaged over all electronic transitions. Bloch's solution for the close-collision energy loss differed from that of Bohr and Bethe. Bohr had assumed for close collisions that the atomic electron was free, and Bethe had represented such collisions as plane waves. Bloch showed that the Bohr classical solution was valid for hard close collisions, while the Bethe solution was valid for weak scattering. Bloch then provided a solution which reduced to the Bohr solution for hard collisions and almost reduced to the Bethe solution for weak collisions.

These original studies required that the particle velocity be much greater than that of the electrons bound in the atoms. Later extensions of the Bethe-Bloch theory attempted to find semi-empirical ways to correct it for energetic inner-shell atomic electrons (see, for example, 72c).

For details of the Bethe-Bloch stopping theory there are extensive review articles, e.g. (62b,63b,78h, and 80e). We shall use the Bethe-Bloch approach only in extending our calculations to relativistic ion velocities. This chapter will mostly concentrate on low energy non-relativistic stopping theory (energies below 10 MeV/amu) and it will not repeat these reviews.

After this work in the 1930's the next major step in electronic stopping theory was the consideration of the target as a collection of interacting electrons, i.e. a plasma, and to consider the energy loss to collective effects such as dynamic polarization and energy loss to plasmons. The classic papers to treat the target electrons as a plasma were by Fermi (40c) and Fermi and Teller (47b). These papers begin with Maxwell's equations and establish some estimates of the polarization of the target electronic medium. They consider the problems of binding energies, the dispersion of the electron oscillators and damping constants on collective motion. One result is that they find a typical energy loss of per unit distance for slow particles is proportional to the ion velocity

$$\frac{dE}{dx} \sim v \, \rho^{1/3}$$

where v is the particle velocity and ρ is the electron density.

Electronic interactions of a particle with a plasma were then extensively treated by Lindhard (54a), Neufeld and Ritchie (55a) and Fano (56c). We review below primarily the results of Lindhard who presented generalized methods to treat the response of a free electron gas to a perturbation and he derived an explicit function for the interaction.

The Lindhard treatment is a many-body self-consistent treatment of an electron gas responding to a perturbation by a charged particle. It naturally includes the polarization of the electrons by the charged particle and the resultant charge-screening and the plasma density fluctuations. It treats smoothly both individual electron excitation and collective plasmon excitations without separate 'distant' and 'close' collision processes. Finally, when used with the local-density-approximation it can be directly applied to any target and, for example, the effects of chemical bonding or crystal structure on stopping power are simply evaluated.

Lindhard's approach to the interaction of a particle with a free electron gas makes the following assumptions:

- The free electron gas consists of electrons at zero temperature (single electrons are described by plane waves) on a fixed uniform positive background with overall charge neutrality.
- The initial electron gas is of constant density.
- The interaction of the charged particle is a perturbation on the electron gas.
- All particles are non-relativistic.

The electronic stopping of a charged particle in the local density approximation may be stated as:

$$S_e = \int I(v,\rho) \, Z_1^2 \, \rho \, dV \qquad (3-1)$$

where S_e is the electronic stopping; I is the stopping interaction function of a particle of unit charge with velocity, v, with a free electron gas of density, ρ; Z_1 is the charge of the particle, ρ is the electronic density of the target, and the charged particle integral is performed over each volume element, dV, of the target. (We use this form of a stopping equation because it simply expands to the form needed for heavy ions). The electronic density of a target atom is normalized so that its atomic number $Z_2 = \int \rho \, dV$ with the integration over the atomic volume. Each of the three components of Eq. (3-1) will be discussed below.

With these assumptions, Lindhard derived the interaction function, I, of Eq. (3-1) as:

$$I = \frac{4\pi e^4}{mv^2} \cdot \frac{i}{\pi \omega_o^2} \int_o^\infty \frac{dk}{k} \int_{-kv}^{kv} \omega d\omega \left[\frac{1}{\varepsilon^\ell(k,\omega)} - 1 \right] \qquad (3-2)$$

where the longitudinal dielectric constant, ε^ℓ, is derived to be

$$\varepsilon^\ell(k,\omega) = 1 + \frac{2m^2\omega_o^2}{\hbar^2 k^2} \sum_n \frac{f(E_n)}{N} \times$$

$$\qquad (3-3)$$

$$\frac{1}{\left\{ k^2 + 2\vec{k}\cdot\vec{k}_n - \frac{2m}{\hbar}(\omega - i\delta) \right\}} + \frac{1}{\left\{ k^2 - 2\vec{k}\cdot\vec{k}_n + \frac{2m}{\hbar}(\omega - i\delta) \right\}}$$

where e and m are the charge and mass of an electron; ω_o is the classical plasma frequency defined as $\omega_o^2 = 4\pi e^2 \rho/m$; E_n is the energy and $\vec{k}_n$ the wave vector of the electron in the n'th state; $f(E_n)$ is the distribution function and is an even function of $\vec{k}_n$, and δ is a small damping factor. Simple polynomial fits to a numeric evaluation of Eq. (3-2) can be found in Ref. 79d.

The physical properties of Lindhard's particle-plasma interaction theory can be shown in several ways. In Figure (3-1) is shown the interaction term I, of Eq. (3-2) and (3-3), versus a free electron gas density. Each curve has a flat section at low electron densities where the ion is going much faster than the mean electron velocity. Each curve bends down where the ion velocity becomes equal to the Fermi velocity, v_F, of the free electron gas, defined as

$$v_F = \left(\frac{\hbar}{m} \right) \left(3\pi^2 \, \rho \right)^{1/3} \tag{3-4}$$

For example, in figure (3-1) the top curve is for particles with velocity of 100 keV/amu $= 6 \times 10^8$ cm/sec. The Fermi velocity for an electron density of 10^{24}/cm^3 is about 3×10^8 cm/sec, which is where the curve is inflecting. At higher electron densities, some of the electrons of the free electron gas can respond adiabatically because of their higher velocities and the interaction is reduced. For any single electron density there is a maximum of interaction strength which occurs for particles with a velocity about equal to the electron Fermi velocity.

In Figure (3-2) we show Iρ, the interaction term, I, times the charge density, ρ. If the particle has a unit charge, this plot shows the differential energy loss per unit path length, in units of eV/cm, for a particle in a uniform free electron gas. For a given charge density, the energy loss per path length has two competing components: a decreasing interaction strength as electron density increases, and an increasing number of electrons per unit volume.

The Local Density Approximation in Stopping Power Theory

The theoretical electronic energy loss of a proton in a solid (in contrast to a free electron gas) is calculated using the local-density approximation. In essence, this approximation assumes that each volume element of the solid is an independent plasma. The stopping power is calculated for a particle in a plasma of each volume element's density, and the final stopping power is computed by averaging over these values which are weighted by their distribution in the solid. Referring to eq. (3-1):

$$S_e = \int I(v,\rho) \, (Z_1^*(v))^2 \, \rho \, dV \tag{3-5}$$

where I is the interaction of the particle with velocity v in a plasma of density ρ. The charge of the proton, Z_1^*, has as asterisk to indicate this may be a value different from the atomic number because the ion may not be fully stripped. By integrating over ρdV we weigh each density interaction by the probability of that density occurring in the solid.

An extended comment might be made to explain the local-density-approximation to those completely unfamiliar with it. This is a widely used method to evaluate the theoretical mean response of a solid to a perturbation. For our application, we consider the solid to be an electronic plasma with fluctuations in density. We first calculate the interaction (energy loss) of an energetic particle immersed in a

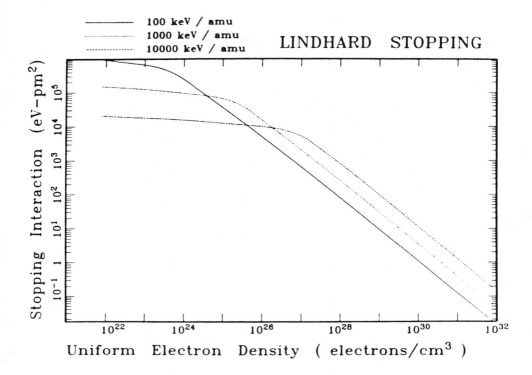

Figure (3-1) The stopping interaction of a charged particle with a free electron gas
of various densities. The stopping interaction derived by Lindhard is shown in the text
as equation 3-2. Each curve is flat for the section where the particle is much faster
than the electrons in low density electron gases. At about the point where the particle
velocity equals the Fermi velocity of an electron gas the interaction curve inflects. For
greater density electron gases the interaction becomes less since some of the electrons
are moving faster than the particle and these collisions become more adiabatic.

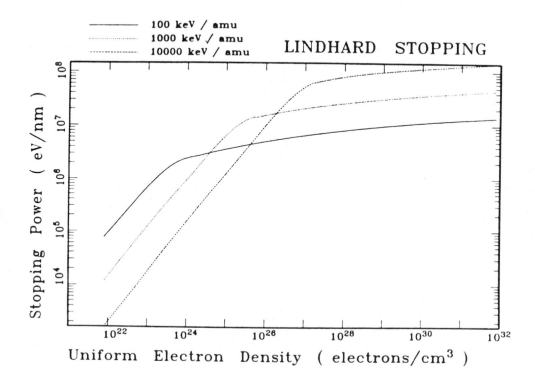

Figure (3-2) The stopping power or energy loss per unit path length of a particle in a free electron gas is the product of the interaction strength shown in figure (3-1) times the electron density. These two factors compete since as the electron density increases, the number of electrons per unit volume increases but the interaction strength decreases. The result is a linear increase of stopping power with electron density for dilute gases, and then a leveling off for more dense gases. The inflection-point for a particle is where its velocity equals the Fermi velocity of a free electron gas.

uniform plasma sea with the same electronic density as any single volume element of the solid. A basic assumption is now made that the averaged interaction of a single particle to a uniform plasma is *identical* to the averaged interaction of a single volume element of plasma to a particle whose spatial location is uniformly probable. This equivalence allows the evaluation of the mean interaction of a single particle with a single volume element of plasma. This process is then repeated for the interaction of the particle with every volume element of the solid target to obtain the mean interaction of the particle with the solid.

Two of the more important assumptions in using this approach with Lindhard stopping theory for energy loss calculations are:

- The electron density in the target varies slowly with position.
- Available electron energy levels and transition strengths of the atoms of the solid are described by those in a free electron gas.

The basic physics of the local density approximation applied to stopping theory may be seen in plots of the integrand of the stopping cross-section equation (3-1), as evaluated for atomic targets, see figures (3-3) to (3-5). These plots show three curves which link the various parts of the stopping process. Figure (3-3) shows various shapes of the charge density of Cu with the Thomas-Fermi atom shown as a dotted line, the isolated atom Hartree-Fock atom shown as a dashed line and the solid-state Cu atom shown as a solid line. (see section 2 for a review of how these various atomic models were developed). The plots use for an ordinate the factor $4\pi r^2 \rho$ where ρ is the electron density. With this factor the area under the curve equals 29, the atomic number of Cu. Clearly the Thomas-Fermi atom has no shell-structure but is a reasonable average value. The isolated atom Hartree-Fock curve shows pronounced shell structure but it has a long tail extending out many Angstroms since it is not contained. Finally, the solid-state structure is contained within 1.4 Å, with the electrons from 1.2-1.4 Å being averaged over the various bonding angles in the Cu face-centered-cubic crystal.

Figure (3-4) shows as a dotted line the same solid-state Cu distribution as shown in (3-3). It shows as a dashed line the value of the particle-plasma interaction term, I, of equations (3-2) to (3-3) and of figure (3-1). For each radius the density of the Cu atom is taken and the equivalent I is calculated as indicated in figure (3-1). Since figure (3-4) specifies that the ion has a velocity of 10000 keV/amu (about 20 times the Bohr velocity) it is moving much faster than most of the electrons in the solid, and so its interaction is relatively independent of the electron velocity. In figure (3-1) the density of electrons in Cu is mostly on the flat section of the curve labeled 10000 keV.

Finally, the term ρI is plotted as a solid line in figure (3-4). The area under this curve is the integrand of eq. (3-1) and hence is the stopping cross-section of a moving particle in Cu. This solid line shows how the energy loss is distributed among the various Cu electrons with all but the innermost electrons absorbing energy about proportional to their density.

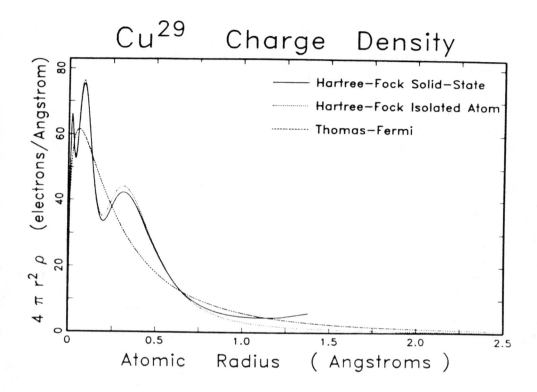

Figure (3-3)   The description of atoms in solids has progressed from a Thomas-Fermi description (1932-1955) to Hartree-Fock isolated atoms (1957-1974) to complete Hartree-Fock descriptions of atom in solids (1976-1982). The Thomas-Fermi atom had no shell structure, but it allowed analytic solutions to the complex problems of stopping and range theory, see for example the LSS theory (62a). The use of Hartree-Fock atoms which are not simply expressed has led to full numeric treatment of the interaction of particles with matter.

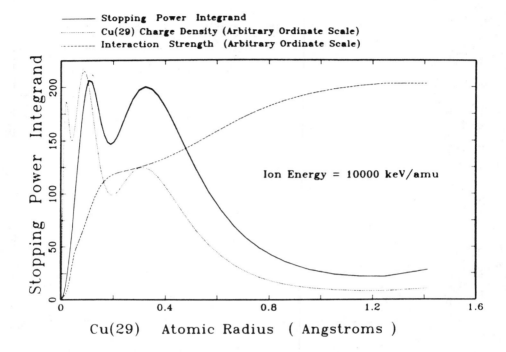

Figure (3-4) The nature of the interaction of a high speed particle with a solid is illustrated by considering a target of copper (Cu). The dotted line is the charge distribution of electrons, ρ in a Cu atom, and is identical to the solid-state curve of figure (3-3). The dashed curve is the interaction strength, I, of a high velocity particle (10^4 keV/amu) with the charge distribution. It is small only for the inner k-shell electrons. The product of I_ρ is the stopping power, see equation (3-5). This product is shown a solid line. It shows that the energy loss is evenly distributed across the Cu electrons except for the inner shell.

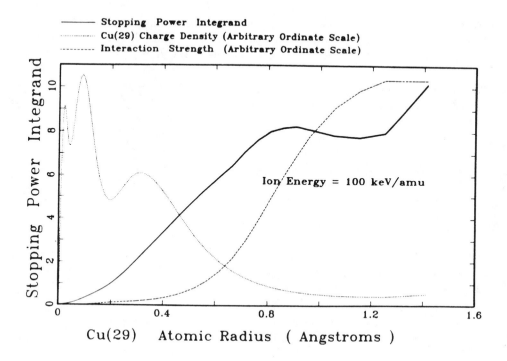

Figure (3-5)   The interaction of a slow particle with copper. The dotted line is the electronic charge distribution of a Cu atom in a solid. The dashed curve is the interaction strength of the slow particle with this distribution. There is almost no interaction with the core electrons, and the interaction is almost exclusively with the conduction electrons. The solid line shows the stopping integrand, equation (3-5) and shows that the electronic energy transferred to the copper atom is mostly to the outer electrons. Since there is little energy transferred to either the K or L shells, it indicates there will be few x-rays produced during deexcitation of the atom.

In contrast, figure (3-5) shows the same curves except they are evaluated for a low velocity particle of 100 keV/amu (about twice the Bohr velocity). The copper density curve (dotted line) is identical to that of figure (3-3). The dashed interaction curve is quite different from that of figure (3-4) for the particle is now at a velocity which is large only when compared to the low density outermost electrons of Cu atoms. For the inner electrons the interaction is almost adiabatic and there is little excitation. The solid line, ρI, is the stopping power and it is clear that the outer shell electrons, about 20% of the total electrons, absorb almost 90% of the energy loss. The inner k-shell electrons absorb almost no energy and one would anticipate that there would be almost no k-shell x-rays emitted from the Cu target.

The first attempt to evaluate electronic stopping cross-sections for protons in solids using the Lindhard stopping formalism and the local density approximation was by Bonderup who used Lenz-Jensen atoms to represent the atoms in the solid (67a). This work was extended to isolated Hartree-Fock atoms by Rousseau et al. (70b) and to actual solid-state charge distributions by Ziegler (78a).

The Electronic Stopping Cross-Sections of Heavy Ions

We have completed the discussion of the electronic energy loss of the very light ions H and He. The energy loss of heavy ions to the electrons of a solid will be presented in three steps: (a) the stopping of very low velocity ions ($v_1 < v_F$) where we shall show that the energy loss is generally proportional to the ion's velocity, (b) the stopping of high velocity ions ($v_1 > 3v_F$) where we can scale proton stopping powers to equivalent heavy ion stopping using Thomas-Fermi scaling rules, and (c) the use of more complex theory to bridge the low and high velocity regions ($v_F < v_1 < 3v_F$). Since the Fermi velocity, v_F, of solids is about equal to the Bohr velocity, v_o, the three velocity intervals will be approximately (a) < 25 keV/amu, (b) > 200 keV/amu and (c) 25-200 keV/amu.

Electronic Stopping of Low Velocity Heavy Ions

We are concerned here with heavy ions with ion velocities, v_1, less than the Fermi velocity, v_F, of the material they are traversing. Since the Fermi velocities of solids usually falls between 0.7 to 1.3 v_o, this means the ion velocity will be less than 30 keV/amu. Here the majority of target electrons move much faster than the ion and collisions with the ion are mostly adiabatic without direct energy loss to collisions.

Firsov proposed a semi-classical model to evaluate this low velocity energy loss (59a). He considered two isolated atoms colliding with their undistorted electronic spheres penetrating each other. The excitation energy of the atomic shells was assumed to be distributed among all the electrons. The electronic energy loss of the ion goes to the final ionization of the target atom at the end of the collision or soon after it. Firsov estimated the effective cross-sections of various inelastic processes as a function of the relative collisional velocity of the two atoms and the impact parameter of the collision. This was done by considering an arbitrary plane located between the two nuclei, and the capture and loss of electrons was defined by the flux of electrons through this plane. Since the two atoms have a relative velocity, v, any captured electron must be accelerated up to this velocity and the ion loses a small energy loss proportional to this velocity. Firsov used Thomas-Fermi atoms to obtain electron distributions of the atoms and found that the electronic stopping cross-section would increase proportional to the heavy ion velocity.

Lindhard and Scharff also considered the problem of low velocity stopping, but used a model of a slow heavy ion in a uniform electron gas (61b,63a). Relative to the ion the electrons have a slight drift velocity. During a collision with the ion a net energy is transferred which is proportional to this velocity. Hence, although Lindhard considered a more dynamic interaction between the target electrons and the ion, a velocity proportional energy loss was found which was similar to the geometric model of Firsov.

These theoretical models have been the basis of many later calculations which have introduced more accurate Hartree atoms into the interactions (see, for example, 68e, 76d, 77m and 80g), but the preponderance of this work still yields velocity proportional stopping cross-sections.

Experimentally, there is general agreement with the prediction that the stopping cross-sections of low velocity heavy ions is velocity proportional. Extensive studies are presented in ref. 65b, 66b, 68d, 72i and 80a. There are two residual problems. First, although the stopping may be proportional to velocity, if a straight line is drawn through the data it may not go through the origin.

A second problem to velocity proportional energy loss theory is that for a few elemental targets there is clearly a different low velocity energy loss dependence. One example is for the semiconductors silicon and germanium. For metal targets the low velocity energy loss is primarily caused by the conduction band electrons which may be accurately described by a free electron gas. But the band-gap in the semiconductors clearly makes these targets different from metals in their theoretical electromagnetic response (60b,70g).

One would expect anomalously lower energy loss to materials with a band gap since there are fewer low-energy excitation levels available. Extensive analysis of experimental low velocity stopping powers in Si and Ge leads to the conclusion that stopping goes as $S_e \propto v^{0.7}$ in these narrow band-gap materials for ions with atomic number less than 19. For heavier ions the data is too sparse to draw any conclusion.

The Electronic Stopping of High Velocity Heavy Ions

One broad empirical rule for the calculation of fast heavy ions in solids is to relate such stopping to the equivalent proton stopping powers. This is called the heavy-ion scaling rule and has the form:

$$\frac{S_{H}(v_1,Z_2)}{S_{H}(v_2,Z_2)} = \frac{S_{III}(v_1,Z_2)}{S_{III}(v_2,Z_2)}. \tag{3-12}$$

This rule states that the electronic stopping of heavy ions, S_{III}, at two velocities is directly proportional to the stopping of protons, S_{H}, at the same velocities and in the same material. S_{H}, in this relationship, is the stopping power per unit ion charge.

The previous discussion about He to H stopping which led to equation (3-8) indicates this relationship between heavy ion stopping and proton stopping can be made simpler:

$$S_{III} = S_{H}(Z_{III}^{\bullet})^2 = S_{H}Z_{III}^2\gamma^2 \tag{3-13}$$

where Z_{III} is the atomic number of the heavy ion and γ is its fractional effective charge. This effective charge term can be estimated from Thomas-Fermi atomic theory, which may be applicable in the region where Thomas-Fermi atoms approximate Hartree-Fock atoms, i.e. where $0.3 \le \gamma \le 0.8$.

The basic scaling relationships in a Thomas-Fermi atom are:

$$\text{Charge density} \equiv \rho \propto Z^2$$

$$\text{Electron binding energy} \equiv E_b \propto Z^{7/3}$$

$$\text{Binding energy per electron} \equiv e_b \propto Z^{4/3} \tag{3-14}$$

$$\text{Electron velocity} \equiv v_e \propto Z^{2/3}$$

where Z is the nuclear charge of the Thomas-Fermi atom.

As discussed in the historical summary of this chapter, scaling laws for heavy ion stopping powers received considerable attention in 1938-41 because of interest in nuclear fission experiments. Lamb suggested that the electron binding energy would be the primary parameter in determining γ (40b), while Bohr suggested that the electron velocity would be critical (40b,41a). Later evidence supported the Bohr view that one might consider the heavy ion to be stripped of all electrons whose classical orbital *velocities* are less than the ion velocity. This Bohr concept was set in explicit form by Northcliffe (60c) as:

$$\gamma^2 = 1 - \exp\left[-v_1/(v_o Z_1^{2/3})\right] \tag{3-15}$$

where Z_1 is the ion atomic number and v_1 is its velocity which is compared to the Bohr velocity, v_o. The $Z_1^{2/3}$ of equation (3-15) comes from the Thomas-Fermi atom relationship shown in equation (3-14), for the velocities of electrons in heavy ions. Equation (3-15) expands in first order to the Bohr relation: $\gamma = v_1/(v_o Z_1^{2/3})$, and it should be valid over the region where the Thomas-Fermi atom approximates heavy ions; $0.3 \le \gamma \le 0.8$. This relationship has been supported by dozens of authors, see for example 65a, 66b, 68d, 72i and 77h. A typical parameterization of this Bohr/Northcliffe relationship is found in 77h:

$$\gamma^2 = 1 - \exp\left[-.92\, v_1/v_o Z_1^{2/3}\right].$$
(3-16)

The accuracy of this is shown by plotting it versus experimental reduced heavy ion stopping powers defined using equation (3-13):

$$\gamma^2 = \frac{S_{1I}}{Z_{1I}^2 S_{1I}}$$
(3-17)

where S_{1I} is the equivalent proton stopping power. An example is shown in figure (3-6) with over 1000 data points from 127 heavy ion/target combinations. The accuracy of the fit is about 10%, which is as accurate as the experiments themselves.

The simple Thomas-Fermi picture is valid for a great range of heavy ion stopping. Reference (80a) analyzed this formula (3-16) in detail and found that the stopping proportionality was valid for $3 \le (v_1/v_o) \le 100$. Below a relative velocity of 3 the Thomas-Fermi atom no longer represents the almost neutral heavy ion, and the simple physics of the Bohr's stripping model becomes inadequate as discussed in the next section. For very high velocities relativistic effects become important.

The Electronic Stopping Cross-Sections of Medium Velocity Heavy Ions

For the energy loss of medium velocity heavy ions in an electronic plasma we have constructed a model based on the ideas of Kreussler, Varelas and Brandt (81c) and Brandt and Kitagawa (82a). The physical assumptions which differ from our previous discussions of particle and proton electronic stopping are:

(a)   The charge state of the ion is determined approximately by assuming all electrons are stripped whose velocities are less than the relative velocity of the ion to the Fermi velocity of the solid.

(b)   For distant collisions (i.e. with electrons outside the ion's diameter) standard stopping theory holds with the ion having an effective charge determined by (a) above.

(c)   For close collisions (i.e. with electrons which penetrate the ion's electronic shells) the energy loss increases because there is less shielding of the nucleus. A comprehensive treatment is made of the reduced shielding and the increased energy loss with first order perturbation theory.

(d)   The effective charge of an ion is based on its charge state in a solid, assumption (a) above, with an additional term to account for the increased interaction of close collisions, assumptions (c) above.

We shall briefly review the physics of this approach without detailing the mathematical derivations which are to be found in (81c) and (82a).

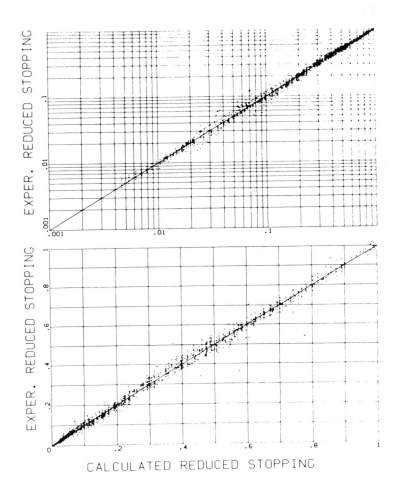

Figure (3-6) Experimental verification of Thomas-Fermi scaling of electronic stopping powers. The ordinate of each point is an experimental stopping power reduced to an effective charge by dividing by the equivalent proton stopping power and Z_1^2, see equation (3-17). The abscissa is the Thomas-Fermi charge state of the ion which assumes all electrons are stripped which have a velocity less than the ion velocity. In Thomas-Fermi atoms the electron velocity is proportional to $V_o Z_1^{-2/3}$. This excellent agreement between theory and experiment is only valid for ions with velocities $v_1 > 3v_o$.

The first major new concept is to calculate the stripping of the heavy ion not by comparing the ion's electron velocities to the ion velocity as suggested by Bohr, but to compare them to the relative velocity between the ion and the electronic velocity of the medium. For ion velocities, v_1, much greater than the electron velocities of the medium the two stripping concepts are similar and hence the previously discussed scaling laws (3-12) will hold.

For lower velocities, $1 \leq (v_1/v_0) \leq 5$, the ion velocity approaches the Fermi velocity of a solid, v_F, and the target inner-shell electrons can no longer be excited and the energy loss is mostly to the conduction electrons. In the local density approximation (discussed at the beginning of this section) this is illustrated in figures (3-4) and (3-5).

The relative velocity, v_r, of the ion to the velocity, v_e, of the valence electrons in a solid is defined as

$$v_r \equiv < | \vec{v}_1 - \vec{v}_e | > \qquad (3-18)$$

$$= \frac{v_e}{6} \frac{(v_1/v_e + 1)^3 - | v_1/v_e - 1 |^3}{v_1/v_e} \qquad (3-19)$$

If we assume the conduction electrons are a free electron gas, then their mean kinetic energy is:

$$\frac{1}{2} m v_e^2 = \frac{3}{5} E_F$$

and we can substitute into (3-19)

$$v_e = \left(\frac{3}{5} \right)^{1/2} v_F.$$

Obtaining for $\qquad v_1 \geq v_F$:

$$v_r = v_1 \left(1 + \frac{v_F^2}{5 v_1^2} \right) \qquad (3-20)$$

and for $\qquad v_1 \leq v_F$:

$$v_r = \frac{3}{4} v_F \left(1 + \frac{2 v_1^2}{3 v_F^2} - \frac{v_1^4}{15 v_F^4} \right). \qquad (3-21)$$

The final degree of ionization of the ion now depends only on the ion velocity, v_1 and the target Fermi velocity, v_F. The explicit relation will be discussed later.

Once the charge state of the ion is known it is necessary to establish its electronic charge distribution, and hence its diameter, based on its degree of ionization. This diameter will be used to separate the close and distant collisions of assumptions (b) and (c) above.

The next step is to calculate the electronic stopping using effective charge theory by adding together the energy loss in distant collisions, which consists of target electrons that see a charge qZ_1, and the energy loss to the electrons of the medium which penetrate the ion's diameter and feel increased nuclear interaction. Using perturbation theory, for *heavy ions* the BK theory produces the simple expression for the effective charge of an ion, γ:

$$\gamma = q + C(1 - q) \ln \left[1 + (2\Lambda v_F/v_o)^2 \right] \qquad (3 - 30)$$

where q is the fractional ionization, $q = 1-N/Z_1$, and N is the number of electrons still bound to the ion of atomic number Z_1. The only undefined quantity is C which is about $1/2$ and weakly dependent on the target. We have found from extensive data analysis of the stopping cross-sections of the same ion in many targets that the best approximation is $C \approx (v_o/v_F)^2/2$, and so the effective charge of the ion is:

$$\gamma = q + (1 - q)\frac{(v_o/v_F)^2}{2} \ln \left[1 + (2\Lambda v_F/v_o)^2 \right] \qquad (3-31)$$

The physical meaning of this equation is that q is the degree of ionization of the ion, and γ is its effective charge. The first term, q, is the effective charge for distant collisions, i.e. for target electrons which do not penetrate the ion's electronic volume. For all degrees of ionization some electrons of the target will penetrate the ion's electronic cloud and *increase* the energy loss and this is the second term in equation (3-31). The sum of these two is the effective charge, γ.

The final knot to tie in evaluating the electronic stopping of heavy ions is how to evaluate the ion's degree of ionization, q, which will depend on the ion's velocity and the target material. If we argue that we can use the Bohr stripping criteria (based on Thomas-Fermi atoms) and upgrade it to strip all electrons whose *relative* velocity, v_r/v_o, is less than one then:

$$q \sim \frac{v_r}{v_o Z_1^{2/3}}, \qquad (3-34)$$

Brandt felt that this simple expression could be refined in a manner similar to his earlier work (75i) evaluating the Bohr stripping relation in which he numerically evaluated the Thomas-Fermi electronic distribution in order to predict more accurately the effective charge of an ion. This previous work still based ion stripping on the ion electron velocities relative to the absolute velocity of the ion and did not consider any dependence on the electron velocities in the target. Brandt later suggested his comments on the Bohr stripping criterion could be extended to relative velocities (81d), but he did not derive a new expression before his sudden death. We use in our treatment the expressions for Bohr stripping as revised by Northcliffe (3-16), and have revised the definition of the relative velocity from v_1/v_o to v_r/v_o:

$$q = 1 - \exp\left(\frac{-cv_r}{v_oZ_1^{2/3}}\right). \qquad (3-35)$$

where "c" is a constant and v_r is calculated using (3-32) or (3-33). The constant should be about unity, and a first estimate would be to use directly the value found previously for high velocity heavy ions, c = 0.92, as discussed in equation (3-16). This would make the ionization level of heavy ions to be :

$$q = 1 - \exp\left(-\frac{.92v_r}{v_oZ_1^{2/3}}\right) \qquad (3-36)$$

The accuracy of the above approach is shown in figure (3-7) for several thousand heavy ion stopping powers. The two plots show heavy ion stopping power measurements, S_{HI}, reduced to obtain the corresponding effective charge by using equation (3-17): $\gamma^2 = S_{HI}/(Z_1^2 S_{H})$, where Z_1 is the ion atomic number and S_H is the equivalent proton stopping power. We then can use equation (3-31) to solve for the ionization fraction, q, from the effective charge, γ. The abscissa is what might be called the "effective ion velocity" relative to the target electron velocity. The solid curve is the universal ionization fraction of ions, q, equation (3-36).

The fit between the theoretical and experimental values is remarkably good, showing a data spread not much different from experimental error. The only problem occurs for very low ionization fractions, q < 0.5, where the data clearly does not follow the curve to the origin. This low value of ionization is where the ion is moving slowly through the solid, slower than some of the target electrons, and it is almost neutral. The problem probably lies in the approximations of BK theory, especially the application of perturbation theory to a situation to which it is clearly inappropriate. The outer shell of the ion contains the slowest electrons, and the stripping of these must be done using a more detailed approach. But a significant point is that the data of the slowest ions in figure (3-7) do not disperse. They stay grouped about a tight common pattern which intersects the abscissa at about 0.1 effective ion velocity.

Summary of Electronic Stopping Cross-sections

(1)   For H ions we directly use a fitted function to obtain H stopping cross-sections in each element.

(2)   For He ions we multiply the equivalent H stopping by the He effective charge at that velocity, $S_{He} = S_H (Z_{He}\gamma_{He})^2$.

(3)   For heavy ions we scale proton stopping powers using Brandt-Kitagawa

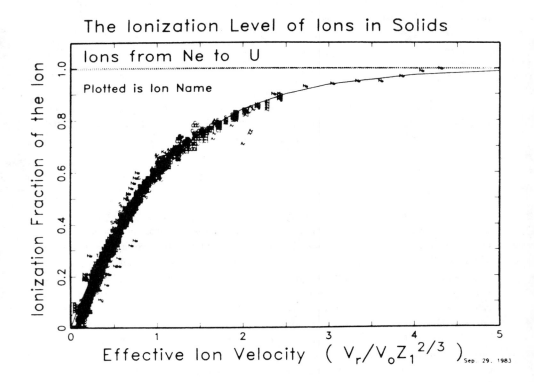

Figure (3-7)   The ionization level of heavy ions in solids, evaluated using the Brandt and Kitagawa (BK) formalism. The abscissa is called the ion's *effective velocity* and it combines the Thomas-Fermi stripping criteria, figure (3-6), with Brandt's proposal that electrons are stripped whose velocity is less than the relative velocity, v_r, between the ion and the solids Fermi velocity. The ordinate is the fraction of charge remaining on the ion as determined by reducing experimental data to an effective charge using equation (3-17) and then solving for the ionization level using equation (3-31). The solid line is the Northcliffe-Bohr estimate, equation (3-36). The agreement is remarkably good, deviating only at the lowest effective velocities, and where the ion is almost neutral.

theory.

The calculation of heavy ion $(Z_1 > 2)$ electronic stopping takes the following steps:

(a) The relative velocity of the ion, v_r, is calculated with equations (3-32) or (3-33) which depend only on the ion velocity, v_1, and the target Fermi velocity, v_F

(b) The fractional ionization of the ion is calculated with equation (3-39).

(c) The screening length of the ion, Λ, is calculated as a function of the ions charge state.

(d) The effective charge, γ of the ion is then calculated with equation (3-31).

(e) The final stopping power is then found using equation (3-13): $S_e = S_H (Z_1\gamma)^2$, where S_H is the proton equivalent stopping power.

(4) For very low velocity ions, $v_1 \sim v_F / Z_1^{2/3}$, we use velocity proportional stopping.

The major assumptions of our approach to calculating electronic stopping powers which were discussed in this section are:

- The electron density in the target varies slowly with position.

- Available electron energy levels and transition strengths of the atoms of the solid are described by those in a free electron gas.

- Target band-gap effects can be reflected in the target's Fermi velocity.

- The degree of ionization of the ion depends only on the relative velocity of the ion to the Fermi velocity of the target.

- For distant collisions the electronic energy loss is described by Lindhard's free electron gas theory incorporated into a local-density-approximation for the particle-solid interaction.

- For close collisions the electronic energy loss for heavy ion is corrected by Brandt and Kitagawa theory.

---

4 - TRIM CALCULATIONS OF RANGE DISTRIBUTIONS

Section Summary

The formalism for a Monte Carlo computer program which simulates slowing down and scattering of energetic ions in amorphous targets has been described in detail by Biersack and Haggmark (80d), and the program is shown with extensive comments in reference (84a). This program is called TRIM for the expression: Transport of Ions in Matter. It was developed for determining ion range and damage distributions as well as angular and energy distributions of backscattered and transmitted ions. The computer program provides particularly high computer efficiency, while still maintaining a high degree of accuracy. This is achieved mainly by applying an analytic formula for determining nuclear scattering angles, and by suitably expanding the distance between collision at high energies. This Monte Carlo program is used to calculate the range distributions of a variety of ion/target combinations, and they are shown with precise experimental profiles.

Introduction

The computer simulation of the slowing down and scattering of energetic ions in materials has been used recently in studies of ion implantation, radiation damage, sputtering, and the reflection and transmission of ions. The Monte Carlo method as applied in simulation techniques has a number of distinct advantages over present analytical formulations based on transport theory. It allows more rigorous treatment of elastic scattering, explicit consideration of surfaces and interfaces, and easy determination of energy and angular distributions. The major limitation of this method is that it is inherently a computer-time consuming procedure. Thus there is often a conflict between available computer time and desired statistical precision. In the Monte Carlo computer program presented here, we attempt to alleviate this problem by using techniques which reduce computer usage by at least an order of magnitude and at the same time sacrifice little accuracy.

Several ion transport procedures based on the Monte Carlo method have been reported, see for example, Ref. (63e,74e,72j). Aside from considering crystalline or amorphous targets, their major differences lie in their treatment of elastic or nuclear scattering. Only Oen, Robinson, and co-workers treat this scattering in a precise manner by numerically evaluating the classical scattering integral for realistic interatomic potentials. Other authors base their formalisms on either the momentum approximation extended to large angles or fitted, truncated Coulomb potentials to obtain analytical representations of the scattering integral. Since energetic ions undergo many collisions in the process of slowing down, the method used to evaluate the scattering integral is of critical importance in terms of its relative computer efficiency. Therefore, we have made use of a new analytical scheme which very accurately reproduces scattering integral results for realistic potentials.

As with other simulation programs, our method consists of following a large number of individual ion or particle "histories" in a target. Each history begins with a given energy, position, and direction. The particle is assumed to change direction as a result of binary nuclear collisions and move in straight free-flight-paths between collisions. The energy is reduced as a result of nuclear and electronic (inelastic) energy losses, and a history is terminated either when the energy drops below a pre-specified value or when the particle's position is outside the target. The target is considered amorphous with atoms at random locations, and thus the directional properties of the crystal lattice are ignored. This method is applicable to a wide range of incident energies-approximately 0.1 keV to several MeV, depending on the masses involved. The lower limit is due to the inclusion of binary collisions only, while the upper limit results from the neglect of relativistic effects. Also, nuclear reactions are not included. The efficiency for dealing with high energy particles has been increased by introducing an energy dependent free-flight-path between collisions is longer at high energies and is steadily reduced in the course of slowing down.

The nuclear and electronic energy losses or stopping powers are assumed to be independent. Thus, particles lose energy in discrete amounts in nuclear collisions and lose energy continuously from electronic interactions. For low energies, where nuclear scattering and energy loss is particularly important, the program utilizes the above mentioned analytic scheme based on solid-state interatomic potential as described in Section 2. The electronic energy loss has been described in Section 3.

For the sake of computer efficiency, effects of minor influence on range distributions, such as the "time integral" or electronic straggling, which are usually neglected in analytic and Monte Carlo formalisms, have been carefully tested to check whether they can be approximated or neglected. In most cases for energies below 1 MeV, the electronic straggling was found to be of little importance for the projected range profiles. Therefore, in the present program, it is normally neglected but is available as an option in the form of an impact parameter dependent electronic energy loss. The time integral τ was found to be of little influence in all cases except for the very lowest energies, i.e., below 1 keV. It is therefore included only in an approximate manner with best accuracy at low energies.

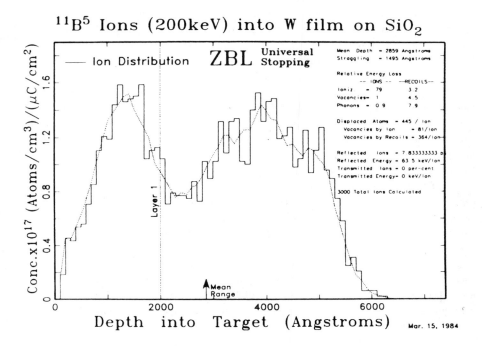

Figure (4-1) An example of a Monte-Carlo calculation of the ion penetration of a solid. The figures of section 4 are various plots of the same calculation showing different information. The ions are boron with atomic mass 11, at 200 keV, which are incident on a target consisting of a layer of tungsten (W) 2000 Angstroms thick, on a thick substrate of SiO_2. This unusual combination is chosen to illustrate some special ion implantation phenomena. The above plot shows the depth distribution of the boron ions into the target. The profile is in the form of a histogram made with solid lines. The dashed line is a fit through the profile and it will be put on other plots so the various effects can be related to the implantation profile. The units of the ordinate have been chosen to easily çonvert from the peak concentration to the ion implantation units of Coulombs/cm². For example, if the boron peak in the W layer is desired to be at a concentration of 10^{20}, then the implantation dose required is 10^{20} /1.6x 10^{17} = 600 micro-Coulombs / cm². The data table on the right part of the plot shows some of the details of the calculation such as the mean depth and the straggling. Also shown is how the ion energy is distributed into ionization, vacancy production and phonon production. Below this is the the number of vacancies produced by each incident ion, and how many of these are direct vacancies caused by an ion-atom collision and how many result from the collisional cascade produced in the solid. Finally, it is noted that over 7% of the incident ions are reflected back out of the target.

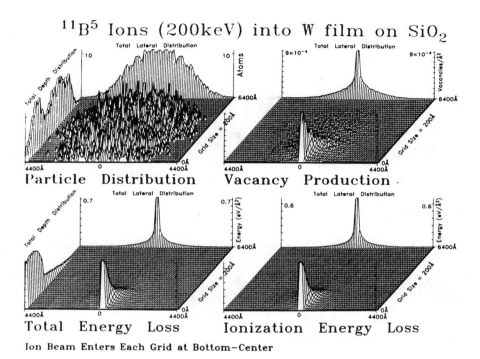

Figure (4-2) This plot shows a 2-dimensional view of the ion penetration of the solid. In each of these plots the ion is assumed to be incident into the bottom-center of the plot and going into the page. The upper left hand plot of this group shows the final distribution of boron ions. Along the left border of this plot is the summation of the distribution and hence is the same curve as shown in Figure (4-1). Along the back border is the ion distribution summed along the other axis and this gives the summed lateral distribution of the ions in the solid. Note that the particle distribution extends to the surface (depth = 0) and this also indicates that there probably is some back-scatter from the surface. The upper right plot shows the vacancy production in the target, with a sharp peak where the beam enters the target, and this then spreads out as the ions scatter transversely. Along the back of the plot is the lateral distribution of vacancies. The lower right plot shows the distribution of energy going into ionization of target atoms. This plot is very similar to the vacancy distribution, but later plots will show great differences in detail. The lower left plot is the distribution of total energy loss of the ions into the target.

$^{11}B^5$ Ions (200keV) into W film on SiO_2

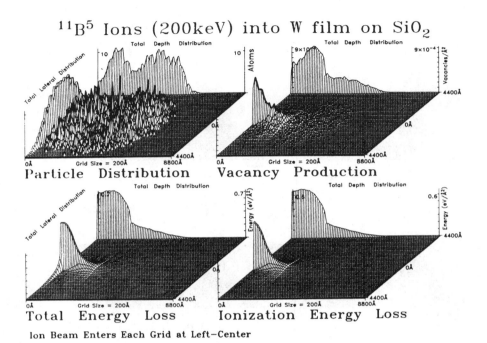

Ion Beam Enters Each Grid at Left–Center

Figure (4-3) This plot shows a 2-dimensional view of the ion penetration of the solid. It is identical to the plot of figure (4-2) only the plots have been turned ninety degrees and the beam here enters from the left-center of each grid. The interesting new plots here are the summed distributions along the back of each grid. These are the summations which show what is happening as a function of depth. The upper left plot shows a very unusual 2-peaked distribution from this single implantation. This result is from the ions forming a normal peak in the W surface layer, and the ones which get through this layer suddenly find the oxide layer to have a completely different energy loss mechanism and a totally independent peak is formed here. This is verified by the three energy loss distributions in the other plots. In each of them there is a very great energy loss in the surface metal layer, then an abrupt drop to a much lower level when the ions are moving in the oxide.

Lateral Distribution from a Mask Edge

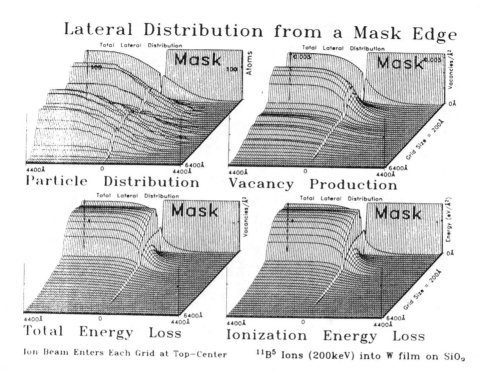

Particle Distribution   Vacancy Production

Total Energy Loss   Ionization Energy Loss

Ion Beam Enters Each Grid at Top-Center $^{11}B^5$ Ions (200keV) into W film on SiO_2

Figure (4-4) This plot shows a 2-dimensional view of what happens if a mask covers part of the solid's surface. In these views the beam is entering the surface of the solid at the far side of each plot, and hence the ions end at the near side of the plots. The mask is assumed to cover half of the surface as shown. These plots can then be used to see to what degree the mask changes the implant distribution. Along the back edge of each plot is again a summed distribution showing the total lateral effects. In the upper left plot of the particle distribution you can see that right at the mask edge the total ion concentration is exactly one-half that of the bulk. This is a maxim of ion implantation that just below a mask edge the number of ions is exactly one half the number found in unmasked regions. Similarly the vacancy production and the energy loss is one-half the normal values under a mask edge. You can follow the line of the mask edge down into the solid because we have placed a discontinuity in the plot at this point. This is not real, only something to add legibility.

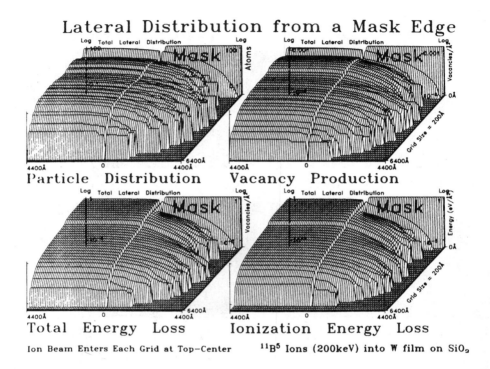

Lateral Distribution from a Mask Edge

Particle Distribution   Vacancy Production

Total Energy Loss   Ionization Energy Loss

Ion Beam Enters Each Grid at Top-Center $^{11}B^5$ Ions (200keV) into W film on SiO_2

Figure (4-5)   This plot shows a 2-dimensional view of what happens if a mask covers part of the solid's surface. It is identical to figure (4-4) except that the vertical scale is logarithmic. Although most of the ions fail to scatter far laterally, in this logarithmic view it can be seen that a few ions scatter laterally almost as far as they go longitudinally. For applications such as doping semiconductors or exposing photoresist there is sensitivity to 1% effects, and these plots are more useful than the linear plots shown previously.

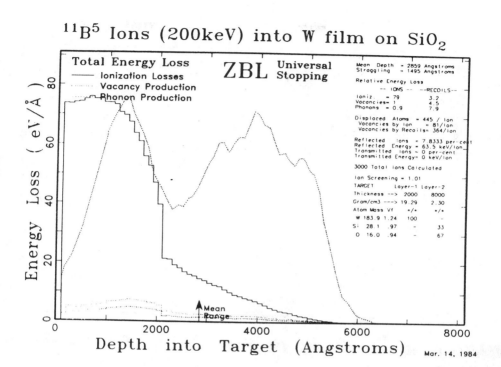

$^{11}B^5$ Ions (200keV) into W film on SiO$_2$

Figure (4-6)   This plot shows how the initial ion energy is distributed among various types of energy loss. About 80% of the energy loss goes into electronic excitation of the target atoms, shown as a solid line. The thin dotted curve with a peak in the middle of the plot is the final ion distribution and it is the same as in figure (4-1). The ionization energy loss shows an abrupt discontinuity where the target changes from tungsten to oxide, indicating that the electronic coupling between the ion and the two types of solids is very different. The energy loss into target vacancy production and into phonon production are each of the order of 10% of the electronic energy loss.

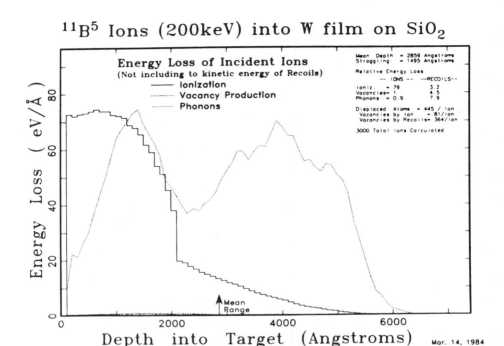

$^{11}B^5$ Ions (200keV) into W film on SiO_2

Figure (4-7) This plot shows how the incident ions lose their energy. It does not include the energy transferred to target atoms which will be involved in collision cascades (these are in figure 4-8). About 96% of the direct deposition of energy into the target by the ions is by electronic excitation. The amount of energy into phonons and vacancy production is minimal. This is because boron has a small nuclear charge, and there is little nuclear stopping.

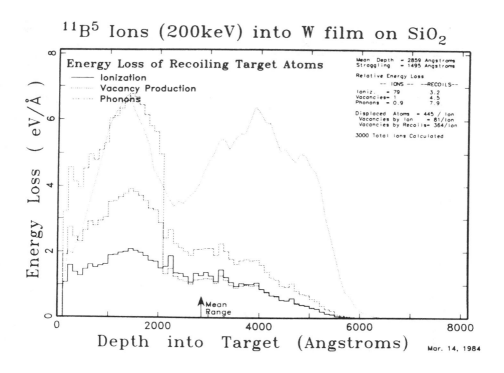

Figure (4-8) This plot shows the energy loss in the target due to collisional cascades. The ion has a hard collision with a target atom and transfers a significant amount of energy to this atom. Since this atom has a more charged nucleus than the ion and is going much slower, it has a high probability of colliding with another target atom, a process which is so likely that it is called a collisional cascade. As shown in the figure the energy loss by ionization is the least important for these recoils. The largest energy loss is to phonons in the metal layer, and to vacancy production in the oxide substrate. This difference occurs because the amount of energy a boron atom can transfer to a heavy tungsten atom is very small relative to that to a Si or O atom. A slow moving tungsten atom will probably collide with other target atoms and shake up the lattice with phonons but it may not have enough energy to displace the lattice atoms from their sites.

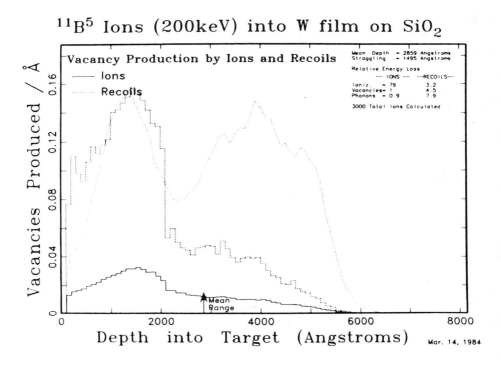

Figure (4-9)   This plot shows on the same scale how the vacancies in the target are produced. The lower solid line are the number vacancies produced directly by the incident ions. The upper dotted curve shows the number produced by the recoiling target atoms. It can be seen that on average each target atom knocked out of its site will produce about 4 additional vacancies. This ratio only slightly changes between the metal and the oxide layers.

TEXT REFERENCES

00a Mme. Pierre Curie, Comptes Rendus 130, 76 (1900).
03a J. J. Thomson, "Conduction of Electricity Through Gases," Cambridge University Press, (1903).
09a H. Geiger and E. Marsden, Proc. Roy. Soc. 82, 495 (1909).
11a E. Rutherford, Phil. Mag. 21, 212 (1911); ibid 21, 699 (1911).
12a J. J. Thompson, Phil. Mag. 6-23, 449 (1912).
12c C. G. Darwin, Phil. Mag. 23, 901 (1912).
13a N. Bohr, Phil. Mag. 25, 10 (1913).
15a N. Bohr, Phil. Mag. 30, 581 (1915).
30a H. A. Bethe, Ann. Physik 5, 325 (1930).
31a N. F. Mott, Proc. Camb. Phil. Soc. 27, 553 (1931).
32a H. A. Bethe, Z.f. Physik 76, 293 (1932).
32b A. Sommerfeld, Z.f. Physik 78, 283 (1932).
32c W. Lenz, Z. F. Physik 77, 713 (1932); H. Jensen, Z. f. Physik 77, 722 (1932).
32d C. Moller, Ann. Physik 14, 531 (1932).
33a F. Bloch, Ann. Physik 16, 287 (1933).
33b F. Bloch, Z.f. Physik 81, 363 (1933).
34a H. A. Bethe and W. Heitler, Proc. Roy. Soc. A146, 83 (1934).
38a W. F. G. Swann, J. Frank. Inst., 226, 598 (1938).
40b W. E. Lamb, Phys. Rev. 58, 696 (1940).
40c E. Fermi, Phys. Rev. 57, 485 (1940).
40d N. Bohr, Phys. Rev. 58, 654 (1940).
41a N. Bohr, Phys. Rev. 59, 270 (1941).
41b J. Knipp and E. Teller, Phys. Rev. 59, 659 (1941).
47a G. Moliere, Z. Naturforschung A2, 133 (1947).
47b E. Fermi and E. Teller, Phys. Rev. 72, 399 (1947).
48a N. Bohr, Mat. Fys. Medd. Dan. Vid. Selsk. 18, No. 8 (1948).
48b A. Bohr, Mat. Fys. Medd. Dan. Vid. Selsk 24, No. 19 (1948).
49a P. Gombas, "Die Statistische Theorie des Atoms und ihre Anwendungen," Springer-Verlag, Austria (1949).
53b J. Lindhard and M. Scharff, Mat. Fys. Medd. Kgl. Dan. Vid. Selsk. 27, No 15 (1953).
54a J. Lindhard, Mat. Fys. Medd. Dan. Vid. Selsk., 28, No. 8 (1954).
55a J. Neufeld and R. H. Ritchie, Phys. Rev. 98, 1632 (1955).
56c U. Fano, Phys. Rev. 103, 1202 (1956).
57a O. B. Firsov, Zh. Eksp. Teor. Fiz. 32, 1464 (1957).
57b O. B. Firsov, Zh. Eksp. Teor. Fiz., 33, 696 (1957).
57c N. Bichsel, R. F. Mozley and W. A. Aron, Phys. Rev. 105, 1788 (1957).
58a O. B. Firsov, Zh. Eksp. Teor. Fiz. 34, 447 (1958).
58b O. B. Firsov, JETP, Vol. 7, 308 (1958).
58c W. Whaling, "Handbuch der Physik, Bd. XXXIV, 13, Springer-Verlag, Berlin
60b R. M. Sternheimer, Phys. Rev. 117, 485 (1960).
60c L. C. Northcliffe, Phys. Rev. 120, 1744 (1960).
61b J. Lindhard and M. Scharff, Phys. Rev. 124, 128 (1961).
62a F. W. Martin and L. C. Northcliffe, Phys. Rev. 128, 1166 (1962).
62b J. D. Jackson, "Classical Electrodynamics," Chapt. 13, Wiley, New York (1962, 1975).

63a   J. Lindhard, M. Scharff and H. E. Schiott, Mat. Fys. Medd. Dan. Vid. Selsk. 33, No. 14 (1963).
63b   U. Fano, Ann. Rev. Nucl. Sci. 13, 1 (1963).
63d   F. Herman and S. Skillmann, "Atomic Structure Calculations," Prentice-Hall (1963).
63e   M. T. Robinson and O. S. Oen, Appl. Phys. Lett. 2, 30 (1963).
64a   J. Lindhard and A. Winther, Mat. Fys. Medd. Dan. Vid. Selsk. 34, No. 4 (1964).
64c   J. P. Biersack, Hahn-Meitner Report, HMI-B37 (1964).
65a   W. Booth and I. S. Grant, Nucl. Phys. 63, 481 (1965).
65b   J. H. Ormrod, J. R. MacDonald and H. E. Duckworth, Can. J. Phys. 43, 275 (1965).
66b   C. D. Moak and M. D. Brown, Phys. Rev. 149, 244 (1966).
67a   E. Bonderup, Kgl. Danske Vid. Sels. Mat. Fys. Medd. 35 No. 17 (1967).
67c   P. T. Wedepohl, Proc. Phys. Soc. 92, 79 (1967)
68a   J. Lindhard, V. Nielsen and M. Scharff, Mat. Fys. Medd. Dan. Vid. Selsk. 36, No. 10 (1968).
68b   S. Lindhard, V. Nielsen, M. Scharff and P. V. Thomsen, Mat. Fys. Medd. Dan. Vid. Selsk. 33, No. 10 (1968).
68d   T. E. Pierce and M. Blann, Phys. Rev. 173, 390 (1968).
68e   K. B. Winterbon, Can. J. Phys. 46, 2479 (1968).
70a   L. C. Northcliffe and R. F. Schilling, Nucl. Data Tables, 7, 233 (1970).
70b   C. C. Rousseau, W. K. Chu and D. Powers, Phys. Rev. A4, 1066 (1970).
70e   H. Bichsel, Amer. Inst. of Phys. Handbook, 3rd Ed. (1970).
70f   W. S. Johnson and J. F. Gibbons, "Projected Range Statistics in Semiconductors," Stanford University Bookstore, Stanford, CA, 1970 (now out of print).
70g   W. Brandt and J. Reinheimer, Phys. Rev. 2b, 3104 (1970).
72b   H. D. Betz, Rev. Mod. Phys. 44, 465 (1972).
72f   W. K. Chu and D. Powers, Phys. Lett. A40, 23 (1972).
72i   M. D. Brown and C. D. Moak, Phys. Rev., B6, 90 (1972).
72j   K. Guttner, H. Ewald and H. Schmidt, Rad. Eff. 13, 111 (1972).
74a   J. F. Ziegler and W. K. Chu, Atomic Data and Nucl. Data Tables, 13, 463 (1974).
74e   M. T. Robinson and I. M. Torrens, Phys. Rev. B9, 5008 (1974).
74h   E. S. Mashkova and V. A. Molchanov, Rad. Eff. 23, 215 (1974).
75a   J. D. Jackson, "Classical Electrodynamics," Chapt. 13, Wiley, New York (1962, 1975).
75b   P. Sigmund, Chapt. 1, "Radiation Damage Processes in Materials," ed. by C. H. S. Du Puy, Noordhoff, Leyden (1975).
75e   D. K. Brice, "Ion Implantation Range and Energy Deposition Distributions, Vol. 1, High Energies," Plenum Press, New York (1975).
75f   K. B. Winterbon, "Ion Implantation Range and Energy Deposition Distributions, Vol. 2, Low Energies," Plenum Press, New York (1975).
75g   J. F. Gibbons, W. S. Johnson and S. W. Mylroie, "Projected Range Statistics: Semiconductors and Related Materials," 2nd Edition, Halsted Press, Stroudsbury, PA, USA (1975).
75k   B. M. Latta and P. J. Scanlon, Phys. Rev. 12A, 34 (1975).
76d   B. M. Latta and P. J. Scanlon, Nucl. Inst. and Meth. 132, 133 (1976).

77a   H. H. Andersen and J. F. Ziegler, "Hydrogen Stopping Powers and Ranges in All Elements," Vol. 3 of series "Stopping and Ranges of Ions in Matter," Pergamon Press, New York (1977).

77c   W. D. Wilson, L. G. Haggmark and J. P. Biersack, Phys. Rev. $\underline{15B}$, 2458 (1977).

77f   W. D. Wilson, L. G. Haggmark and J. P. Biersack, Phys. Rev. $\underline{15}$, 2458 (1977).

77h   J. F. Ziegler, Appl. Phys. Lett. $\underline{31}$, 544 (1977).

77m   D. J. Land, J. G. Brennan, D. G. Simons and M. D. Brown, Phys. Rev. $\underline{A16}$, 492 (1977).

78a   J. F. Ziegler, "Helium Stopping Powers and Ranges in All Elements," Pergamon Press, New York (1978).

78h   M. Inokuti, Y. Itikawa and J. E. Turner, Rev. Mod. Phys, $\underline{50}$, 23 (1978).

79d   G. J. Iafrate and J. F. Ziegler, Jour. Appl. Phys. $\underline{50}$, 5579 (1979) plus errata available from the authors.

79g   P. Loftager, F. Besenbacher, O. S. Jensen and V. S. Sorensen, Phys. Rev. $\underline{A20}$, 1443 (1979).

80a   J. F. Ziegler, "Handbook of Stopping Cross Sections for Energetic Ions in All Elements," Pergamon Press, New York (1980).

80b   U. Littmark and J. F. Ziegler, "Handbook of Range Distributions for Energetic Ions in All Elements," Pergamon Press, New York (1980).

80d   J. P. Biersack an L. G. Haggmark, Nucl. Inst. and Meth., Vol. 174, 257 (1980).

80e   S. P. Ahlen, Rev. Mod. Phys. $\underline{52}$, 121 (1980).

80g   S. A. Cruz, C. Vargas and D. K. Brice, Nucl. Inst. and Meth., $\underline{170}$, 208 (1980).

82a   W. Brandt and M. Kitagawa, Phys. Rev. $\underline{25B}$, 5631 (1982).

82b   J. P. Biersack and J. F. Ziegler, "Ion Implantation Techniques," Springer-Verlag, p. 122 (1982).

83a   J. P. Biersack and P. Mertens, "Charge States and Dynamic Screening of Swift Ions in Solids," p 131, Oak Ridge Rpt. No. CONF-820131, Oak Ridge (1983).

83b   W. N. Lennard, H. R. Andrews, I. V. Mitchell, D. Phillips and D. Ward, ibid, p 136 (1983).

84a   J. F. Ziegler, J. P, Biersack, U. Littmark, "The Stopping and Range of Ions in Solids," Vo. 1, Pergamon Press, New York (1984).

ION IMPLANTATION DAMAGE IN SILICON*

Siegfried Mader

IBM Thomas J. Watson Research Center
Yorktown Heights, New York

This chapter describes aspects of ion implantation damage which are important for Si process technology. Primary damage consists of atomic displacements and amorphization of Si (except for B implantation). Annealing restores crystallinity and induces electrical activation of implanted dopant ions. It can also cause the formation of residual defects with well defined crystallographic nature. During prolonged annealing, these defects change their sizes and configurations. Their role in device processing will be discussed.

1. INTRODUCTION

The most widespread application of ion implantation is localized doping of semiconductor wafers during planar device fabrication. The more conventional doping methods, which are alternatives to implantation, employ thermal indiffusion of dopant atoms from source layers or from the vapor phase. Compared to in-diffusion, implantation allows a more precise control over the amount of impurities which are introduced into the substrates. This is because we can -- in principle -- measure the electrical charge deposited by an ion beam directly; whereas, for in-diffusion we have to rely on highly temperature dependent thermodynamic driving forces and kinetics. However, the price is high: implantation cannot be had without radiation damage. Ions which are propelled into the substrate crystal collide with substrate atoms and displace them from their lattice sites in large numbers. The success of semiconductor device fabrication has in part been due to our ability to grow perfect crystals of the substrate material. Thus, it is not

*   Portions of this chapter were published in "Ion Implantation Techniques," ed. by H. Ryssell and H. Glawischnig, Springer-Verlag, Berlin, Heidelberg, New York (1982).

63

surprising that the success of ion implantation doping depends on the restoration of the damaged crystals, at least in critical junction regions.

In this chapter we will describe some aspects of implantation damage and its annealing in Si. In modern technology one uses junction depths of 0.5μ or less and the trend is to shallower junctions. For doping with high concentrations (of the order of $10^{20}/cm^3$) it has become customary to select implant depths which are smaller than the desired junction depth and to thermally diffuse the implanted profile to the final position. This predeposition and drive-in scheme avoids junction positions in the regions of primary implant damage. But it does not guarantee a restoration of crystal perfection. Frequently secondary or residual defects evolve from the primary damage. Lower doping concentrations ($<10^{18}/cm^3$) can be implanted directly into the desired depth position; they are less likely to produce residual defects.

In the following sections we will first list methods for the characterization of damage and defects and briefly describe the primary damage structure. Then we deal with the events during thermal anneal and with the evolution of residual defects.

In 1972 Gibbons reviewed damage production and annealing [1], and in 1973 Dearnaley et al. published a monograph with a large section on semiconductor applications [2] These references contain most of the relevant concepts. In the intervening years the literature proliferated with many detailed observations and measurements. This can be seen from bibliographies on implantation, e.g. [3], which are books in themselves. Convenient summaries of the damage and defect aspects of our subject can be found in recent conference proceedings of the Materials Research Society [4] and of the Royal Microscopical Society [5].

2.   CHARACTERIZATION OF DAMAGE AND DEFECTS

A very suitable tool for characterization of implantation damage is transmission electron microscopy (TEM). There are two reasons for the good match between object and methodology: 1. the scale of distances between damage features is of the order of nm to μm and most of them are in the range of TEM resolution; 2. the defects are present in shallow layers just below the surface of the implanted crystal. These layers are accessible to TEM by simply etching away the substrate wafer.

TEM images arise from variations of Bragg diffraction at distorted crystal lattice planes. The methods for interpretation of diffraction contrast are well established [6]. They are particularly useful for characterization of crystallographic features of secondary defects which can form during heat treatment of the primary damage structure.

X-ray topography also images distortions of lattice planes. Its limited lateral resolution renders it less sensitive for implantation damage. Optical microscopy of defects decorated by chemical etches (Sirtl etch, Jenkins-Wright etch) also suffers from limited lateral resolution. Closely spaced defects do not give rise to individually distinguishable etch figures and a dense defect structure may remain unnoticed. The situation is somewhat better for surface examination with a scanning electron microscope (SEM) after a very light chemical etch.

Another well established and frequently used method for characterizing implant damage is Rutherford backscattering (RBS), [7.]. Its popularity among practitioners of implantation stems -- in part -- from the similarity of tools. For both, implantation and RBS, one needs accelerators and ion beams, in the latter case one uses H^+ or He^+ and energies in the MeV range.

RBS is particularly useful for exploring primary implantation damage which is less accessible to TEM. Backscattering can measure the number of atoms which are displaced from regular lattice sites. It is unsurpassed for detecting the presence of an amorphous layer and for measuring its thickness. It can also determine profiles of implanted ions which are heavier than Si and it can ascertain whether they occupy substitutional lattice sites. In a wafer with residual defects after heat treatment RBS analysis shows dechanneling yields which are caused by crystallographic defects. Correlations between these dechanneling effects and microscopic defect structures are emerging [8], [9].

Point defects are created in abundance during implantation and a powerful method for their analysis is electron paramagnetic resonance (EPR), compare [10]. Amorphous Si gives rise to a fingerprint in the EPR spectra, this was used to study implantation induced amorphization, e.g. in [11].

3. PRIMARY IMPLANTATION DAMAGE

Detailed theories and tabulations exist for the ranges of implanted ions in solid targets [12], [13], [14]. But the understanding of what happens to the Si target itself is less detailed. Brice [13] tabulated depth distributions of the energy which is deposited in the crystal.

An ion slowing down and coming to rest in the target loses energy by electronic stopping and by nuclear collisions. Both loss mechanisms contribute to the primary implantation damage. Collisions with energy transfers above a threshold of about 15 eV displace Si atoms from their lattice sites and create a Frenkel pair. The recoiled Si atoms can act as projectiles for secondary collisions and generate displacement cascades. Electronic excitations enhance the diffusion of point defects which can exist in several charge states. The travel of recoiled atoms, thermal diffusion in the beam heated target and electronic enhancement effects contribute to migration and agglomeration of point defects during the implantation process.

Accumulation of displacement damage can lead to amorphization of the Si structure. There is a critical dose for the formation of an amorphous layer which increases with decreasing mass of the implanted ions and with increasing target temperature. In a target at room temperature 5.10^{14} P^+/cm^2 or 10^{14} As^+/cm^2 produce amorphous layers, whereas B^+/cm^2 ions do not amorphize a Si target.

The nuclear stopping power increases with increasing projectile mass. For light ions, like B^+, this translates into a trail of well separated primary recoils in the wake of the implanted ion. But heavy ions produce closely spaced recoils and their secondary cascades touch each other or overlap. The results are amorphous zones with dimensions of the order of 10 nm. When these zones completely fill the volume of the implanted layer, this layer becomes amorphous. It can be shown that this happens when the energy deposited into collisions exceeds 6.10^{23} eV/cm^3 or 12 eV for each target atom [11].

Amorphization begins in a layer at the depth of the maximum collision energy deposition (slightly less than the projected range R_p) and spreads towards the surface and towards deeper positions in the target. The interface with the single crystal target is not a well defined plane due to the statistical nature of ion penetration. Beyond the interface we expect a considerable concentration of Si interstitials which had diffused out of the damage clusters during implantation.

4. DEFECT REACTIONS

Before we describe actual defect structures in implanted Si and their changes during heat treatment we digress here to introduce some aspects of dislocation geometry and interactions. Readers who are familiar with dislocations should continue with Section 5. We limit ourselves to those dislocations which occur as residual damage after implantations that had created an amorphous surface layer. In actual device fabrication these defects are important because they can degrade device yield; the last section contains examples.

The defects form and change from one configuration into another one in an evolutionary sequence. The sequence is schematically shown in Fig. 1a. During heating the amorphous layer grows back onto the single crystal substrate, the regrowth kinetics is discussed in Section 5. Small dislocation loops form at the location of the initial amorphous- crystalline interface. There we had expected a considerable concentration of Si interstitials. Interstitial atoms are indicated in a simple lattice model in Fig. 1b, together with four lines. In Fig. 1c the lattice is separated along these lines and displaced in the normal direction. The openings are filled with extra atoms in perfect register with the lattice. These configurations are so-called prismatic interstitial dislocation loops. They can grow by addition of more interstitials along their perimeters, which are -- in this case -- edge dislocations. Absorption of point defects induces climb motion of edge dislocations. The connecting thread through the evolution of residual defects is continued climb, first by absorption of interstitials and later by their emission (or absorption of vacancies).

Dislocations are characterized by their Burgers vectors which equals the displacement of the crystal on one side of the loop plane with respect to the other side. Here they are perpendicular to the loop planes. When two dislocations meet they can combine to form a new one with a Burgers vector which is the sum of the two reacting ones. An important reaction in the face centered cubic structure is the Lomer reaction which reduces the energy of the reacting partners. Such reactions occur when the dislocations in Fig. 1c climb and find themselves close to the perimeter dislocation of a loop with a different inclination. The resulting dislocations are edge dislocations whose extra half planes either extend to the wafer surface (shown at the left and right of Fig. 1d) or to the interior of the crystal (center of Fig. 1d).

Climb motion by absorption of interstitials continues. The center dislocation in Fig. 1d reaches the surface and disappears, while the two other ones climb deeper into the crystal. They form the lower segments of interstitial half loops which are shown in Fig. 1a. Eventually the climb force reverses and the interstitial half loops climb back to the surface and disappear also. The rates of expansion and shrinkage depend on the actual implantation conditions and species, more will be said in Section 6.

The situation in the diamond cubic Si structure is, of course, more complicated than the schematic diagrams of Fig. 1. We have to allow for a larger multiplicity of orientations for the Burgers vectors of dislocation loops.

(a)

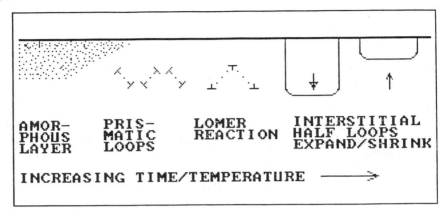

(b)

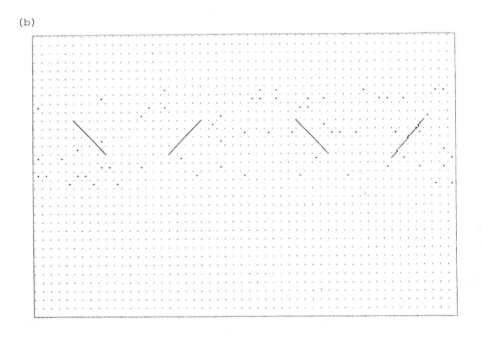

Fig. 1. (a) Evolution of defects. (b) Interstitials. (c) Prismatic interstitial dislocation loops. (d) Edge dislocations.

(c)

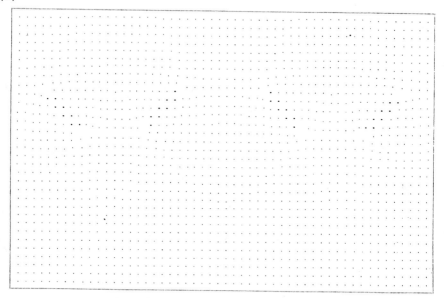

(d)

In the Si structure the possible Burgers vectors are a/2 <110>, and there are six different <110> directions. They can be visualized as edges of a tetrahedon or an octahedron. For wafers with a (001) surface orientation the <110> directions are shown at the left of Fig. 2a with the aid of a half-octahedron. There are four inclined Burgers vector directions, they are labelled a to d; and two directions are parallel to the surface, they are labelled e and f.

Nearly all of the prismatic interstitial loops which form at the initial stage have inclined Burgers vectors. This is because the strains associated with interstitial loops can relax more if the loop plane is as parallel to the surface as possible. As in Fig. 1, we now allow the loops to grow, to touch, and to react with one another. This gives rise to a new structure of larger and winding loops, examples will be shown in Section 6. They now have Burgers vectors e and f parallel to the surface. Since the loops are still located in a thin layer parallel to the wafer surface, they must now have an appreciable shear component. This is a surprising stage in an evolution which is driven by growth of loops through absorption of interstitials. We will elaborate on this below, according to an analysis in ref. [25]. In the center of Fig. 2a, two loops with Burgers vectors e and f are shown (omitting the winding outlines). They each have two edge dislocation segments with extra half planes extending to the surface and to the interior. With continuing absorption of interstitials the latter ones climb to the surface and disappear, the former ones climb deeper into the crystal, while the connecting segments move into the positions of the lateral boundaries of the interstitial half loops.

In order to understand the transition from prismatic loops to shear loops, we have to examine all possible reactions between the four different prismatic loop orientations. For this purpose we draw in Fig. 2b an array of loops with Burgers vectors indicated as line segments; they are projections of loop normals when the array is viewed from above through the wafer surface. The Burgers vector of a loop in the m-th row and n-th column, B(m, n), is chosen at random by a computer. For example, by comparison with Fig. 2a, B(1,1) = b. Note that several neighboring loops happen to have identical Burgers vectors, e.g., B(3,2) = B(4,2) = c.

Now we let the loops grow and touch each other, Fig. 2c. Neighboring loops with identical Burgers vectors will immediately merge into oval or larger shapes. Other neighbors can undergo a Lomer reaction, e.g.,

$$B(1,5) - B(1,4) = d - b = f$$

$$B(2,5) - B(2,4) = c - a = f$$

(a)

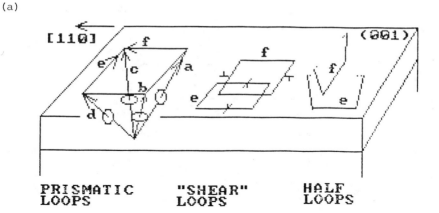

(b)

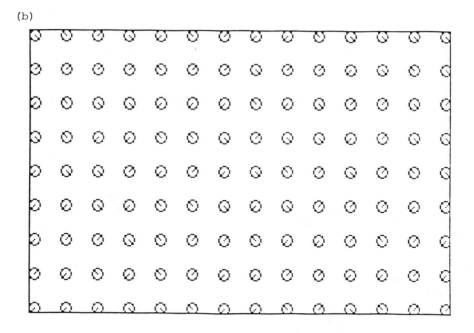

Fig. 2. (a) Defect evolution in (001) wafers. (b) Array of prismatic loops. (c) Loops before Lomer reaction. (d) Loops after Lomer reaction. (e) Shear loops with Burgers vector f. (f) Interstitial half loops.

(c)

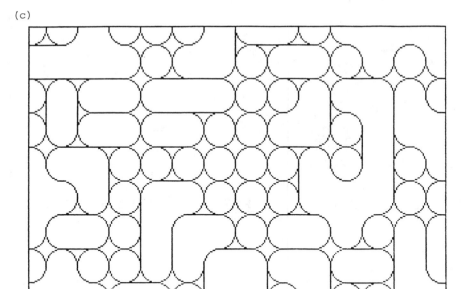

(d)

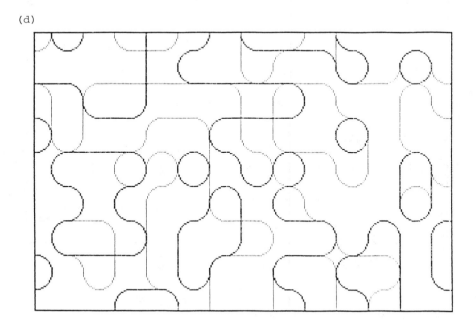

Figure 2c,d

(e)

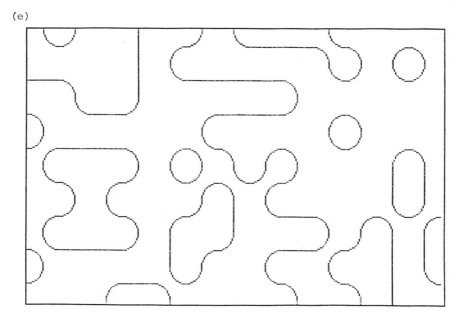

(f)

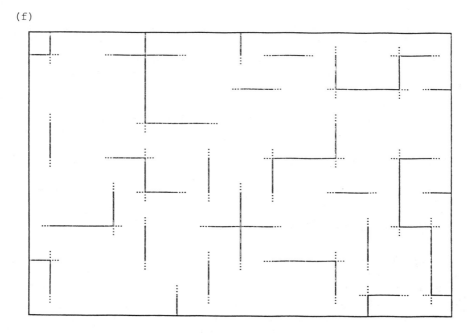

Figure 2e,f

(The minus sign arises from the opposite sense of the dislocation line directions along the reacting segments). There is a third possibility which does not lead to an immediate reaction, e.g.,

$$B(4,2) - B(4,1) = c - d$$

But this is equal to the sum $(e + f)$ and we let such a pair convert into two dislocations e and f.

Next we connect all the segments e and f -- allowing sharp corners to round off, as the line tension of dislocations would want to do -- and arrive at the pattern of Fig. 2d. The whole array of prismatic loops is now converted into a configuration of winding loops and small loops inside larger loops, all with Burgers vectors e and f parallel to the wafer surface. For clarity the set with Burgers vector f is isolated and redrawn in Fig. 2e. The reaction rules can easily be coded for a computer and all the diagrams were, in fact, drawn by an IBM Personal Computer. We examined a large number of random arrays (such as Fig. 2b) and always found a complete conversion into shear loops (such as Fig. 2d), without loose ends.

Finally, to simulate the stage of half loops in Fig. 2a, we eliminate from Fig. 2d all edge dislocation segments with extra half planes extending into the wafer. They climb to the surface and disappear. We retain the ones with extra half planes extending to the wafer surface. (They are the ones for which the Lomer reaction had returned a positive value for e or f). Since they climb deeper into the crystal, we extend them somewhat beyond their lengths in Fig. 2d. The resulting pattern is shown in Fig. 2f.

Figs. 2e and 2f are strikingly similar to TEM pictures of As implanted Si after 850°C and 1000°C heat treatments, as shown in Figs. 7 and 8. This similarity supports the concept of an evolutionary sequence of residual defects by dislocation climb in the direction of interstitial absorption.

5. THERMAL ANNEALING

In order for a dopant atom to be electrically active as donor or acceptor it has to be incorporated into a lattice site of the Si structure. The resistivity of as-implanted material is very high because only a few implanted atoms happen to come to rest in a substitutional site. Furthermore, after implantation, there are more structural defects than dopant atoms and most of these defects have deep trapping and compensating states. Therefore, thermal annealing is necessary to remove the defect states and to incorporate all implanted ions in lattice sites.

A large arsenal of annealing methods is available. One way of categorizing these methods is the time of exposure to high temperature. This time ranges from microseconds for pulsed laser anneal, milliseconds for CW laser anneal, seconds for hating with incoherent light and thermal radiation, to minutes and hours for conventional furnace anneal. The short time methods have been very actively investigated in the past few years and most of the results are conveniently available in conference proceedings [15] [16]. However, in actual device fabrication, where reproducibility and throughput are important, furnace anneal dominates.

In pulsed laser anneal, the surface of the wafer melts locally while the remainder of the wafer remains at ambient temperature. The very rapid resolidification usually restores a perfect crystal. There is not enough time for any process which requires diffusion in the solid, such as precipitation of a supersaturated dopant concentration.

The methods of millisecond to second duration heat the implanted layer without melting and allow activation by thermal diffusion. The diffusion distances are short, of the order of a few nm, and implanted profiles are not significantly broadened. Amorphous layers grow back into crystalline structures. But the duration of the diffusion tends to be not sufficient for complete restoration of crystalline perfection. The interstitials which had accumulated in the crystalline region adjoining the amorphous layer cannot diffuse out of the crystal, but they form agglomerates. Figure 3 shows an example in an As implanted wafer. It had been exposed for 10 sec. to a graphite heater at 1250°C in an apparatus similar to that described by [17]. There are many small interstitial prismatic dislocation loops.

The remainder of this section deals with furnace anneal. As is to be expected, the effects of heat treatments are different for implantations which had amorphized the substrate and for B implantations.

Consider first amorphized layers. They regrow epitaxially onto the single crystal substrate with well defined growth kinetics. The interface velocity V was elegantly determined by Csepregi et al. [18], [19], using RBS. It can be described by

$$V = V_o \exp(-2.35 \text{ eV}/kT)$$

where V_0 depends on the crystallographic orientation of the interface and on the type of doping. For an isothermal anneal at 550°C the interface velocity

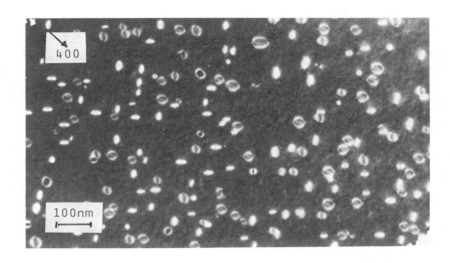

Fig. 3. Interstitial prismatic dislocation loops after regrowth of amorphous Si. Implantation 5.10^{14} As^+/cm^2 at 140 KeV, annealing: 10 sec. exposed to graphite heater at 1250°C.

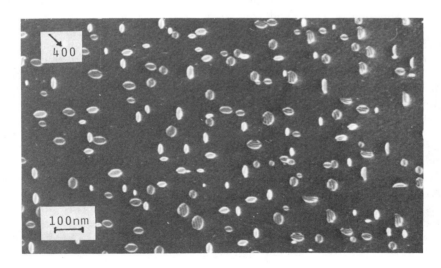

Fig. 4. Frank loops with extrinsic stacking faults after regrowth of amorphous Si. Implantation: 2.10^{14} Sb^+/cm^2 at 150 KeV, annealing: 15 min. at 950°C.

is 8.5 nm/min. for (100) and about 1 nm/min. for (111) interfaces in the case of undoped (Si implanted) wafers and 60 times higher for P and As implanted wafers. These kinetics are obeyed over a very large range of temperatures and growth rates including the regime of very rapid heating by CW laser anneal [20].

In a typical furnace anneal at 900°C to 1000°C the amorphous layer regrows during heat-up. For (001) wafers a single crystal layer is restored except at the position of the original amorphous-crystalline interface. In this region we have found -- in all cases examined in our laboratory -- extrinsic defects, i.e., defects which can be ascribed to agglomeration of interstitials. They usually have the configuration of dislocation loops similar to Fig. 3; but other configurations occur also. Figure 4 shows Frank loops where the excess Si atoms are incorporated on {111} planes and form stacking faults. In TEM pictures they show characteristic contrast of parallel fringes. They were observed after annealing of an Sb implantation.

In interfaces with {111} orientations additional imperfections occur: microtwins on inclined {111} planes. They propagate through the whole thickness of the regrowing layer. Their nucleation was discussed by Washburn [21].

The driving force for the regrowth of an amorphized layer is very large and dopant atoms are swept into lattice sites even for peak concentrations which are larger than the solubility at the regrowth temperature. After a regrowth anneal of e.g. 30 min. at 600°C the implanted dose is electrically active.

Electrical activation can be checked by measuring the sheet resistance ρ_s. In a first approximation ρ_s is inversely proportional to the electronic charge q, average mobility μ and implanted dose N:

$$\rho_s = 1/q \cdot \mu \cdot N$$

The mobility μ depends strongly on the concentration of doping atoms and values of ρ_s have been computed using known mobility data [2]. After regrowth of an amorphous layer, ρ_s measurements usually yield values in the range expected by such computations. This indicates that the defect structures of Figs. 3 and 4 do not significantly degrade the mobility.

The situation is different for B implants (and P implants below the critical dose for amorphization). For low doses ($<10^{13}$/cm^2) one observes a monotonous increase in the fraction of activated dopant atoms with increasing annealing temperature. However, with higher doses an initial increase of activated fraction reverses in the temperature range of 500°C to 800°C and complete activation can only be achieved above 900°C.

During annealing of B implants a large and complex variety of crystallo-graphic defects are formed. In the temperature range of reverse annealing rod shaped defects dominate. They are elongated in $<110>$ directions. An example is shown in Fig. 5. With increasing temperature the rod structure coarsens and, in addition, elongated interstitial dislocation loops and Frank loops appear, Fig. 6. The structures of the rod defects are not completely known, there appear to be several types. Wu and Washburn [22] showed that their shrinking is controlled by the diffusivity of B in Si. This implies that they contain B or are boron precipitates. The existence of precipitates and their dissolution during heat treatment above 800°C further implies that, in this case, the concentration in solid solution -- and therefore the electrical activity -- is limited to the equilibrium solubility of B at the annealing temp-erature. Similar results were recently reported for P [23].

So far we considered heat treatments mainly from the viewpoint of electrical activation. An equally important concern is the broadening and in-diffusion of the profiles. For doping by predeposition and drive-in the primary function of the anneal cycle is to diffuse the junction to a desired depth position. The necessary times and temperatures can be preselected with good accuracy from diffusion models or process models [24].

During drive-in the defects described so far react with one another and form a variety of different arrangements. This will be discussed in the next section. Eventually, with prolonged drive-in, the defects tend to shrink and to disappear. Shrinking occurs by dislocation climb which is primarily con-trolled by self diffusion of Si. Since the activation energy of self diffusion is always larger than that of impurity diffusion the ratio of the rate of defect elimination to the rate of dopant in-diffusion increases with increasing temperature. Therefore, it is desirable to choose as high a temperature as possible -- within the limits of achieving the junction depth of the device design -- in order to remove as many defects as possible.

6. RESIDUAL DEFECTS

After an implanted profile is fully activated and diffused to the desired depth position, it is quite possible that defects are still present in the doped layer. They are referred to as residual defects.

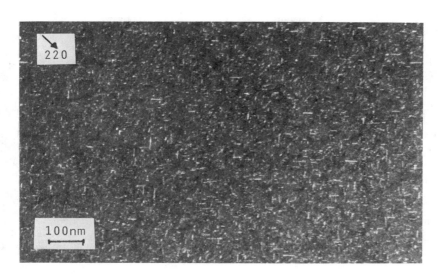

Fig. 5. *Microstructure after B implantation: Rods elongated in <110> directions. Implantation: 5.10^{14} B^+/cm^2 at 150 KeV, annealing: 15 min. at 750°C.*

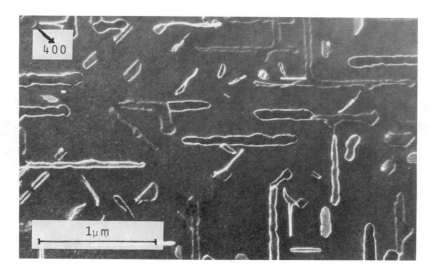

Fig. 6. *Microstructure after B implantation. Elongated interstitial dislocation loops and Frank loops. Implantation: 5.10^{14} B^+/cm^2 at 150 KeV, annealing: 20 min. at 950°C.*

A layer with dislocation loops or stacking faults is not an equilibrium structure because the defects carry with them strain energy, core energy, and stacking fault energy. After sufficiently long annealing times we would expect that they had disappeared (or reorganized themselves into a cross grating of misfit dislocations in those cases where doping changes the Si lattice constant). This is indeed the case. Residual defects after a specific drive-in cycle are just intermediate states of a dislocation structure which is evolving towards its equilibrium. We will try to describe this evolution for the case of As implantation in (001) wafers.

During heat-up in a furnace small interstitial prismatic dislocation loops form, similar to the ones in Fig. 3. Most of them have Burgers vectors which are inclined to the wafer plane. During further heat treatment the loops grow until they touch one another. Two touching dislocation segments then undergo a Lomer reaction which reduces their line energy. It was shown in Section 4 and Ref. [25] that this reaction transforms all of the prismatic loops into two sets of larger loops. They now have long meandering outlines and Burgers vectors parallel to the wafer plane. This structure is shown in Fig. 7.

With continued heat treatment a second stage occurs where the edge dislocation segments climb normal to the wafer surface in the direction defined by absorption of interstitials (or emission of vacancies). This leads to the formation of edge dislocation half loops with straight segments parallel to the surface and end segments turning up to the surface, Fig. 8. The extra half planes are contained between the surface and the dislocation lines. The edge segments initially climb deeper into the crystal to penetration depths of 100 to 200 nm. Then they reverse their advance, climb back to the surface and disappear [26].

The second stage is greatly retarded for implantation through a screen oxide. It appears that recoiled oxygen atoms precipitate along the dislocation lines and stabilize the structure with winding loops, Fig. 9. On the other hand, the second stage is enhanced and the edge segments penetrate deeper for drive-in in oxidizing atmosphere. Frequently a drive-in cycle begins with a short oxidation which is intended to cap the crystal surface and to prevent out-diffusion of the implanted impurities.

Expansion and shrinking of interstitial loops (and half loops) is controlled by non-equilibrium concentrations of point defects. Several factors contribute to the imbalance of point defects during thermal anneal. Not all of them are due to the implantation process. In fact, only the small interstitial loops in the wake of the amorphous regrowth can be ascribed to the primary implant damage. They had nucleated at a position where we had

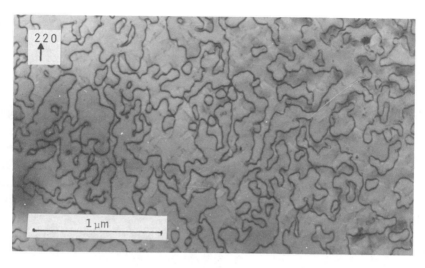

Fig. 7. Winding dislocation loops with Burgers vector parallel to the wafer surface. Implantation: 1.10^{16} As$^+$/cm^2 at 40 KeV, annealing: 1 hr. at 850°C.

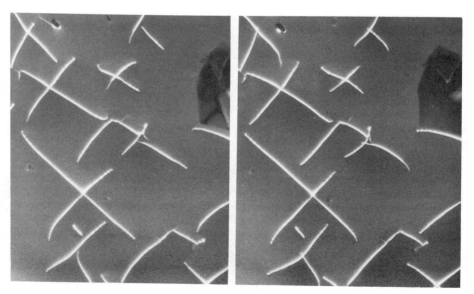

Fig. 8. Half loops of edge dislocations, stereo pair. Implantation: 5.10^{15} As$^+$/cm^2 at 50 KeV, annealing: 10 min. at 1000°C. Average depth penetration of half loops: 150 nm.

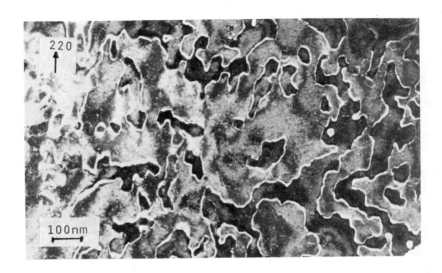

Fig. 9. *Winding dislocation loops decorated with small precipitate particles. Implantation: 5.10^{15} As^+/cm^2 at 50 KeV through 25 nm SiO_2; drive-in: 60 min. at $1000°C$.*

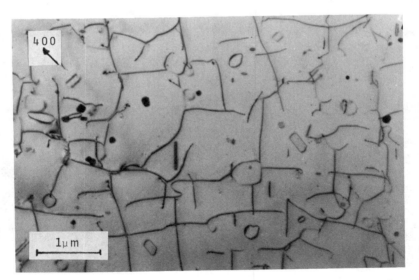

Fig. 10. *Dislocation half loops developing into a cross grating of misfit dislocations. Implantation: 3.10^{15} P^+/cm^2 at 200 keV, annealing: 2 hrs. at $900°C$ and 30 min. at $950°C$.*

expected to find a super-saturation of Si interstitials as a result of the implantation process.

The growth of these loops into the structures of Figs. 7 and 8 takes place while simultaneously the implanted profile broadens and diffuses into the crystal. The diffusivity of dopant atoms is always larger than the self diffusion of Si. This leads to a pile-up of extra atoms (or an undersaturation of vacancies) in front of the steepest gradient of the profile, which tends to enlarge interstitial dislocation loops at that location. (This concept is similar to the one discussed by Goesele and Strunk [27]). When diffusion proceeds into a perfect crystal no dislocation loops from because, presumably, the imbalance of point defects is not large enough to overcome the nucleation barrier. But in the present case nucleation had already occurred as a consequence of implantation damage and in-diffusion merely enlarges existing loops. Once the diffusion front has passed beyond the depth position of the loops or half loops the climb force reverses and aids the shrinking and elimination of the defects.

Oxidizing heat treatment creates an additional super-saturation of Si interstitials, as is known from the formation of oxidation stacking faults and oxidation enhanced diffusion [28]. This enhances the growth phase of our half loops and tends to retain dislocations in the crystal even after all of the volume of primary damage has been consumed by oxidation.

With phosphorus implantation the half loops of the second stage tend to join together and form a cross-grating or a network of edge dislocation. This structure remains in the crystal; it is stabilized by the lattice misfit between the phosphorous doped layer and the substrate. Figure 10 shows the development of this structure. Its formation is strongly enhanced by annealing under oxidizing conditions, as was shown by Tamura [29].

Double implantation, e.g., of P and Sb, greatly reduces the density of residual dislocations for equivalent doping concentrations and junction depths. P and Sb introduce lattice misfits of opposite signs which compensate each other. In addition, the double implants modify primary damage and in-diffusion characteristics [30].

Residual defects after drive-in do not always have the configuration of complete dislocations. Figure 11 shows defects in a layer doped by B implantation. They are Frank loops with extrinsic stacking faults, and they occur between the position of the primary damage and the driven-in junction. Since stacking faults contain extra atoms on {111} planes they also accommodate lattice misfit between the B doped layer and the substrate.

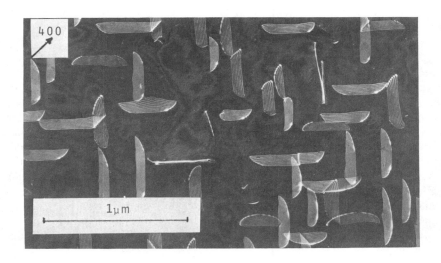

Fig. 11. *Extrinsic stacking faults in layer between 200 nm and 400 nm below surface (within doped layer, below primary damage). Implantation: $5.10^{15} B^{+}/cm^{2}$ at 40 keV, annealing: 30 min. at 970°C.*

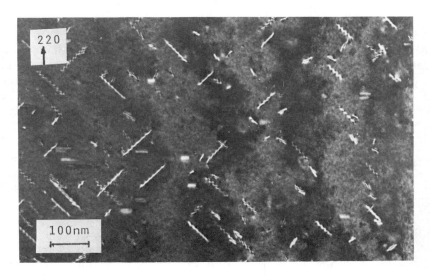

Fig. 12. *Rod defects elongated along $<110>$ directions, formed during H^{+} bombardment at 50 keV in wafer at 575°C, about 400 nm below surface. Dose rate $10^{13} H^{+}/cm^{2}$ sec.*

We conclude this section with a few observations made after implantation into hot targets. In this case the implanted species was H^+, and the purpose was not doping of the substrate but production of Frenkel pairs to induce radiation enhanced diffusion [31]. In targets with temperatures above $500°C$ rod shaped defects were found, shown in Fig. 12. They are elongated in <110> directions and dissociated into narrow ribbons on {113} habit planes. On these planes they form extrinsic stacking faults. These defects are yet another configuration of agglomerated Si interstitials. They have been observed in a variety of cases where displacement damage occurs [32].

For radiation enhanced diffusion to be effective the point defects must have an opportunity to diffuse distances of the order of a micron before recombining or being trapped at immobile agglomerations. This happens at target temperatures above $600°C$. The diffusion distances can be qualitatively appreciated in Fig. 13, where the H^+ beam had produced a super-saturation of point defects in the volume from which the TEM foil was prepared. Interstitial dislocation loops had nucleated and grown to sizes of about $1\ \mu$ diameter, most of them larger than foil thickness in Fig. 13. The scalloped outlines of the edge dislocations reflect the high super-saturation of interstitials in this particular case, compare [33].

We have seen that Si interstitials can agglomerate in a variety of configurations: prismatic dislocation loops, Frank loops with stacking faults on {111} planes, and {113} stacking fault ribbons. One might ask what happens to the other partner of a displacement collision, the vacancy. Vacancy agglomerates have been rarely observed in semiconductors. They occur when displacements are spread out in space and not concentrated in amorphous zones. They form three dimensional cavities without distorting the Si lattice. Therefore, they are far less conspicuous in TEM images, but they can be made visible by out-of-focus contrast [34]. An example is shown in Fig. 14 which is taken from the same H^+. bombarded wafer as Fig. 13. The cavities can grow to sizes of 10 to 20 nm and they tend to develop facets on {111} planes. In our example the distances between cavities (and thus the diffusion distances of vacancies) in Fig. 14 is much smaller than the distances between edge dislocations which absorb interstitials in Fig. 13.

7. EFFECTS OF RESIDUAL DEFECTS

Implantation damage and residual dislocations act as nucleation sites for the precipitation of impurities in a wafer. They can be put to good use as gettering agents [35]. With backside implantation one can remove fast diffusing impurities from the device side of a wafer [36].

Fig. 13. Edge dislocation absorbing interstitials during H^+ bombardment at 150 keV in wafer at 720°C, about 1.5μ below surface. Dose rate $3.5.10^{13}$ H^+/cm^2 sec.

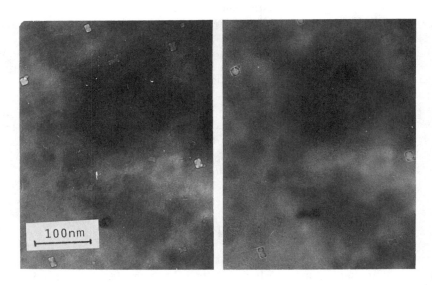

Fig. 14. Cavities (vacancy agglomerates) in same wafer and same depth as Fig. 13 (a) 220 nm under focus, (b) 220 nm over focus.

This property becomes, of course, detrimental when a dislocation threads through a junction and attracts unwanted impurities into the depletion region. Even undecorated dislocations degrade a device, although one can never be completely sure that a dislocation had not attracted some impurities.

The electrical behaviour of dislocations on a microscopic scale is not completely understood. They act as recombination centers which is evident from their SEM contrast in electron beam induced current (EBIC) mode [37]. When they cross a junction they cause a soft reverse characteristic and leakage currents with superlinear dependence on the applied reverse voltage. This was very clearly shown for the case bipolar transistors with implanted bases by Ashburn et al. [38] and for implanted emitters by Bull et al. [39].

But why would dislocations be near the depletion region after a properly executed drive-in? In the previous section we had shown that conditions can be chosen such that dislocations climb away from the junction or disappear altogether. This is true for experiments with blanket wafers. However, when devices are fabricated implantations have to be restricted to local areas. They are defined by windows in masking films or by oxide patterns recessed into the Si surface. The pattern edges usually are sources of mechanical stresses in the substrates. Dislocations react to mechanical stress by glide motion. Even transient dislocations, such as interstitial half loops during As drive-in, can respond to mechanical stress and glide away from the implanted areas into regions where they are cut off from the driving forces for climb which would have eliminated them in the implanted area.

An example is the so-called side wall penetration in implanted emitters [26], [39 to 41], shown in Fig. 15. The area on the left had been implanted with As while the area on the right had been masked by 80 nm SiO_2 and 160 nm Si_3N_4. After drive-in, the implanted area is free of defects while half loops are present at the edge of the window and under the masked area. Slip traces show that the dislocations had indeed moved out of the implanted area.

In this particular case a mechanical stress analysis was possible [26]. The masking film had an intrinsic tensile stress. After etching the window, the edge transmits a stress distribution into the substrate. The shear stress components of this distribution on planes parallel to the substrate surface exert glide forces on dislocations which are present during drive-in anneal. Inside the implanted areas the dislocations are attracted to the mask edge and under the masking film they are pushed away from the edge.

Figure 16 shows an example of an As implanted area which was bounded on both sides by a thick oxide recessed into the substrate surface by 0.5μ.

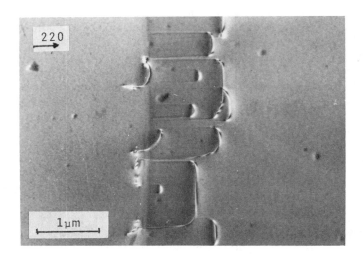

Fig. 15. Penetration of dislocation from As-implanted area (left) into masked area (right). Implantation: 5.10^{15} As^+/cm^2 at 50 keV, annealing: 60 min. at $1000°C$.

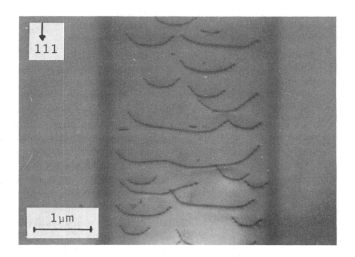

Fig. 16. Dislocation half loops in region bounded by recessed oxide on left and right. TEM foil tilted $35°C$ to show half loop outlines. Implantation: 2.10^{15} As^+/cm^2 at 70 keV, annealing: 45 min. at $1000°C$.

Drive-in under slightly oxidizing conditions did not completely eliminate the dislocation half loops. But they are all confined well within the doped region, no glide or penetration took place. In this case the defining oxide exerted a small compression on the implanted volume. An examination of the associated shear stress components shows that they do not induce glide of dislocations of the type present in Fig. 16.

The two examples show the importance of mechanical stresses for understanding the effects of residual dislocations which are free to respond to shear stresses. Stacking faults (such as the ones in Fig. 11) are enclosed by sessile dislocations. They cannot move when they are exposed to mechanical stress.

So far, in this section, we emphasized that defects should not be in the depletion region of a junction. Residual defects which are away from the junction, closer to the implanted surface, are not always electrically harmful. They can even improve the junction quality by local gettering. However, there is one application where defects must be completely eliminated. This is when the implantation doped wafer is to be a substrate for deposition of an epitaxial layer. Dislocations which terminate at the surface, as in Figs. 8 and 16, replicate themselves through a growing epitaxial layer. Moreover, out-diffusion of dopant material is greatly enhanced along the dislocation lines. If the buried doped layer is the subcollector of a planar transistor, the out-diffusion along the dislocations can reach into the emitter region and give rise to electrical pipes. Buried layers, therefore, have to be more thoroughly annealed than the minimum drive-in of a very shallow junction. Fortunately this is compatible with the use of buried layers, where the desire of a small sheet resistance translates into a deep junction before epitaxial deposition.

ACKNOWLEDGEMENTS

I am grateful to my colleagues J. R. Gardiner, C. T. Horng, I. E. Madgo, B. J. Masters, A. E. Michel, and R. O. Schwenker for many discussions and for interesting specimens from a variety of projects.

REFERENCES

1.   J.F. Gibbons, Proc. IEEE, *60*, 1062 (1972).

2.   G. Dearnaley, J.H. Freeman, R.S. Nelson and J. Stephen: "Ion
        Implantation," North-Holland, Amsterdam 1973.

3.   A.H. Agajanian: "Ion Implantation in Microelectronics," IFI Plenum,
        New York 1981.

4.   "Defects in Semiconductors," ed. by J. Narayan and T. Y. Tan,
        North-Holland, New York 1981.

5.   "Microscopy of Semiconducting Materials, 1981," ed. by A.G. Cullis
        and D.C. Joy, The Inst. of Phys., Bristol 1981.

6.   "Modern Diffraction and Imaging Techniques in Material Science,"
        ed. by S. Amelinckx, R. Gevers, G. Remaut, and J. Van Landuyt,
        North-Holland, Amsterdam 1970.

7.   W.K. Chu, J.W. Mayer, and M.A. Nicolet: "Backscattering Spectro-
        metry," Academic Press, New York 1978.

8.   G. Foti et al., ref 5, p. 79.

9.   D.K. Sadana and J. Washburn, ref. 5, p. 301.

10.   J.W. Corbett et al., ref. 4, p. 1.

11.   J.R. Dennis and E.B. Hale, J. Appl. Phys. *49*, 1119 (1978).

12.   J.F. Gibbons, W.S. Johnson and S.W. Mylroie: "Projected Range
        Statistics," Halstead Press, Stroudsburg 1975.

13.   D.K. Brice: "Ion Implantation Range and Energy Deposition Distri-
        bution," IFI Plenum, New York 1975.

14.   U. Littmark and J.F. Ziegler: "Handbook of Range Distributions for
        Energetic Ions in All Elements," Pergamon Press, New York 1975.

15. "Laser and Electron Beam Processing of Materials," ed. by C.W. White and P.S. Peercy, Academic Press, New York 1980.

16. "Laser and Electron-Beam Solid Interactions and Materials Processing," ed. by J.F. Gibbons, L.D. Hess, and T.W. Sigmon, North-Holland, New York 1981.

17. R.T. Fulks, C.J. Russo, P.R. Hanley and T. I. Kamins, Appl. Phys. Lett. *39*, 604 (1981).

18. L. Cspregi, E.F. Kennedy, T.J. Gallagher, J.W. Mayer and T.W. Sigmon, J. Appl. Phys., *48*, 4234 (1977).

19. L.Cspregi, E.F. Kennedy, J.W. Mayer and T.W. Sigmon, J. Appl. Phys. *49*, 3906 (1978).

20. S.A. Kokorowski, G.L. Olson, and L.D. Hess, J. Appl. Phys. *53*, 921 (1982).

21. J. Washburn, ref. 4, p. 209.

22. W.K. Wu and J. Washburn, J. Appl. Phys. *48*, 3742 (1977).

23. D. Nobili, A. Armigliatu, M. Finnetti and S. Solmi, J. Appl. Phys. *53*, 1484 (1982).

24. D.A. Antoniadis and R.W. Dutton, in "Process and Device Modelling for Integrated Circuit Design," ed. by F. van de Wiele, W.L. Engl and P.G. Jespers, Noordhoff, Leyden 1977, p. 837.

25. S. Mader and A.E. Michel, Phys. Stat. Sol. (a) *33*, 793 (1976).

26. S. Mader, J. Electron. Mater. *9*, 963 (1980).

27. U. Goesele and H. Strunk, Appl. Phys. *20*, 265 (1979).

28. S.M. Hu, ref. 4, p. 333.

29. M. Tamura, Phil. Mag. *35*, 663 (1977).

30. A. Schmitt and G. Schorer, Appl. Phys. *22*, 137 (1980).

31. B.J. Masters and E.F. Gorey, J. Appl. Phys. *49*, 2717 (1978).

32.   T.Y. Tan et al., ref. 4, p. 179.

33.   M. Kiritani, Y. Machara and H. Takata, J. Phys. Soc. Japan, *41*, 1575 (1976).

34.   M.R. Ruchle in "Radiation-induced Voids in Metals," ed. by J.W. Corbett and L.C. Ianniello, AEC Information Services 1972.

35.   T.E. Seidel, R.L. Meek, and A.G. Cullis, J. Appl. Phys. *46*, 600 (1975).

36.   H.J. Geipel and W.K. Tice, Appl. Phys. Lett. *30*, 325 (1977).

37.   A. Ourmaza et. al. ref. 5, p. 63.

38.   P. Ashburn, C. Bull, K.H. Nicholas, and G.R. Booker, Solid-St. Electron. *20*, 731 (1977).

39.   C. Bull, P. Ashburn, G.R. Booker and K.H. Nicholas, Solid-St. Electron. *22*, 95 (1979).

40.   M. Tamura, N. Yoshihiro and T. Tokuyama, Appl. Phys. *17*, 31 (1978).

41.   T. Koji, W.F. Tseng, J.W. Mayer and T. Suganuma, Solid-St. Electron. *22*, 335 (1979).

EXPERIMENTAL ANNEALING AND ACTIVATION

JOZSEF GYULAI

Central Research Institute for Physics
H-1525 Budapest, P.O.Box 49, Hungary and
Dept. of Physics, Technical Univ. Budapest
H-1521 Budapest, Hungary

ABSTRACTS

Damage annealing and its consequences in electrical activity of implanted and incorporated dopants are reviewed. In the Introduction, apart from a general and more detailed outline, references are given to most books, conference proceedings and review papers, where majority of the results can be found.

Best results are achieved, when annealing low dose (about 10^{12} ions/cm^2) and high dose ($>10^{15}$ ions/cm^2) implants. In case of high dose implantation, substrate crystal orientation, thermal history, solid solubility influence the final state.

Sections II and III include short reviews of device aspects of amorphization, dopant diffusion, focused ion beams (FIB) and Ion Beam Synthesis, for the latter mainly new results on fabrication of implanted SOI (Silicon On Insulator) structures. A brief review of beam annealing is also given.

In Sec. IV and V, some results in newly emerging areas, as damage and activation for extreme low and high energy implantations, are briefly summarized.

Section VI deals with implantation and implantation damage as influencing factors of chemical reactions on implanted surfaces, e.g. oxidation and silicide growth. To certain extent, gettering by implantation induced defects is also a related topic (Sec. VII).

Damage aspect of two cases are discussed as "Indirect Beam Effects", where atoms or layers originally present on the surface are important, not the implanted species itself (Ion Mixing and Recoil implantation, Sec. VIII).

In the last sections (IX), annealing strategies and techniques are treated. Here, emphasis was put on dopant behavior (redistribution, electrical activity) during different rapid annealing cycles. Perspectives of applications are also summarized.

A detailed list of references concludes the chapter.

I. INTRODUCTION

This chapter is intended to summarize results, which made ion implantation useful in semiconductor technology. In other words, what are the methods to overcome drawbacks of a technique, which leads to a state far from equilibrium and to the presence of lattice defects, as reviewed in the chapter on radiation damage by Mader.

The idea that after ion bombardment a thermal treatment is necessary to bring the system closer to its equilibrium state, is as old as the idea of ion implantation itself. However, details of this "treatment" has kept research active for two decades and still new ideas emerge. The increasing demands of semiconductor technology have initiated research to explore application of different radiations, such as lasers, particle beams or, simply, high-intensity light sources quoted as Rapid Thermal Annealing (RTA) or RTP ("P" for Processing).

Today's semiconductor technology (planar technology) with a layer-by-layer preparation technique, is a natural area, where the cleanliness and the reproducibility of implantation is appreciated.

This layer-by-layer technology is sensitive to surface problems, as forthcoming surfaces will be converted to become interfaces and this interface may inherit structural imperfections and impurities. In this series of treatments, ion implantation interacts with the top few hundred nanometers of the matter, adds extra atoms, removes some more by sputtering and leads to a new and complex defect structure. This "disorder" has to be converted into the desired order on an atomic scale.

Recent demands on miniaturization bring in new complexity. First of all, the redistribution of the implanted species during this post-implantation treatment in scope of VLSI demands is considered detrimental. Furthermore, the lateral structure of the device causes new sources of structural problems, mechanical stresses and misfits. In most cases it seems desirable that constitutional and, thus, unavoidable distortion of bonds at functional interfaces should be localized to minimum, i.e. to a few lattice spacings and let the material possess its normal state elsewhere. This is why the interest "expanded" to understanding of mechanical properties during post-implantation treatments e.g. /Miller et al., 82a/.

Needless to say that for the technology it is optimum that this post-implantation treatment should be part of the procedure,

as a whole. At the end, the device should reach a state,
equilibrium or not, but certainly a stable enough state for a
long-time operation at the desired temperature.

It is a general statement that the study of ion implantation
is a study of radiation damage effects and annealing. Thus, the
topic of this chapter is included into all basic text-books on
implantation (Mayer et al., 70a; Dearnaley et al., 73a; Townsend
et al., 76a; Carter and Grant, 76b; Ryssel and Ruge, 78a; Poate
and Mayer, 82b; Feldman et al., 82c; Ryssel and Ruge, 86a), into
review articles (Namba and Masuda, 75a; Eisen and Mayer, 76c;
Tokuyama et al., 78b; Davies and Howe, 80a; Wolf, 79a; Thompson,
81a; Poate, 82d; Mayer and Gyulai, 83a; Ziegler, 84a). Invited
talks and papers are referred to, which were presented on
conferences, first in Ion-implantation in Semiconductors (Eisen
and Chadderton, 71a; Ruge and Graul, 71b; Crowder, 73b; Namba,
75b; Chernow et al., 77a) and later in Ion Beam Modification of
Materials (Gyulai et al., 79b; Benenson et al., 81b; Biasse et
al., 83b; B.M. Ullrich, 85a; S.U. Campisano et al., 87a;) then,
parallel to IBMM, the conferences on Ion Implantation: Equipment
and Techniques, IIET, (C.M. McKenna et al., 81c; H. Ryssel and H.
Glawischnig, 82e; J.F. Ziegler and R.L. Brown 85b; M.I. Current
et al., 87b). Volumes edited by the Materials Research Society
contain also relevant reviews and original papers, especially on
research on new annealing techniques (lasers, electrons and ions,
sometimes quoted as directed energy deposition methods; Gibbons
et al., 81d; Narayan and Tan, 81e; Appleton and Celler,
82f;Picraux and Choyke, 82g; Kear et al., 82h; Narayan et al.,
83c; Mahajan and Corbett, 83d; Ludeke and Rose, 83e). Devoted to
the non-conventional annealing, a proceedings volume, edited by
Anderson et al.,(80b) and the book edited by Poate and Mayer
(82b) is to be mentioned. In the past years, there were numerous
reviews on laser annealing (e.g. Foti, 79c, Gyulai and Revesz,
79d). A recent review by Kear et al. (81f) and the chapter
written by Lau and Mayer (82i) should be mentioned. This latter,
however, is rather intended to film growth than to implantation
damage annealing. English translation of a book by Smirnov et al.
(83f) contains, among others chapters on implantation, laser and
e-beam annealing, a detailed review on implantation in InSb and
on radiation effects (electrons, X-rays) in technology.

A newly developing area, which has an impact on applications
of ion implantation, is the Silicon-On-Insulator (SOI) technique,
as we will return to it later. As reviews of the field we refer
to works by Cullen (83g), P.L.F. Hemment et al. (87c).

The astonishingly large number of basic publications referred here indicates that the only possibility for a review is to focus to one segment of the problem, possibly onto the most important and representative one. Here, a more-or-less consistent picture can be drawn.

Thus, we'll not only confine ourselves to silicon, but mostly deal with effects of medium and high dose implantation. Other semiconductors, like GaAs, though being in focus these days, will not be treated here, even if in many new achievements the ion implantation was the key factor. Channel implants for MESFETs (Metal Schottky-Gate Field Transistors), contacts, Schottky barrier adjustments have drawn more attention to date.

Advancements on silicon technology could help to· overcome difficulties in preparation of high quality layers on GaAs, but the sensitivity on stoichiometry of the substrate still poses a major drawback in the use of this material. Among others, experiments to compensate for matrix atom deficiency by implantation, are promising (Krautle, 81g). Non conventional annealing techniques have also been explored (review by Williams, 83h).

In the silicon field, we will not include implantation doping with low doses, as for applications like channel implants for MOSFETs (Metal-Oxide-Semiconductor Field Effect Transistors), where doses as low as $*10^{10}$ atoms/cm^2 might be involved, the implantation has become a simple industrial routine. Problems can be expected because of statistical nature of ion implantation, where the threshold voltage of a submicron gate is set by only a few hundred boron or arsenic atoms and the electric field under the gate cannot be treated as a continuum anymore.

II. LOW AND MEDIUM DOSE IMPLANTS

The low to medium dose regime is characterized by the fact that the individual ion tracks do not overlap. This situation can be seen in transmission electron microscope as individual amorphous regions or by channeling of energetic ions in Rutherford backscattering mode (Backscattering Spectrometry, BS).

(Throughout this chapter, BS will be used as a unified name for Rutherford Backscattering, RBS, and for channeling effect measurements with light ions, mostly with MeV He$^+$ particle (78c).

*On channeled BS spectra taken in aligned direction with atomic
rows or planes, a partial loss of the channels (i.e. loss of
"transparency" of the crystal) indicates the situation referred
to as "non-overlapping amorphous" regions. Here, depth
distribution of the damaged region can also be deduced.)*

A series of experiments were made to visualize the
individual ion tracks and to understand their regrowth properties
(80c; 80d; 81h) Here, implantations were made at 40-50K
temperature in order to prevent self-annealing during ion
bombardment, but sample preparation needed procedures up to 350K,
i.e. this is the starting temperature of the measurements. As an
example, Fig.1 shows the evolution of bismuth tracks for
increasing ion energy and decreasing mean

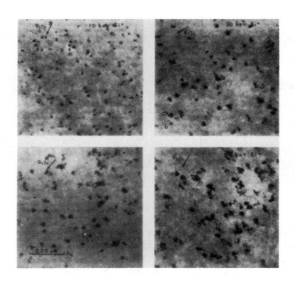

*Figure 1. Bright field electron micrographs of damaged regions in
Si bombarded with $3x10^{11}$ ions/cm^2 bismuth. (Mean deposited energy
for 10 keV Bi is 1.84 eV/atom; for 30 keV, 0.54 eV/atom; for 45
keV, 0.33 eV/atom; for 80 keV 0.16 eV/atom). Shown are the
Burgers vectors. (From Howe and Rainville, 81h.)*

deposited energy, which is responsible for the amount of damage. While for lower energy (10 keV Bi^+; large specific energy deposition) the efficiency of creating a cascade is close to unity, for higher energies "subcascades" develop (Fig.2). As no structure has been revealed within the cascades, it is likely that they are the "amorphous" regions.

Annealing properties were investigated on a relative scale, i.e. the decrease of defect density from its value at 300K during 10-minutes isochronal anneals. Figure 3 shows that the annealing properties also depend on the implanted species. As stated in (80d), the damage caused by heavier projectiles anneals at higher temperatures. Conclusions on annealing of an early work (71c), on 200 keV Sb implants, implanted at 80K, also fits to the Sb-curve. Implantation of molecular ions causes simultaneous and also overlapping cascades, which anneal at higher temperatures.

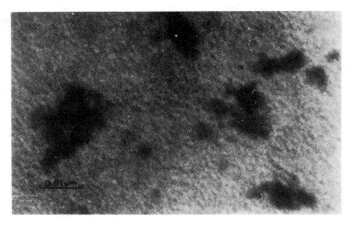

Figure 2. "Sub-cascades" for 118 keV Bi^+ in silicon, $3x10^{11}$ ions/cm^2 (from Howe and Rainville,81h).

More recent results (81h) show that there is no explicit dependence of the damage annealing on the mean deposited energy (Fig.4), i.e. all Bi implants anneal in the same temperature. It is to be noted that individual amorphous regions anneal below 300°C, while continuous amorphous layers anneal between 500-600°C.

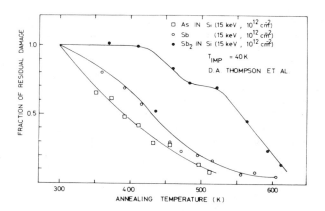

Figure 3. Annealing of defects produced by low temperature implantation during ten-minutes isochronal annealing steps. (From Thompson et al.,80d.)

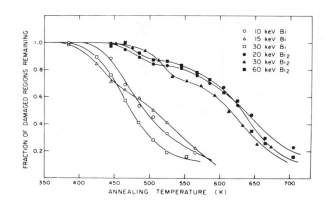

Figure 4. Annealing of defects produced by low temperature Bi implantation with different energies during ten-minutes isochronal annealing steps. (From Howe and Rainville,81h.)

   The purpose of heat treatment for low dose implants is, of course, to i) bring dopants into a lattice position, where they contribute to electrical conduction on a normal fashion (not as trapping centers), ii) reorder the lattice that carrier scattering or trapping should be minimal (good mobility).

   This procedure will result in that the dopants become "electrically active". It is also common to speak about the "electrically active fraction" of the implanted dose. Certainly, full electrical activity would be ideal, but, in order to keep process temperatures low, a compromise is often made and implantation doses include an over-dose.

   Electrical activation at low temperatures can be enhanced by ionizing irradiation, which produces the necessary amount of vacancies for the solid state reactions bringing dopant atoms to lattice positions. The use of 100-500 keV electron beams or UV light has been shown to increase electrical conductivity of phosphorus implanted (40-90 keV; 1.2×10^{14} - 2×10^{15} atoms/cm^2) silicon by a factor of 2-3 (78d).

   A discussion of low temperature electrical activation of implanted phosphorus is given in a review of Tokuyama et al., (78b), based on the experiments of Miyao et al. (76d) and T. Mitsuishi et al. (77b).

   Figure 5 shows the dose dependence of electrically active fraction of 50 keV phosphorus implanted into (111) silicon. Three regions can be distinguished. At low doses, most of the carriers are located deep in the crystal (Fig.6). This comes from the channeled part of the beam. Since channeled ions stop in a less defective crystal, the 480°C anneal was enough for electrical activation. The slight decrease of the curves with increasing dose comes from the fact that the amount of channeled ions is believed not to be dependent on the dose.

   Returning to the analysis on Fig.5, an intermediate dose defines the region II. Here activation becomes effective even at this low temperature anneal. The peak position here is about 2/3 of the projected range. The peak carrier concentration, however, changes by orders of magnitude for only a factor of two increase in dose. This is believed to be connected with the influence of the amount of disorder increasingly formed with such doses. If damage clusters overlap, electrical activation becomes easier (77c), as the cluster includes more dopant atoms, but compensating defects form more or less the same way and in the same amount, as for single amorphous clusters.

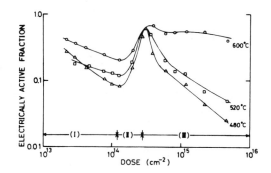

Figure 5. Dose dependence of electrically active fraction of phosphorus (50 keV, into (111) Si), for 15 min. annealing at 480, 520 and 600°C. (From Tokuyama et al., 78b.)

In this dose region, often has been found that during isochronal annealing cycles (15-30 min. anneals on sequentially increasing temperatures with an increment of 50 – 100°C) the electrical activity is not monotonic function of the

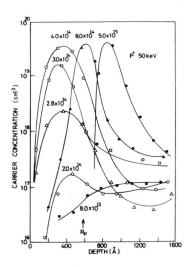

Figure 6. Carrier concentration profiles of 50 keV phosphorus implanted into (111) Si and after 15 min. anneal at 480°C.(From Tokuyama et al.,78b.)

annealing temperature. A drop in the number of carriers is sometimes found around 500-600°C and an anneal to at least 900°C is necessary to achieve nearly full activation.

This "reverse annealing" is often encountered. Its origin is mostly believed to be connected with an exchange reaction, where the substitutional atom goes to interstitial, leaving a vacancy behind. This was suggested in case of In (190 keV, 10^{13} and 10^{14} ions/cm^2) in (83i). In that paper, a correlation between sheet resistivity and substitutionality was also shown.

III. ELECTRICAL ACTIVITY FOR HIGH DOSE IMPLANTATION

Region III on Figs.5 and 6 shows again a different character. With doses of 4×10^{14}/cm^2, and an annealing at 480°C, the only change was the parallel shift of the carrier concentration curve toward greater depths. No increase in peak concentration occurred, therefore , doping efficiency drops for this temperature. Higher temperatures are needed to achieve close-to-unity dopant efficiency.

The electrical activity at high doses is rather complicated. Near the solid solubility, many dopants partially coalesce into precipitates. Arsenic tends to form dimers, etc.

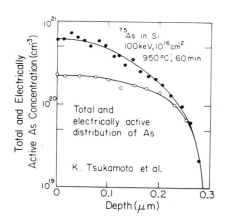

Figure 7. Comparison of total and electrically active As in Si. (From Tsukamoto et al., 80e.)

A results published by Tsukamoto et al. (80e) compares the
total arsenic distribution with the measured electrically active
dopant concentration (Fig.7).

It would be lengthy to review, what we mean by "amorphous",
because there is a reasonable difference in microstructure
depending on the technique how amorphous layers are produced. The
word "amorphous" is used in case of implantation, when no long-
range order exists which, usually, is checked by diffraction and
associated with the disappearance of channeling in RBS.

It was shown by Peto et al. that the electronic structure of
the implantation amorphized silicon is characteristically
different from the evaporated silicon (83j).

Amorphization dose depends on species, energy, sometimes on
dose rate, substrate temperature, substrate thermal properties,
on impurity concentration, as major factors. (E.g. heavy doping
with boron increases amorphization dose for Ne ions in Si RT
implants by more than a factor of ten, to $3 \times 10^{16}/cm^2$
(Dvurechenski et al., 86b). Dose rate effects are more important
for GaAs, mainly because of poorer heat conduction and larger
variety of possible defects.

Especially, if an oxidizing ambient is present, which is
usually the practical case, annealing of high dose phosphorus
implants results in formation of a network of edge-type
dislocations (Fig.8, 78e).

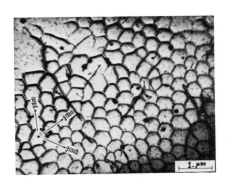

*Figure 8. Formation of
dislocation network on
(111) Si during P diffu-
sion under an oxidizing
ambient (Full cycle:
implantation 50 keV
$10^{16}/cm^2$; 550°C 24h
+ 900°C 30 min dry N_2
1050°C 30 min dry O_2).
TEM micrograph of edge-
type dislocation network
(From Tseng et al., 78e).*

III.A. Epitaxial Regrowth of Implanted Layers and Effect of
       Dopants

        Results of the experiment by Muller et al. (75c) on
annealing of 200 keV, $5x10^{14}$ and $10^{15}/cm^2$, room temperature
arsenic implant into silicon of different crystal
orientations, as (111), (100) and (110), draw attention to the
close relationship between implantation damage anneal and solid
phase epitaxy. Here, a 1000°C, 30 min. anneal left a large amount
of disorder in the (111) silicon, while the (100) and (110)
regrew perfectly. Electrical measurements showed that the number
of carriers was also lower in the layer of (111) orientation by
20% and 30% compared with those on (100) and (110), annealed at
900°C. Carrier mobilities did not show any marked difference.
        Clear picture was achieved, when multiple energy, self-
implantations at LN_2 temperature were used to produce amorphous
layers. The regrowth (in a vacuum furnace to prevent oxidation)
showed marked difference for (100) and (111) substrates (Fig.9).
TEM revealed stacking faults and microtwins in the (111) oriented
regrown layers (76e).
        Regrowth rates are dramatically influenced also by dopants
present (implanted) within the amorphous layer. Phosphorus,
arsenic and, especially, boron enhances growth rates up to an
order of magnitude (77d), while oxygen, carbon and implanted
noble gas atoms decrease growth rates (77e; 78f; 78g).
        The understanding of growth rate enhancement was more recent
and is reviewed later in this section.
        The deteriorating effect of oxygen, carbon and other species
can be explained on a simple chemistry or solubility arguments.
If, for example, SiO_2 islands are present at the crystalline-
amorphous interface, growing crystalline faces must embrace these
islands and when growing fronts meet on the opposite side of the
SiO_2 block, defect-free growth is impossible. Another reason that
continuous and laminar growth is inhibited for these species is
their solubility in silicon. The amount of the implanted species
exceeding solid solubility may stay there in a so-called
metastable state, but has a tendency to form precipitates. These
precipitates for gases are simply bubbles, as revealed also by
TEM and, similarly to the mechanism proposed for oxide blocks,
they also act as centers for polycrystalline growth.
        The influence of dopant atoms fell into the focus of
attention more recently. Earlier Csepregi et al. (77d) suggested

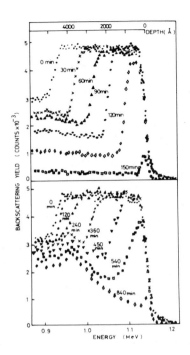

Figure 9. Aligned backscattering spectra for 2 MeV $^4He^+$ ions incident on "self-implanted" Si with multiple energy at LN_2 temperature, preannealed at 400°C 1 hr, and annealed at 550°C. Upper curve for (100), the lower for (111) oriented silicon. (From Csepregi et al., 76e.)

that the growth is governed by broken bonds and vacancies. Sigmon (81i) drew attention on that charge state of vacancies will be influenced by the Fermi level at the amorphous/crystalline interface.

The effect of carrier compensation in SPE growth was experimented in (82j). Arsenic, arsenic + boron and arsenic + phosphorus implants were used. Together with the boron, silicon ions were also used to enhance amorphization. All implants were made at LN_2 temperature. Figure 10. shows two examples, where compensation occurs. The slower growth rate is to be noted, when the growth front passes the compensated area.

It is worth to mention that the effect of the Fermi level on lattice location of dopants (As+B, 10^{16}/cm^2 doses) was demonstrated earlier (79e).

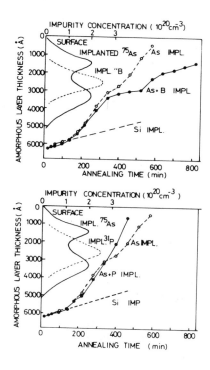

Figure 10. Amorphous layer
thickness vs annealing time
at 475°C for (a) As and B
implanted, and (b) As and
P implanted (100) Si.
Calculated impurity profiles
are also shown.
(From Suni et al., 82j.)

III.B. "Perfect Doping" and VLSI

If a carefully performed SPE growth restores perfect
crystallinity, a useful property of this rather low temperature
treatment might be a challenge to industry practices. First, the
doping implantation would be done into an amorphous region, thus
no channeled tails would form and simple range theories apply.
Most advantages can be expected for light ions, as boron.
Secondly, only a small spatial redistribution of the dopant is
expected during low-temperature SPE growth. These two features
would satisfy VLSI demands. The first reordering experiments with
implanted amorphous layers doped with P, B or As with a further
implantation step, already showed highly substitutional

character, and thus reasonable electrical activity of the dopants (77d).

The technique has a further virtue in satisfying VLSI demands. Namely, the implant profiles in the amorphous layer would obey simple range theories and will not exhibit channeled tails.

Ishiwara et al., (83k) implanted boron or phosphorus into UHV evaporated amorphous silicon and into self-implanted (room temperature implantation) silicon layers.

The behavior of boron was perfect from electrical point of view. Following low temperature annealing, the electrical activity even surmounted 100%. This behavior is shown on Fig.11.

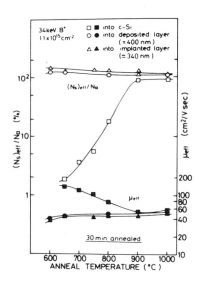

Figure 11. Annealing characteristics of carrier concentration and mobility from sheet resistivity and Hall effect measurements on implanted amorphous and deposited amorphous layers doped with boron. For comparison, measurements of crystalline silicon (c-Si) are also displayed.
(From Ishiwara et al., 83k.)

More recent results (83l) are demonstrating advantages of the technique called also as "dual implantation". TEM micrographs show that only few dislocations form during annealing (600°C 1 hr and 900°C, 10 min.) in the dual- implanted layers, in contrast with the boron implanted one, where high density of dislocation loops are present. Though there is an advantage for preamorphization at LN_2 temperature over the room temperature Si-

implantation, this latter is fully satisfactory, as an important
point in practical applications.

Figure 12 shows a comparison of boron profiles for the
regular case and for dual implantation. As mentioned above,
neither channeled tail, nor appreciable redistribution can be
observed. Comparison can be made with MOSFETs fabricated
parallel, using single (boron) implanted source and drain
regions. The subthreshold I-V characteristics are displayed on
Fig.13. Superiority of the dual-implant is attributed to the
lower sheet resistance of source and drain regions.

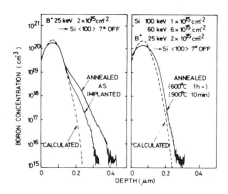

*Figure 12. Boron depth profiles
(SIMS) for single boron (a) and
dual (Si^++Bi^+) implants. Note
the minute redistribution and
the absence of channeled tail
in the latter. (From
Tsaur and Anderson, Jr., 831.)*

Recently, not only silicon self-implantation was tried to
preamorphize silicon, but other group IV elements, as non-
doping impurities were explored. In a detailed work of Seidel
(85c and 85d), amorphization with Ge ions was found as superior.
In this case, only some extended defects were found after
annealing (see Sec.III.C on BF_2^+).

In Sec. III.D, an experiment will be reviewed (86c
Angelucci), where problems involved with preamorphization, i.e.
with dual implantation become obvious. First, the formation of
extended defects, second, the enhanced diffusion for certain
species. Mainly because of this latter, amorphization with the
dopant itself becomes attractive.

Similar technique was applied to improve crystalline quality
of Silicon-On-Sapphire (SOS). The good match of lattice constants
at high temperature is partly destroyed because of different

thermal expansions of the two materials. (Upon cooling a large,
-10.2 kbar, compressive stress develops.)

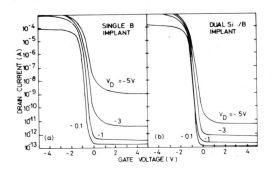

*Figure 13. Subthreshold I-V characteristics of simple boron doped
and dual implanted p-channel MOSFETs.   (From Tsaur and
Anderson, Jr., 83k.)*

    First successful attempt to overcome this, using a multiple
energy, deep and channeled self-implant at LN_2 temperature was
done by Lau et al. (79f), who have converted the silicon near the
interface amorphous, but preserved a good enough crystal on the
surface. Following amorphization, a sequential anneal   produced
an SPE growth to the interface.
    Minority carrier lifetime increased 500 times, the great
reduction of leakage current and noise of IC-s was demonstrated
(79g). The idea became a useful technique because of shrinking
Si layer thicknesses "regular" energies were enough (81j and k).

III.C. Simultaneous Amorphization and Doping

    Simplicity of the procedure justifies experiments, where
amorphization will be done by the dopant itself, and regrowth and
activation will be a single step. In this case, limited
redistribution of the dopant is still maintained, but the mutual
position of dopant and damage distribution can be tailored only
within limited range, i.e. by choosing proper energy and species
to have large enough fraction of the ion energy deposited into
nuclear processes. (Nuclear stopping is necessary to produce
defects and amorphization in semiconductor crystals.)

In case of n-type doping, arsenic closely satisfies these criteria, but p-type dopants with similarly good properties do not exist: different "tricks" are needed to reach high concentration of defects.

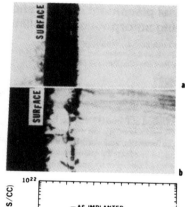

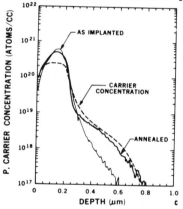

Figure 14. Cross-sectional TEM micrographs for phosphorus implanted into (111) Si, 120 keV, $10^{16}/cm^2$ at 100°C.
(a) unannealed,
(b) 837°C annealed,
(c) P profiles for (a) and (b) and carrier concentration from (b).
(From Sadana et al.,83m).

A detailed analysis of a common industry practice is given in (83m), namely, the case of high-dose (10^{15}-$2\times10^{16}/cm^2$) 120 keV phosphorus into boron-doped p-type (111) Si. The annealing was made in the range 750-900°C in N_2 atmosphere. If heating by the beam occurred during implantation, either the amorphous layer was thinner (about 100°C), or did not form at all (about 300°C). In this latter case, color bands were even formed. Figure 14 shows the case when moderate heating occurred. The original

amorphous layer reordered by the annealing. Apart from the
reasonable amount of residual disorder (microtwins), the SIMS
profiles show an interesting feature. The phosphorus
redistribution occurred only in the partially amorphous region.
There, the carrier concentration coincided with the measured
profile. In the regrown amorphous region, the electrical activity
was only partial. In contrast to the case found with antimony
(see III.D., 74a), here the diffused-in part has not reached
solid solubility.

The most important p-type dopant is, of course, the boron.
The boron being a light element, no amorphization will occur even
for high doses at RT implantation. It was already in 1971, when
Muller et al. (71d) have proposed to use molecular ions of BF_2^+
instead of elemental boron ions. The idea comes from
characteristics of most implanters that as a source of boron
ions, BF_3 gas is used. Then the ion current of BF_2^+ molecular
ions has much higher intensity that of elemental B^+. It was shown
already in (71d) that fluorine ions will add to the lattice
damage and thus will help to amorphize the surface. This will
result in a better annealing characteristics of the layers. Those
days, as a drawback was considered that the molecule hitting the
Si surface will fall apart to its constituents and the total
energy will be split. The boron is taking 22%, the other two will
take 39% each, therefore, the penetration depth will be reduced.
It is to be noted that the energy taken by the ions will result
in a good part overtapping penetration, thus the damage of the
fluorine will closely coincide with the boron distribution.

With shrinking device dimensions the "disadvantage" turned
into a major advantage. In implantation machines, at least a few
ten kV must be used to extract reasonable current from the ion
source. This can be too large for modern applications resulting
in too deep penetration. The BF_2^+ technique automatically will
help by reducing ion energies. It has only two drawbacks: the
fast diffusivity of boron and the occurrence of channeling tail.

Two approaches are used to avoid these difficulties: (1) to
preamorphize the silicon and (2) to use rapid thermal annealing,
RTA, to reduce dopant diffusion. An optimalization of
implantation and annealing conditions is made by Wu et al. (86e)
using both of these approaches. It was found that
preamorphization results in a leakage current of three orders of
magnitude higher than without it. The leakage current arises from
post-annealing defects in the n-region. The paper recommends
post-RTA annealing instead of preamorphization. This low-

temperature annealing is believed to help stress relaxation. Similar observations were made by Brotherton et al. (86f), who have found a narrow band of extended defects at the original amorphous/crystalline interface and deep-level donors beyond it. If considerations on metallurgical junction depth are included, it is a difficult optimization task to amorphize deep enough to eliminate channeling and shallow enough to avoid defects in the implanted p^+ region, which are responsible for generation currents.

Application of RTA to anneal BF_2^+ implanted layers is expected to reduce boron redistribution. This is, in fact, true, but there are still special problems. A critical concentration effect was found by Powell (84b). As Figs. 15 and 16 represent, there is a big difference for 2×10^{15} and $6\times10^{15}/cm^2$ BF_2^+ doses, when subjected to furnace and RTA heating. The enhanced diffusion for higher doses is connected with higher damage concentration. SIMS profiles correlate well with carrier concentration measured by spreading resistance probe.

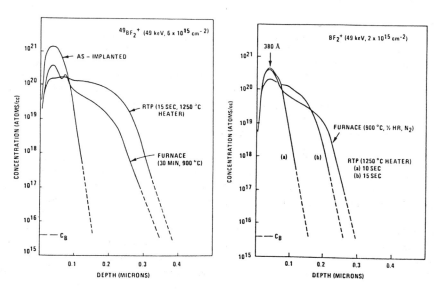

Figures 15 and 16. SIMS depth profiles of boron in BF_2^+ implanted Si annealed in furnace and by RTA for two doses. Note the inverted effect of the two types of annealing (after Powell, 84b)

If preamorphization is applied, the use of Ge ions is recommended by Seidel (85c). In this case, the formation of the so-called "spanning dislocations" can be avoided. Both (85c) and Narayan et al. (85e) found simultaneously that with RTA the boron moves deeper, the fluorine, in contrast, towards the surface.

Delfino and Lunnon (85f) have compared BF_2^+ implantation with BCl^+ with the hope of increasing amorphization effect. Even though Cl has four times higher efficiency in amorphization, the channeling tail was not better reduced. Diodes were comparable for both implants.

When speaking of channeling tails for BF_2^+ implants, care should be taken to implantation conditions. Hansen et al. have shown (86g) that molecular ions, which dissociate after having passed mass separation in the beam line of the implanter, will be accelerated to full voltage in the accelerating tube as individual ions. This can falsify experimental results.

Elements, as As and P were also experimented in molecular ion form. Phosphorus in form of $PF^+, .., PF_5^+$ ions was implanted (87d) and it was found that a saturation effect is found above triply overlapping subcascades. In case of As (86h), As_2^+ molecular implant resulted in an enhanced metallurgical junction, which is deeper than the electrical, as activation is poorer for molecular implantation.

The conclusion is that machine originated and VLSI considerations suggest to use molecular ion implantation, but different problems and trade-offs have to be solved.

III.D. Implantation and Diffusion

Damage annealing in case of implantation doping is accompanied by redistribution of the dopants, i.e. by diffusion. Up to recent years, when VLSI demands applied for shallow pn-junctions, implantation had been intentionally followed by a diffusion "drive-in" process. This was beneficial, because the junction was finally formed in a region, where no direct implantation damage was present. The diffusion in case of an implanted "predeposition" runs differently compared with other predeposition techniques, as the implanted distribution should be considered as a source (limited amount and non-delta function distribution), furthermore, defect pattern is more complicated with implants resulting in non-constant diffusion coefficient.

If diffusion is enhanced in a "real-time" mode by defects produced during implantation, or during irradiation of any kind (neutrons, gammas, protons etc.) after implantation doping, the so-called Radiation Enhanced Diffusion condition is met (Ryssel and Ruge, 86a, Sec. 3.9.3).

For constant diffusion coefficient, first Ryssel made calculations (73c) resulting in a Gaussian distribution modulated by erf terms. For non-constant diffusion coefficient, we refer to the works of Fair (77f). There, different diffusion coefficients are associated with different charge states of vacancies.

Deepest part of the distribution can form "supertails" made up of primary ions trapped into channels then diffused in and stopped at defects (74b).

Radiation enhanced diffusion mostly results in a more abrupt tail region of the distribution (Fig.17), which can even be desirable in certain applications (73c).

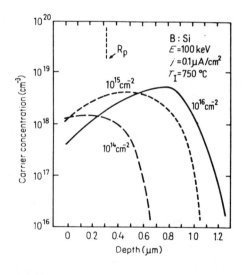

Figure 17. Radiation enhanced diffusion of boron in Si as a function of dose implanted at 750°C

(after Ryssel, 73c)

Experiments on other dopants (Sb, Ga, In) were carried out by Gamo et al.(71e) and Namba et al. (72a).

Depth dependent lattice location after diffusion, in case of some group II and VI implant atoms, was known longsince (Gyulai et al. 71f). However, it was only a recent success of

Servidori et al. (87e) to demonstrate that also diffusion of dopants can be different at different depths corresponding to defect state of that region. In their experiment, B, P, As and Sb were implanted into (100) Si and, following an anneal between 700 and 900°C, rocking curves were obtained in a triple-crystal X-ray diffractometer together with TEM investigations. With this sensitive technique, not only the usual dislocations and clusters were found below the original amorphous-crystalline interface, but after the epitaxial growth, a vacancy-rich surface layer and a deeper region enriched in interstitials were also found. In these layers, corresponding to their defect status, different implants diffuse differently. Thus, B and P show enhanced diffusion in the deep region, where interstitials are abundant, in contrast with As and Sb, which exhibit only a small enhancement. These latter dopants, however, show a somewhat enhanced diffusion in the vacancy-rich surface layer, while the boron there is retarded.

Experiments performed by Angelucci et al. (86c) are in accord with the above findings. They have shown that in case of Si^+ pre-amorphization this effect can severely change profiles of boron and phosphorus, while As and Sb profiles change only negligibly. This effect poses a limitation to application of dual implantation, i.e. to reach "perfect" doping.

Quantitative data for changes in junction depth for dopants and for different annealing/diffusion conditions are shown in Table 1.

The transient behavior of dopant diffusion is to be noted. This means that the defects exert effect at the beginning of the cycle, once they are annealed, even prolonged diffusion does not influence the junction depth.

Another conclusion can be drawn that RTA is, in general, a recommended technique, when shallow junctions are desirable.

As already shown in Sec. III.C, in case of boron, to meet VLSI demands, another idea has to be explored, namely, the BF_2^+ implantation.

Recently, even an exact analytical solution was found to describe diffusion for an arbitrary initial implanted profile with a diffusivity being a general function of dopant concentration and sample temperature (Moroi and Hemenger 87w) provided the gradients of these functions disappear at the front surface of an ion-implanted semi-infinite wafer. The calculation is a generalization of the forms by Ghez et al. (84c).

Table 1. Changes in junction depth for dopants and for different diffusion conditions (from Angelucci et al. 86c)

Annealing conditions	Changes in junction depth (nm)			
	B	P	As	Sb
1100°C, 8s	40	30	0	0
1100°C, 20s	40	30	0	0
900°C, 3min	70	35	25	0
900°C, 10min	65	35	30	0
750°C, 30min	125	75	25	25

Correct description of diffusion following implantation, is extremely important in simulation programs describing IC technology. Especially, when considering shrinking dimensions of the devices that allows only the smallest redistribution of the dopants. Thus, main conditions under which calculations were made is the i) high-temperature short-time anneals (Rapid Thermal Annealing, RTA, e.g. 84d) and ii) low-temperature longer-time annealing (84e). Recently, calculations based on realistic variables were published on the effect of implantation damage on two-dimensional boron diffusion, i.e. calculations describing effects found at mask edges (87q).

III.E. Diffusion and Solid Solubility

High dose implantation can produce atomic concentrations well above equilibrium solid solubilities. For a long time, it was a common belief that in such metastable state no substitutionality could be found for implanted atoms.

Diffusion as a thermal equilibrium process, indeed, is subjected to solid solubility, as diffused Sb profiles from a metastable surface peak show a kink (74a; Fig.18). (In that paper the heavily n-type peak was recommended to produce n^+n junctions in one step.) Electrically active profile for this case (76f)

showed carrier concentrations well above solid solubility
(Fig.19).

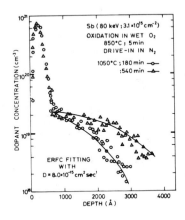

*Figure 18. Diffusion of Sb from
an implanted and wet-oxidation
capped layer. The kink at about
600 nm corresponds to equilibrium
solid solubility limit.
(From Gyulai et al.,74a.)*

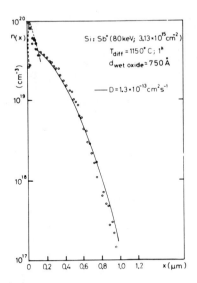

*Figure 19. Carrier concentration
profile of Sb in (111) Si,
diffused under an oxide cap.
Note carrier concentration in
the peak being by a factor of
two above solid solubility.
(From Mayer et al., 76f.)*

The "collapse" of this metastable state during prolonged heat treatment was demonstrated (79h). At 1150°C anneal, the substitutional fraction in the peak area stays constant and first all interstitial atoms diffuse (obeying the law of thermal equilibrium) into the depth (Fig.20).

For higher dose (>10^{16}/cm^2) and higher energy (200 keV) Sb implants with almost full substitutionality were produced (82k) in furnace (no cap layer was formed). Cross-sectional TEM micrographs show antimony precipitations for 850°C anneal at about a depth of 200 nm. Concentrations up to 1.3×10^{21}/cm^3 were achieved. The results have also given an evidence of the formation of non-equilibrium metastable alloys by solute trapping. Upon heating, Sb in this high concentration peak area tends to precipitate.

Si(Sb) 80 keV 10^5/cm^2, 850°C 1min wet O$_2$, T$_{ANN}$=1150°C in N$_2$

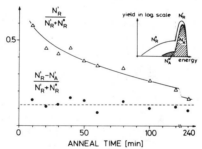

Figure 20. *Stability against "collapse" of a metastable state during prolonged heating. Dashed line shows that the total amount of substitutional antimony stays constant up to a point, where all interstitial atoms have already diffused into the depth. (From Kotai et al.,79f.)*

The same paper (82k) deals also with the behavior of In (125 keV). Again, highly substitutional indium (5×10^{19}/cm^3) was produced. The low melting point of In caused this species to diffuse very fast along grain boundaries in the imperfectly regrown layers, especially, in a 20 nm polycrystalline layer at the surface. Another successful experiment to overcome solid

solubility limit for gallium, again a p-type dopant, was reported
(83n). For low fluences, the substitutional concentration
exceeded 2.5 times solubility. For implants, where the peak
concentration is 1.8 at%, the regrowth again stops at about 20 nm
from the surface. The incorporated Ga here is 1.5 times higher
than the maximum equilibrium solubility limit.

For highest fluences (7.9×10^{15} Ga/cm^2, equivalent to 4.0 at%
peak concentration), only little regrowth occurs.

To conclude this section, a summary graph from a review by
Williams and Short (83o) is shown in Fig.21 for low temperature
(600°C) regrowth on (100) Si.

III.F. Ion Beam Synthesis

Extreme high doses (10^{16}-10^{18}/cm^2) can lead to growth of
different compounds or, at least, when implanted, they represent
the necessary amount of constituents to form chemical bonds
during an anneal thereafter. Usually, to get the new compound
with good structural properties, a heat treatment by some means
is necessary.

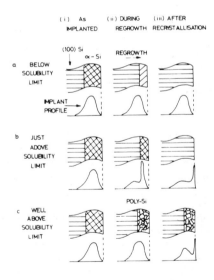

*Figure 21. Schematics for SPE
with high concentration of
dopant and for 600°C regrowth
temperature in (100) Si.
(From Williams and Short, 83o)*

First attempts to form another chemical compound on top of a crystal by implanting proper species dates back to early days of implantation, however, these experiments could have been performed only in a few laboratories, where implanters providing high enough currents were available. Watanabe and Tooi (66a) and Pavlov and Shitova (67a) published early data. Oxygen ion implantation (oxide) isolation of solar cells used in space, was also an early application (Gusev and Guseva 70b).   After a rather long dormant period, Izumi et al. (78h) and Zimmer and Vogt (83p) were demonstrating that good quality devices can be fabricated in a thin crystalline layer of silicon above a buried SiO$_2$ layer produced by hot implantation of oxygen followed by an anneal. The redistribution of oxygen during a post-implantation anneal caused both interfaces to become extremely sharp (83r and 84f) This rather surprising effect had a major impact on usefulness of the technique.

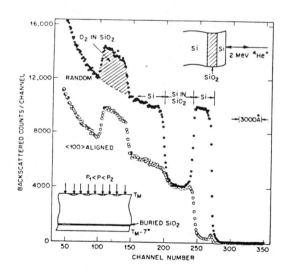

Figure 22. Random and aligned RBS spectra taken after a 30 min annealing at 1405°C. The heating with a regulated power (P) between two values, P_1 and P_2, keeps back surface molten (T_M), (after Celler et al., 86i)

Crystalline quality of the cover layer depends on temperature of implantation and annealing. The first must be optimalized (see later), the second should be the closest to melting temperature (T_M). Celler et al. (86i) have reached the ultimum here: using a self-control of wafer temperature by its optical constants changing at melting. Thus irradiating the back side of the wafer, they could keep the implanted layer at $T_M-7°C$ (Fig.22). With this treatment, excellent quality was achieved.

Already in one of the first reviews (Pinizzotto, 83s), it was made clear that the technique will have problems of cost: it involves about a factor ten difference compared with regular technology.

Cost efficiency is being seriously improved by a new generation of dedicated implanters coming on the market. Special applications (radiation resistant devices) might also help the technique to become practical.

Microstructure of these implanted SOI (Silicon-On-Insulator) layers is still under intense investigation. It was found that islands form during a post-implantation anneal not higher than 1250°C. Silicon islands form in the oxide and SiO_2 islands form in the adjacent silicon (86j).

There is some controversy what the amount of dislocations in the top silicon is concerned. While (86j) states that it was decreasing with increasing annealing temperature, Mogro-Campero et al. (86k) found no such correlation. (The two agree on disappearance of islands above 1250°C anneal.)

Though implantation of oxygen is self-controlled, as it reaches saturation at stoichiometric doses and high-temperature annealing, it is practical to move towards lower doses. Called "subcritical" or "substoichiometric" doses were explored by Stoemenos et al. (86l) and a layer with dispersed islands was found for doses just below critical ($1.3 \times 10^{18}/cm^2$ for 200 keV compared with $1.4 \times 10^{18}/cm^2$ forming a continuous layer at 1150°C). For substoichiometric doses, the implantation temperature turned out to be more critical, than the annealing temperature (White et al., 87f, see Sec.V). In this experiment, to avoid pit formation at high temperatures, 1% O_2 was added to the Ar ambient to maintain a slow oxide growth producing a thin protective layer.

Another candidate for implanted SOI fabrication is the nitrogen. In contrast to oxygen, there is no saturation at stoichiometric concentrations at high-temperature implantation. Again, an utmost high-temperature post-anneal is needed to form a nitride layer. There is a difference in the redistribution behavior of the nitrogen for doses higher than $10^{18}/cm^2$ and below. In the high-dose case, there is very little redistribution in contrast to the dose around $0.7 \times 10^{18}/cm^2$ (85h).

The evolution of a high-dose nitrogen implant with annealing shows cellular morphology after 1200°C. In order to get the desired amorphous Si_3N_4, prolonged anneals at 1250°C or higher, are needed. The high temperature anneal is required not only to reach structural perfection, but to eliminate doping effect of nitrogen, which mainly is localized in the vicinity of the buried layer (Davies et al., 86m).

Figure 23. XTEM micrograph for a nitrogen implanted sample (200 keV, $9.5 \times 10^{17}/cm^2$ at 520°C, annealed for 30 min at 1405°C. Region 1: high quality single crystal Si, Region 2: Si_3N_4 protrusions, Region 3: essentially single crystalline Si_3N_4, Region 4: substrate (after Reeson et al., 87g)

Top quality buried nitride to date was fabricated with 200 keV N^+ to a dose of $0.95 \times 10^{18}/cm^2$ at 520°C and annealed at 1405°C for 30 min encapsulated under a 300 nm SiO_2 (87g). At this temperature, the Si_3N_4 transforms into an essentially single crystalline form with some protrusions towards the surface (Fig.23).

It is an extremely attractive success of Hemment et al. (87c) that good quality doubly implanted layers were fabricated either with deeper nitrid and shallower oxide, or two oxides above each other (Fig.24).

The as-implanted nitrogen profile can be well described with a model developed by Sobeslavsky et al. (87h). The model includes volume swelling and changes in ion ranges due to formation of Si_3N_4. Fitting to RBS spectra is remarkable (Fig.25)

In characterizing SOI layers XTEM and RBS was already
mentioned. Another very promising technique is the spectroscopic

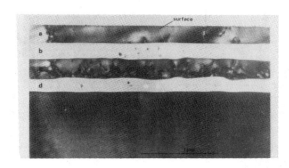

*Figure 24. XTEM micrigraph of a doubly implanted (350 keV and 200
kev, both 0.9x10^{18} O$^+$/cm^2, anneal 1405°C, 30 min (after Hemment
et al., 87c)*

ellipsometry (87i), which, in addition, is non-destructive. With
this technique, as a consequence of the dependence of absorption
coefficient, thus changing light penetration on photon energy,
the characterization of a 1.7 μm thick Si-SiO$_2$ layer is possible
with a spectral range of 1.36-4.3 eV.

To conclude this topic, a few examples will be mentioned to
show some recent applications.

A j-MOS transistor was fabricated on wafers with buried SiO$_2$
(j-MOS on SIMOX). The cross section of the device is shown on
Fig.26 (86n)

The electrical properties of the transistor are superior in
comparison with SOS j-MOS transistors, especially, if breakdown
voltage and off-state leakage current is concerned.

Another device application is a recent high-voltage CMOS
SIMOX by Nakashima et al. (86o). In this application, the oxygen
profile has a secondary peak on the slope of the distribution
towards the surface silicon layer. This layer is called Electric
Field Shielding (EFS). This layer helps in applications for high-
voltage devices by taking off some of the steepness of the oxygen

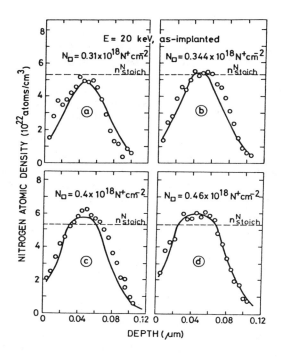

Figure 25. Comparison of calculated (-) and RBS (oo) profiles of 20 keV N⁺ with different doses (after Sobeslavsky et al., 87h)

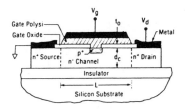

Figure 26. Cross section of a j-MOS SIMOX (implanted buried SiO_2) transistor (after MacIver and Jain, 86n)

distribution. Thus, the drain-to-source breakdown voltage was raised to 180 V.

The device structures also enable to make electrical characterization of the material, as was done by Vu and Pfister (86p) to measure C(V) characteristics, or by Cristoloveanu et al. (86q), who made profiling of inhomogeneous transport properties was made. The conclusion here was that the top of the Si film in the SIMOX structure has high mobility, complete ionization, also shows phonon scattering and has no potential fluctuations and prohibitive stress, similarly to bulk Si. The mobility in the lower part of the film decreases rapidly.

Some materials can grow epitaxially by synthesis + annealing. An example for this is the $Si-Ge_xSi_{1-x}$ system. First, 0.5 at% germanium implanted into silicon was found by Revesz and Mayer to regrow perfectly (79i).

III.G. Focused Ion Beams

In this section an emerging technique will be briefly mentioned without completeness. Direct writing on semiconductors by a focused beam has emerged already in the very early seventies at Hughes Aircraft Laboratories. The technique, however, was treated as a dead-end street by many researchers arguing that if ion currents are high enough for practical throughput, the energy density will produce stable defects and the structure will not be perfect enough for any application.

The idea occasionally came to surface, but it was not long ago that the necessity of direct writing became important. At the same time, beam annealing studies explained those effects (see Sec.III.H), which will also be encountered with focused ion beams (FIB), namely, high fluence effects.

Thus, from point of view of the present chapter, the most important characteristics of FIB is the high current density, 10^{17} ions/cm^{2}s, onto a 10 nm diameter area. Calculation of elementary processes during implantation shows that in this case the next ion impinging into the same area still finds it in the final stage of cooling after the previous cascade. This involves that the next ion hits the surface before the amorphous cluster created by the first ion has been completely annealed.

Thus, on comparison e.g. FIB with boron beam and conventional boron implantation, both subjected to isochronal annealing (Fig.27), FIB will show only i) a small reverse

annealing indicating better amorphousness, ii) a smaller sheet
resistance (84h).

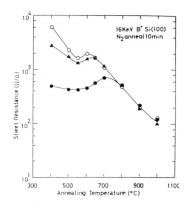

*Figure 27. Isochronal annealing
characteristics of 16 keV,
$1x10^{15}$ boron implanted into
(100) Si; ○ : conventional,
▲ : FIB, 9 cm/s scan speed,
● : FIB, $4x10^{-2}$ cm/s scan speed
(Tamura et al.,84h)*

As a problem, the tail spreading of the FIB implant has to
be solved by a conventional, opposite type dopant implantation
(84i). In this paper, a short channel MOSFET device is described
with a channel length of FIB width.

Critical doses of amorphization for different ions have also
been studied. This dose for boron was about an order of magnitude
lower, while for gallium and arsenic about four times lower than
for regular implantation (T.Tokuyama, 86r).

III.H. Ion Beam Induced Reordering

In recent studies, the topic was called as Self-Annealing
Implantation (SAI) and, if epitaxial growth was involved, Ion
Beam Induced Epitaxial Crystallization (IBIEC). The renewal of a
dormant area (69a and 69b) is justified by the needs of high-
current, high-throughput implantation machines. In those early
studies, it was shown that the residual damage after implantation
is regulated by competitive processes of damage production and
dynamic annealing.

Early conclusions led to a preference of full amorphization
for successive defect-free regrowth (Sec. III.A). However, there
has been a line of studies (73d, 75d and 78i) showing that

epitaxial crystallization of amorphous silicon can be initialized
by irradiation at much lower (200°C) temperature.

Cembali et al. (811 and 821) after showing that relatively
good quality silicon can be produced by SAI with 100 keV, 23
W/cm^2 P^+ (5 s pulsed) beam, that even diodes for solar cell
application can be fabricated.

The amorphization, as shown by early studies (73a), that
damage production depends on energy deposition into nuclear
processes and ion flux, while dynamic annealing on implantation
temperature. Even a study devoted to channeling dependence of
IBIEC (86s) proved that a random direction irradiation is more
effective than a channeled one (as regarded to a crystalline
layer on surface), because of larger energy loss. This concludes
also that the main mechanism of IBIEC is the generation and
migration of point defects within the crystalline material.

A further characteristics of IBIEC stems from its non-
equilibrium nature, as shown by Elliman et al. (87j). Here, Au
atoms were present in the amorphous layer, as a marker for
regrowth. It was found that the regrowth was not inhibited by the
presence of gold up to 1.5 at%, in contrast to thermal epitaxy.

A review by Elliman et al. (87k) summarizes recent results
and leads to conclusions:

i) There exists a critical temperature of implantation
separating the amorphization vs. recrystallization case for
energetic ions with longer penetration than the thickness of the
amorphous layer,

ii) The activation energy of IBIEC, using noble gas ions, is
0.24 eV and the growth was found to be proportional with ion
fluence,

iii) Growth rate at 318°C was 6 $nm/10^{16}Ne/cm^2$; this compares
with 10^{-7} nm/min at that temperature for thermal epitaxy,

iv) Together with growth, the originally rough
amorphous/crystalline interface becomes smoother during IBIEC.

Electrical properties of As implanted SOS after IBIEC with
300 keV Ne needed an additional treatment, CW laser irradiation
in this case, to reach 70-80% electrical activity. However, the
resulting layer after IBIEC+laser treatment had better quality
compared to the control with laser regrowth only (Alestig et al.
871).

It might be premature to foretell the extent of practical
use of the process, but intensive studies are promising. It looks
imperative that the low-loss heat balance of the wafer (including

cooling by radiation) should be kept precisely, which necessitates newly designed target fixtures.

Though related to ion mixing (Sec. VIII.B), it is rather mentioned here that defect processes and RED theory, thus recombination of point defects would account for the temperature dependence of ion mixing in the Si-Nb system. This was shown in a study on dose rate effects for niobium silicide formation by Banwell et al. (86t).

IV. EXTREME LOW ENERGY IMPLANTATION

The extreme low-energy implantation is on the margin of the scope of this review. Low-energy ions bombarding semiconductor surfaces occur in different technology steps, many of them even without any ion optics involved. Especially important are bombardment and defects during process of reactive ion etch (RIE). Implantation with keV ions, however, is not only a negative effect, but can be useful in producing shallow junctions. Here, the topic will not be reviewed, only a few references will be given. Junctions produced with bombardment of 5 keV boron, 1.8 keV phosphorus activated with pulse annealing was reported by Kachurin and Stenina (82m), and 500 eV As implantation resulted in above solid solubility, good quality junctions (83t). Recently Davies (85i) reported excellent abrupt junctions of about 100 nm deep with boron implantation and without preamorphization.

Another approach by Natsuaki et al. (85j) to fabricate 100 nm junctions of boron for CMOS, avoided extreme low energies involving many technical problems, by implanting through WSi_2 layer (Fig.28). WSi_2 as a screen layer has another advantage, namely, effect of a potential knock-on implantation is minimized: tungsten is heavy, silicon is undistinguishable.

A work on tailoring of Schottky barriers with low-energy implantation will be briefly reviewed in Sec.VI.B.

V. HIGH ENERGY IMPLANTATION

The use of high (MeV) energies for implantation is another attractive extension of the technique, because of deeper penetrations of the ions. It is an advantageous property of ion stopping that high-energy ions leave the near-surface region

relatively unaltered, i.e. defect free. Thus, a buried
distribution can be expected. On the other hand, range straggling
is also growing with R_p, therefore, if a narrow distribution is

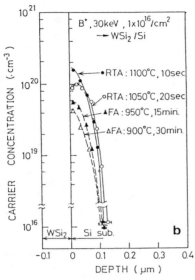

Figure 28. Carrier concentration profiles of boron annealed in furnace (FA) and by rapid thermal annealing (RTA) (after Natsuaki, 85c)

the goal, only some secondary treatment can bring the system
closer to that state. The sharpening of oxygen implanted layers
with heat treatment is an example to this. However, as described
in (87f), when using implantation with two energies, one of them
being 3.5 MeV, oxygen islands in the deeper layer will not
coalesce even with an anneal at 1400°C, while the top layer forms
perfectly (Fig.29), in consequence of the increased straggling.

The high energy implantation, when expecting to be
implementation into technologies, faces obstacles from the
equipment side. To date, only a small market has pushed
development in the direction of rather simple, reduced-cost MeV-
range accelerators with acceptable throughput. The demand was
only recently formulated.

The reader is referred here to a review paper by Cheung
(85g) Though the interest of this chapter is towards annealing,
it is to be mentioned that even data on ranges of heavy ions in
crystalline silicon were not available. Here, we refer to a study
of Wong et al. discussing ranges of 3-11 MeV As, 1-7 MeV P and 1-
4 MeV B ions (87x). It was found that profiles can be adequately

described with Pearson IV distribution. According to Spinelli et al., utmost care has to be taken to avoid channeling (87m).

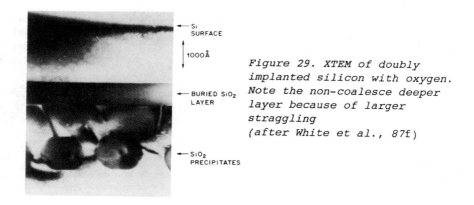

Si
SURFACE

1000Å

BURIED SiO₂
LAYER

Figure 29. XTEM of doubly implanted silicon with oxygen. Note the non-coalesce deeper layer because of larger straggling (after White et al., 87f)

SiO₂
PRECIPITATES

One of the perspective application of high-energy implantation is the threshold voltage adjustment of MOS devices through layered structure of the gate. This is a demand coming from the spreading use of half-finished wafers (gate arrays) by IC customers.

In Section II, low-dose implantation was quoted as a routine. For high energies, it still needs some investigation. Deng et al. (87n) have investigated 750 to 900 keV boron implants in the dose range of 10^{12} to $10^{14}/cm^2$ into an n^+poly - Si(500 nm) - PSG(400 nm) - n^+poly - Si(500 nm) - SiO_2(40 nm) - Si structure. For this, an 850 keV boron has a range of 1.52 μm. For doses $<10^{13}/cm^2$, the expected threshold voltage (V_{TH}) shift was 90% achieved during a 15 sec 750°C RTA. Higher temperatures increase activation efficiency with full activation at 1000°C. Above a certain "damage recovery" temperature, T_R, the V_{TH} will not change. For a dose of $3 \times 10^{13}/cm^2$ T_R = 550°C, for $10^{14}/cm^2$ T_R = 750°C.

In contrast to low energies, where spatial distribution of a threshold voltage adjustment implantation can be considered as a delta-function, here realistic values of range straggling has to be taken and the Poisson equation has to be solved. With a value of 0.15 μm straggling, theoretical V_{TH} values are displayed on Fig.30. If boron peak is getting deeper in the substrate, the experimental V_{TH} values decrease from the theoretical ones. Full activation can only be achieved for low doses. In that case, 750°C 15 sec was enough, however, for high doses, no full activation was found even for 1000°C anneals (87n).

The damage associated with this implantation correlates partly with leakage current. Compared with unimplanted devices, the leakage current increases by three-orders of magnitude, but for an RTA of 15 s 600°C and above (in forming gas), leakage current dramatically drops even below values of the unimplanted ones. In contrast to this, interface trap density is increasing

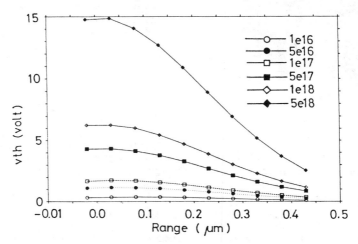

Figure 30. Calculated V_{TH} vs implant range for peak boron concentrations at 850 keV and for different doses, (after Deng et al., 87n)

with near-surface concentration (rather than dose) and it is stable against annealing. Therefore, care must be taken to choose proper energy.

These secondary defects were studied for 2 MeV boron by Tamura et al. (86u) by XTEM. Doses in this study were 2×10^{13}-$5 \times 10^{14}/cm^2$). Fig.31 shows an RTA study of secondary defects. It was found found that all defects lie on (111) planes, therefore movements will be through glide mechanism. It is another new feature of high-energy implantation, that the secondary damage peak lies deeper than the primary one, which is associated with nuclear energy deposition (Fig.32).

Still a low-dose, high energy application, but using both p-type (boron) and n-type (phosphorus) implantation resulted in a twin-wellp$^+$p epitaxial CMOS process, where both wells are formed with implantation (Stolmeijer, 86v). First the p-well by 350 keV 1.5×10^{12}/cm^2 boron, then the n-well by 1 MeV 9×10^{12}/cm^2 phosphorus (doubly charged technique). For the boron the peak is located just under the field oxide to keep V_{TH} of the n-channel transistor high and to minimize effect of the parasitic lateral npn transistor. Then, V_{TH} is set by a low-energy B$^+$ implant. The n-well should be deep to maintain high dose with low surface concentration, again to supress the parasitic pnp transistor. The

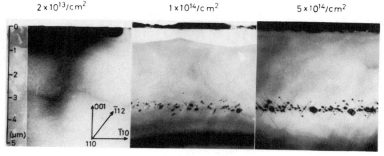

2 MeV, B*, 800 °C, 15 min

Figure 31. XTEM micrographs on secondary defects formed for 2 MeV boron (after Tamura et al., 86s)

technique simplifies CMOS technology by avoiding time consuming steps. The wells with retrograde profiles will minimize latchup sensitivity, together with a high transconductance and high V_{TH}, while susceptibility for punchthrough stays low. (A technical problem, similarly to the molecular ion case, can arise: doubly charged P^{++} ions may change charge state before and inside the accelerating tube resulting in a shallow rising edge of the profile; the solution, again, is good vacuum and near-target mass separation.)

Another technology using 1 MeV n-type retrograde well and self-aligned TiSi$_2$ (Salicide) with high latchup holding voltage was published by Lai et al. (86w). To our topic, again the leakage current is relevant for the reverse-biased polysilicon-gated p$^+$n-well diode. Fig.33 shows the leakage current with a factor of ten jump at 2-3 volts. This refers to defect generated g-r centers at a depth of 0.66-0.76 um, i.e. 0.3 um shallower

than R_p. Nevertheless, the value of this current is only 0.01
fA/μm^2.

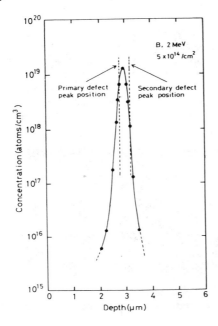

Figure 32. SIMS profile of boron after 800°C 15 s RTA compared with primary and secondary defect peak positions
(after Tamura et al., 86s)

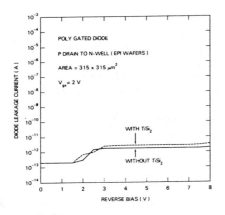

Figure 33. Leakage current for the reverse-biased polySi-gated p$^+$n-well (Salicide) diode
(after Lai et al., 86w)

The location of secondary defects relative to R_p is still under debate. Similarly to the boron case, Tamura and Natsuaki (86x) have found secondary defects for phosphorus centered at 2.1um with an R_p = 1.9 um for 2 MeV P^+, also somewhat deeper than R_p (Fig.34). The defects formed above a critical dose of somewhat below 10^{14} P^+/cm^2, were again unpleasantly stable.

In contrast, Rai et al. (87o) report for 6 MeV Si somewhat shallower damage peak than R_p.

The behavior of arsenic studied by Byrne et al. (82n) and silicon (87p) is similar. In both cases the lower amorphous / crystalline interface is more abrupt and the recrystallization rate is equal for the upper and lower interfaces. Complete regrowth of the amorphous layer occurs at 750°C.

MeV implantation now needs even data accumulation e.g. for ranges. Though spectacular applications are around, it looks as a tough problem to fight extreme stability of secondary defects.

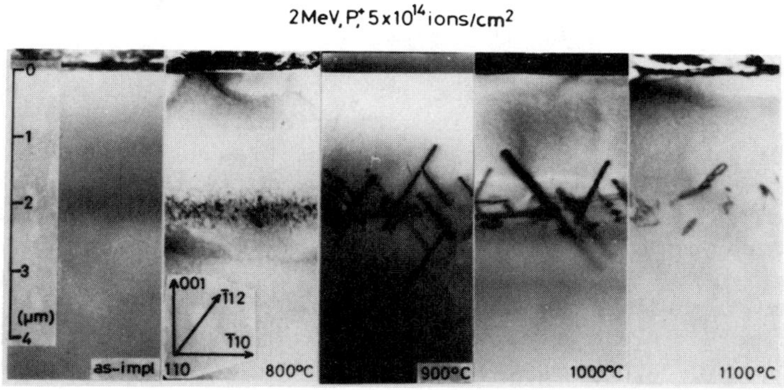

Figure 34. XTEM micrographs of secondary defects for 2 MeV 5×10^{14} P^+/cm^2 for 15 min isochronal anneal (after Tamura and Natsuaki, 86x)

VI. SOLID STATE CHEMICAL REACTIONS ON IMPLANTED SURFACES

Apart from the fact that ionic or metallic bonds might be involved, the basic physics is very similar for SPE growth and for the growth of layers with different composition on a surface.

High-dose ionic synthesis was one extremity. In the present section, growth of SiO_2 or different silicides will be dealt

with, when implanted species and implantation damage is present
in the near-surface region.

The effect of the Fermi level on lattice location (79e) and
on SPE growth (82j,82r) through changing vacancy concentration at
the interface clearly   influences   the probability of atomic
jumps, i.e. chemical reactions. The state of the interface,
therefore, is extremely important.

VI.A. Oxidation of Implanted Surfaces

The most important chemical reaction on the silicon surface
is thermal oxidation.

The influence of implantation on oxide growth rates for some
metals became reasonably well understood (76g). In case of metals
(titanium), the growth rate correlates with differences in
electronegativity.

In semiconductors, the damage effects are also important.
The presence of impurities at the growing front causes new
phenomena (e.g. Deal et al., 67b). Depending on temperature and
impurity species, the segregation coefficient, oxide growth rate
and impurity diffusion coefficient will compete and may results
e.g. in "snow plow" and "push out" effects. The growth rate is
also influenced by the presence of impurities.

If impurities have been implanted, the radiation damage and
high local (even above solid solubility) concentrations are the
new features that cause a difference.

Meyer and Mayer (70c) observed first the enhancement of
oxide growth rates on implanted surfaces. Fritzsche and Rothemund
(73e) reported, e.g. passivating effect of nitrogen and have
called attention on importance of the fact, when the dopant can
form a stable compound with the silicon or with SiO_2.

Strong   redistribution   in   form   of   snow   plow   effect   of
implanted As (Fig.35) and Sb was shown (75e). Further studies
(75f and 77h) investigated dopants i) bonded covalently causing
only damage (Si and Ge), ii) direct oxidizing (O) , and iii)
having some ionic character in Si (B,Al,Ga,P,As,Sb). Strong
inhibition was measured for Ge and Ga, while strong enhancement
was caused by As and Sb. Neither straight correlation was   found
either   with   electronegativity   differences,   nor   definite
connection with damage, structure, apart for the case of Ge and
Si implants, where damage effects must play role.

An important study in this matter was performed by Gotzlich
et al. (79j). Here, most important dopants in IC technology have

been investigated in details. (Boron enhances growth by 1.1 times; Ar, 1.3 times; Sb, 3.5 times and As, 7.5 times for a dose of $10^{16}/cm^2$). The study concluded in an empirical formula, which fitted the phenomenon of enhanced/inhibited oxide growth into a technology simulation program (ICECREAM).

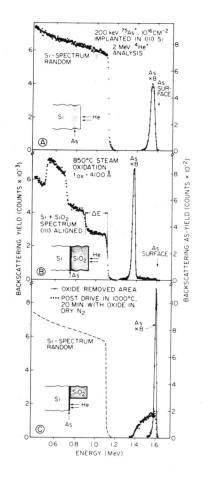

Figure 35. Random backscattering spectra taken on (a) as-implanted, (b) oxidized, and (c) diffused arsenic implanted (111) silicon. (From Muller et al.,75e)

Most recent systematic investigation on oxide growth on implanted surfaces was made by Williams and Christodoulides

(81m). Here, twenty five dopants were checked either with various doses or with a single medium-to-high dose implant at 20-80 keV into (111) Si. Samples were first annealed in Ar at the intended oxidation temperature then oxidized (700-900°C). This paper also summarizes the character of redistribution for dry oxidation.

Definite conclusions are still not possible, apart from the fact that impurities, in general, decrease bond braking energy at the Si-SiO$_2$ interface.

Recently interest turned towards oxidation of surfaces bombarded with fluorine, especially, as a consequence of plasma oxidation in the presence of halides. Implanted (40 keV) fluorine enhanced thermal dry oxidation rate with an accumulation at the Si/SiO$_2$ interface. Some F atoms form SiF$_4$, which is mobile in the oxide and has a tendency to desorb (Kuper et al. 86y).

We quote as an important result, but falling only marginally to the scope of this chapter, the direct oxidation with an extreme low-energy (<100 eV) oxygen beam, which resulted in a thin (4 - 5 nm) high-quality oxide, good for FET gate applications (86z).

VI.B. Silicide Growth on Implanted Surfaces

Another important compound formation on silicon surfaces is the silicide formation. Silicides of near-noble and refractory metals are used in, or as ohmic contacts, Schottky barriers and interconnects. Stable silicide, or ones with predictable reaction properties serve as barriers against Al atom migration, etc. In numerous technological solutions, the silicides grow on previously implanted surfaces. (Though it is a low-dose and low-energy application of implantation, Schottky barrier heights are tailored, e.g. by Sb implantation+anneal, followed by a Ti-W evaporation and low temperature heat treatment, 79k, see also Sec. IV.)

Reactions in presence of dopants (impurities) run analogously to the oxidation case,. i.e. dopants may segregate in either phase, may be snow plowed, etc. If precipitations or stable compounds form at the interface, growth will be inhibited.

First Wittmer and Seidel (78j) described snow plow of implanted arsenic for near-noble silicide growth at rather high temperature (500°C). The snow plow occurs at a temperature as low as 250°C during Pd$_2$Si formation (81n). Figure 36 shows the redistributed arsenic on the silicon side as measured by neutron activation analysis and step-by-step anodic removal. Comparison of cases with and without preannealing (prior to metal sputtering) shows that the reordering of the lattice and with greater part of As in lattice sites, enhances the amount of arsenic being pushed. This is connected with the slower growth rates in the range, where dopant concentration is high. It is important here that the snow plowed arsenic is partly substitutional. Detailed studies of As profiles and substitutionality (82o) show that 50 % of the arsenic is substitutional.

Electrical measurements (83u) point on the usefulness of this snow plow effect, as Pd_2Si contacts can be formed to shallow junctions, because the junction will be displaced instead of becoming leaky.

The behavior of implanted oxygen and nitrogen into Ni-Si system was investigated (81o and p). Both elements are forming a barrier against silicide growth, when initially implanted in the nickel film. A blocking SiO_2 gradually builds up, while nitrogen rapidly moves to the interface and forms a barrier. When dopants are initially in the silicon, they incorporate into the silicide without forming a barrier, though the nitrogen is retarding the growth. A model is proposed that oxygen bonds with the moving species are much weaker than those with stationary species.

It has been shown, again by Scott and Nicolet (83v), that the growth of Pd_2Si is not affected by implanted oxygen.

Titanium silicide is another silicide, which has wide applications, therefore, it has been investigated (83w), when grown on implanted surfaces. Dopants were Sb, As, P and Ar, O. The growth rates were much slower on the implanted sample and the layer thickness formed is about the half of the one grown on non-implanted surfaces (Fig.37).

Atom redistribution showed different pattern than for Pd_2Si. Here, antimony for 600°C started to accumulate on the surface of Ti and the Si-$TiSi_2$ interface. At higher temperatures, Sb outdiffused from the sample. Arsenic, on the other hand, distributed evenly in the silicide up to 1.5×10^{20} atoms/cm^2.

The even redistribution of arsenic is a fast process both in furnace annealing, both with RTA. For furnace annealing, as it was shown by van Ommen et al. (86aa), 800°C anneal is necessary to remove radiation damage. In that paper, slight oxidation was recommended to avoid As loss at the surface. Yu et al. have found (87r) that the first phase formed in the presence of As was Ti_5Si_3. Connected with the formation of the final phase, $TiSi_2$,

the As redistributed into two peaks: one at the interface and one
at the surface, where it is partially lost by sublimation.

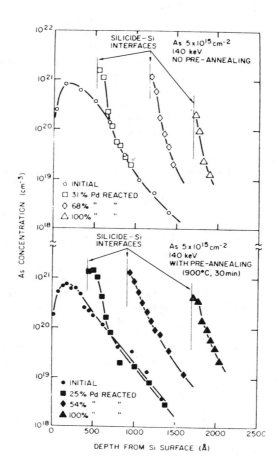

*Figure 36. Redistributed arsenic profiles during Pd₂Si growth at
250°C for 17, 70 and 375 min., respectively. Measurement were
made with neutron activation analysis combined with anodic
stripping. (From Ohdomari et al.,81n.)*

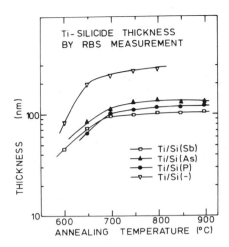

Figure 37. Thicknesses of TiSi$_2$ grown on implanted surfaces. Implantation: Sb, 75 keV, 1.25x10^{16}/cm^2, P and As, 80 keV, 1.25x10^{16}/cm^2; annealing: 20 min. isochronal. (From Revesz et al.,83w.)

Optimization of RTA parameters yielded low resistance (about 14 uohm cm) and uniform silicide layers by redistribution and activation of As (87s).

Ion mixing was found to be successful in formation of ohmic contacts with titanium silicide (Kwong and Alvi, 86bb). Here double diffused phosphorus implants in an ion mixing mode were used to form the n+ layer. In some cases, As was implanted into the already formed silicide as a source material for the n+ layer formation. Contact resistance of 0.95 ohm was attained. The lowest leakage current density was 0.02 fA/μm^2. The technique fits well into self-aligned MOS fabrication.

If pn junction is the goal, boron implantation through the Ti layer followed by RTA yielded a structure of TiN/TiSi$_2$/p$^+$-Si/n-Si with good electrical properties and excellent thermal stability. The spontaneously formed TiN proved to be a good barrier under the Al (85k).

Boron implanted as BF$_2$ into Si prior to Ti evaporation and silicide formation, also moved into the silicide and outdiffusion

was also found. The fluorine coalesced at the interface without
serious effect on contact resistance (851).

Argon and oxygen, while forming precipitates in the Si can
completely block $TiSi_2$ growth. For oxygen, TiSi and Ti_5Si_3 phase,
for argon, only TiSi phase was observed (83w).

Another important silicide is the $MoSi_2$. Mo-Si reaction on
As implanted surfaces results in As incorporation into the Mo
layer, predominantly near the interface. The contact has
rectifying characteristics, as no enhanced diffusion of the As
into silicon occurs at 600°C (Ohdomari et al. 86cc). The cause of
this is a consequence of the moving species being the Si, as no
point defects are created under this condition.

It is out of the scope of this present review to summarize
impurity effects on silicide growth, in general. Examples were
selected, where implanted species had an influence on compound
formation and electrical behaviour of the layers.

In the next paragraph, an application is briefly reviewed,
where damage annealing is believed to cause a direct effect.

VII. IMPLANTATION, DEFECTS AND GETTERING EFFECT

In semiconductor technology "gettering" is a procedure where
defects are intentionally produced at a distance from critical
(functional) interfaces, and by a subsequent heat treatment,
unwanted mobile ion impurities move from the space charge region
to the damaged area, which is, preferentially, on the back
surface of the wafer or lies laterally far away. There they form
stable complexes with the vacancies. The use of ion implantation
to produce the necessary defect density was described first by
Buck et al. (72b) then Seidel and Meek (73f). In (75g), it was
shown that the gettering efficiency is related to the residual
disorder after the treatment at 850-1150°C. Based on the work on
thermal history (76h), and dependence of regrowth on substrate
orientation (75c), Sigmon et al. (76i) have proven that gettering
efficiency will also be related with details of the annealing
cycle and with the orientation of the wafer. For their
implantation (Si, multiple energy, $4x10^{15}/cm^2$, LN_2 implant) and
annealing (1000°C 1 hr) conditions, the gettered gold amounted
$5x10^{13}$ atoms/cm^2 on (111) silicon, but no measurable gettering
was found for (100) Si.

It was demonstrated (75g and 791) that most commonly used
implant atoms are reasonable or good getters, including noble

gases, phosphorus and, as was discussed, $BF_2{}^+$ ions. In the latter paper the minority carrier lifetime in gold gettered layers was also measured.

Figure 38 shows an example (81q), where 10^{14} and $10^{16}/cm^2$ gold was gettered by Ar implantation and annealed at different temperatures. Lower concentrations of gold were measured by DLTS and the higher by RBS. Triangles and squares are results on samples, where high beam currents prevented amorphization during Ar implantation.

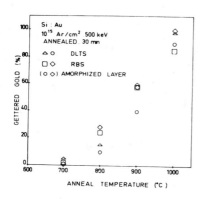

Figure 38. Gettered gold percentage for an initial gold doping of 10^{14} and $10^{16}/cm^3$ as a function of annealing temperature. Denoted by "Amorphized layer" is made with low beam current during Ar implantation. (From Lecrosnier et al.,81q.)

A review on gettering, including III-V compounds was published by Lecrosnier (83x).

Implantation with oxygen can be used for internal gettering of defects to reduce somewhat the temperature to achieve oxygen to precipitate (87t). The implant was placed near the surface and an epitaxial layer deposited at 1030°C served for the capacitor and diode structures. It was shown that the defect region will not extend into the epi layer. Gettering efficiency was good in the buried layer.

VIII. DEFECT ASPECTS OF INDIRECT BEAM EFFECTS

In this section, two effects are briefly summarized. We say indirect, because these effects involve atomic layers initially present on the target. One of them, the recoil implantation, is usually a parasitic effect, the other one, the ion mixing, is a developing area. As this chapter is mainly devoted to damage and its annealing, the review on these indirect effects will not cover all aspects.

VIII.A. Recoil Implantation

For distribution of recoiled atoms we refer the reader to special review articles.

Sputtering and recoil implantation are closely related, therefore, first calculations of Nelson (69c) and Sigmund (69d) considered recoil as forward sputtering. These calculations resulted in shallow depths. Later reviews were published by Moline (77i), Winterbon (79m) and Littmark and Hofer (80f). The problem is of great interest in i) shallow depths required by VLSI, ii) through-oxide implantations, iii) for material removal, "machining" by sputtering, and iv) in case of poor vacuum conditions. More recent range calculations based on the Boltzmann equation by Christel et al. (81r) are referred to, as values for oxygen recoil.

If vacuum conditions are poor during implantation, an unintentional coverage of the wafer will occur leading to recoiled carbon. This recoil, mainly for high-mass projectiles, can be reasonably deep (77i).

The effect of oxygen on regrowth was discussed in (77j). The implanted, recoiled oxygen had a surface concentration of 10^{21} atoms/cm^3 to account for the reduced rate of regrowth. An "anomalous" residual defect was described by Natsuaki et al. (77k), which was produced by recoiled oxygen when heated to 1000°C in dry N_2 atmosphere. The normally present honeycomb-like defects were transformed to an even more complex structure, which resulted in reduction of breakdown voltages.

A good way on damage reduction after a "through-oxide" arsenic implant is given by Hagmann (83y). In a set of samples, combinations of screen oxide strip, preoxidation (800°C, 30 min.) anneal at 1100°C, 75 min., reoxidation and strip, were tried. The check of residual disorder caused by recoil oxygen was made by visual inspection after a Wright etch. Implantation temperature was also a parameter as adjusted by ion current density (between 1 and 8 µA/cm^2). Figure 39 shows results of defect densities for different target temperatures and for different process sequences. Both processes started with growing 22.5 nm screen oxide (at 1000°C) followed by implantation, then a partial (10 nm) strip of the oxide. After this, process A:1100°C, 75 min.

anneal, reoxidation (970°C 95 min.) and strip. Process B: total strip of screen oxide, pre-oxidation (800°C, 30 min.) then as A. Both procedures yield zero defect densities for the smallest flux during implantation. High current densities were satisfactory also when screen oxide has been removed before annealing. It is believed that low energy implantation and high temperature anneal favors oxygen outdiffusion and defect precipitation and annihilation.

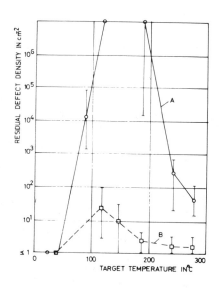

Figure 39. Defect density of 50 keV, $2.1x10^{16}/cm^2$ arsenic implantation and after anneal at 1100°C, 75 min., and reoxidation. A and B are two process sequences, see text. (From Hagmann 83y.)

In a recent work Hagmann et al. (86dd) propose a mechanism to understand defect reducing effect of low temperature oxidation. Oxidation rate is enhanced by point defects generated during implantation. These defects anneal better for higher oxide growth rates. This latter can be increased by removing the screen oxide. In practice, the removal of the screen oxide can only be done approximately, as only buffered HF can be applied. This might be enough, however, to ensure increased oxide growth conditions. Omitting screen oxide at all is less beneficial, because oxygen knock-in is important to produce proper defect structure.

Magee et al. (81s and t) have shown that the oxygen is gettered in residual damage regions, and at temperatures of 1000°C, it outdiffuses into the deposited oxide.

The nature and annealing behavior of this center (83z) was analyzed by ESR. The center was identified as gettered oxygen into dislocation loops or microtwins. If the concentration of oxygen exceeds $10^{20}/cm^2$, this center is stable even at 1000°C.

As a conclusion, low current densities are essential when using through-oxide arsenic implants.

VIII.B. Ion Mixing

Ion mixing is a procedure where energy of the impinging ion is used to produce atomic motion in a layered target. The procedure often ends in formation of phases, compounds. The advantage of the method is in that, say that for $10^{15}/cm^2$ incoming particles (e.g. Xe^+), $10^{17}/cm^2$ target atoms will be involved in the reactions, especially, when the target is at elevated temperatures. As the process is understood now (81u), the energy transfer in elastic collisions produces a recoil mixing and generates defects, which cause atoms to diffuse (radiation enhanced diffusion). Higher target temperature favors this latter.

It is essential that bombarding ions should reach all interfaces of the layered structure to produce mixing. This points on the fact that mixing is not just a heating effect. Furthermore, it is not just a nuclear procedure either.

In a chapter on damage annealing, the ion mixing will not be a major topic, but as a versatile technique leading to stable or metastable phases, should be briefly mentioned and review papers should be referred to, e.g. Tu and Mayer (78k), Mayer et al. (81u), Lau et al. (83aa).

Although for phase forming metallic multilayers, Hung et al. (83bb) have shown the role of radiation damage during mixing: in Al-Ni, Al-Pd and Al-Pt multilayers only the AlNi, AlPd and AlPt phases were present being the most "radiation hard" ones with simplest elementary cell and as most probable atomic mixture, though in thermal process many other phases form and are stable.

The evidence of the role of defects formed during mixing with 250 keV Ar^+ ions in the Si-Ni system was described by Gerasimenko et al. (83cc). Inhibiting effect of these defects for silicon atom movement during mixing causes that for lower fluences $NiSi_2$, for higher fluences Ni_2Si phase was formed, the previous with sharp interface, thus high barrier after an anneal at 300°C.

IX. ANNEALING STRATEGY; RAPID ANNEALING TECHNIQUES

Still in stage of development, but many of the new techniques are prospective in semiconductor processing. Certainly, before accepted in production, research has to prove advantages and point out most fruitful applications. As a rule, solid phase processes are more advantageous, because
   I) Dopant redistribution is negligible,
   II) the structure is brought closer to equilibrium,
   III) defect reactions can be simple,
   IV) lower temperatures are involved, thus compatibility is better with other technology steps,
   V) mechanical stresses are more-or-less limited to constitutional problems,
   VI) mass handling, where applicable favors reproducibility.

Disadvantage can be I) in lengthy operation, II) it may lead to, or has to accept thermally stable defects, III) it is subjected to equilibrium thermodynamic and, thus usually, to solid solubilities, equilibrium segregation coefficients, substrate orientation, etc.

In this paragraph, a short summary will be given on an expanding and potentially important area in semiconductor processing, the use of the so-called rapid annealing. This terminology includes novel heating techniques of any kind.

In recent years hegemony of furnace annealing has changed. Furnaces with well-controlled ramp-up, keep and ramp-down cycles produced structures the closest possible to thermal equilibrium. Aging of devices is the best, when near to equilibrium states were reached during fabrication. Limitations to this strategy are various. Though in at furnace with a parallel-type treatment, i.e. dozens of wafers are processed simultaneously, the throughput is a virtue, but equilibrium excludes the use of so-called quenched (metastable) states. VLSI considerations with shrinking dimensions tend to keep dopants localized in a small volume, demands of keeping layered structures thin and unaltered, have pushed research to find new directions. Laser annealing (LA), (Shtyrkov et al. 75h) looked for some time as a solution,

but up to the present day it has not proven overall
applicability. In fabrication of 3-D devices controlled depth,
i.e. cool-wafer annealing techniques look inevitable, but there
are still enough reserves in today's production technology to
resist non-standard processing techniques. As in laser annealing
the coherent nature of the source did not play, insted of
expensive lasers industry moves in the direction of still
optical, but simpler annealing techniques, i.e. towards lamp
annealing. This is called Rapid Thermal Annealing, RTA or RTP.
Here, thermal treatment lasts for seconds compared to tens of
nanosecs for LA.

RTA is a single-wafer technique, but throughput can be
acceptable. Being optical, heating depends seriously on optical
parameters of the actual structure. In usual arrangements heat
loss by conduction is minimized. This means the wafer being held
hanging at few points. Because of unfavorable optical absorption
of Si for thermal radiation, high-power systems are necessary.
Temperature is limited by restrictions to solid phase processes.

Though in the previous sections results by furnace annealing
or RTA were discussed parallel, an important feature of
successive application of furnace and RTA should be pointed out
here. The 800°C though not very high temperature, in case of
boron, will seriously deepen the junction because of the

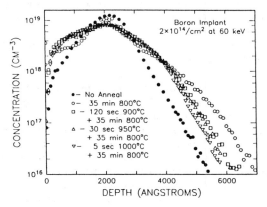

*Figure 40. Reduction of anomalous diffusion during 800°C,
35 min anneal preceded by different RTA cycles (after Michel et
al. 87u)*

occurrence of anomalous diffusion in the tail region. A furnace anneal for 35 min can push the tail by 100-200 nm (87u). If this furnace anneal is preceded by an RTA, the anomalous diffusion can be reduced dramatically (Fig.40).

In this paper, instead of the usual explanation for the anomalous diffusion involving boron interstitials, because of the strong temperature dependence of the reduction, an explanation based on dissociation of the clusters of point defects is favored.

Another aspect of RTA was pointed on by Borenstein et al. (86ee). Here, the problem of transition metal impurities (Fe) present originally in the silicon can cause more problems for RTA than did for furnace annealing. In the latter case deep levels of iron were detected for many silicon material.

Redistribution of arsenic, though during higher temperature RTA, was investigated by Nylandsted Larsen et al.(86ff). The electrically active arsenic had a peak concentration of $2.5 \times 10^{20}/cm^3$ for almost all temperature annealing, but the total arsenic profile was higher till 60 s RTA at 1090°C, when the two profiles coincided without As loss.

A clear example on complex nature of redistribution of dopants during RTA was shown by Natsuaki et al. (85j), where implantations were made through WSi_2 layer. Following RTA, the picture depends on the species: boron tail diffuses, phosphorus shows accumulation at the interface, the same holds for arsenic with addition of some outdiffusion (Fig.41).

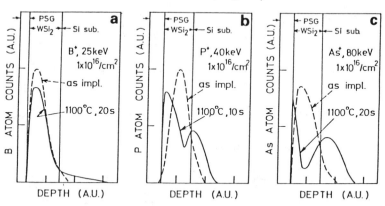

Figure 41. Redistribution of impurities by RTA for implants through a 100 nm WSi_2 overlayer (after Natsuaki et al., 85c)

There are different ways to categorize these annealing techniques.

I) Duration   Characterization

    Picosec (Laser) Pulse
                                       - LPE
                                       - Small redistribution of
                                          dopants

    Nanosec Laser   - LPE
        Electron   Pulse   - Large redistribution
        Ion   - High solubility
                                       - Cold substrate
                                        (3-d IC formation)
                                     - Metastable phases

    Scanned CW Beam   - SPE mostly
                                     - No redistribution
                                     - Little substrate heating

    Strip heater   - LPE
    (also halogen, arc, and   - Large raedistribution
    electron)   - Hot wafer (No 3-d, only
                                        layer preparation)

    Larger Area Heater   - SPE or LPE
    (1-10 sec)   - Litle/large re-
                                        distribution, resp.
                                     - Layer preparation
                                     - Entire substrate
                                     - heated (No 3-d)

    Furnace   - SPE
                                       - Good activation, large
                                     redistribution

II) Purpose

    - Preparation of   - Laser pulse
      undoped substrate   - Strip heater
      layer on insulator   - Large area heater (LPE)
                                 - Electron line heater

    - Preparation of doped   - Large area heater (LPE)
      substrate layer on
      insulator

    - Activation of dopants   - Halogen source or large
                                 area lamp, (SPE)

    - Metastable supersaturated   - Laser pulse
      doping

    - 3-d growth doping   - 1st, Laser strip heater
                                 large area (LPE)
                               - 2nd, Controlled depth
                                 LPE: laser, eleactron,
                               ion pulse

Numerous reviews have been written recently on this topic. In a course on implantation, the review by Hill (83bb), Hess et al.,(83dd) and the special issue of Journal of Crystal Growth, edited by Cullen (83g) should be mentioned.

Summary graphs by Hill (82p and 83dd) are exceptionally tutorial. Figure 42 shows areas in technology, where beam processing can add new dimensions. (Details on figure caption.) From implantation and damage annealing point of view, the most important are (3) annealed implantation damage with no dopant redistribution (Sec.III A-C), (5) supersaturated solid solution of dopant with redistribution (7) reacted polysilicon and metal to form silicide. Secs.VI.B (81u, 82d) and 82q), this point should be extended by reacted metal-metal layers, again Secs.VI.B

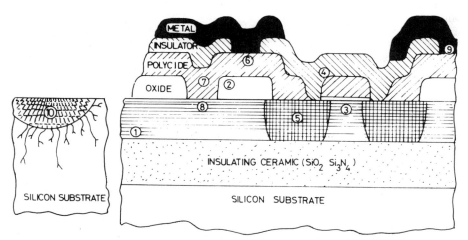

Figure 42. Diagram illustrating the applications of beam processing to integrated circuit structures (1) regrown single crystal on amorphous dielectric, (2) contoured oxide edge, (3) annealed implantation damage w/o dopant redistribution, (4) contoured step in polysilicon/polycide, (5) supersaturated solid solution of redistributed dopant, (6) smoothed polysilicon to give high quality oxide for capacitors, (7) reacted polysilicon and metal (silicide), (8) cleaned surface for Schottky contact, (9) ohmic contact, (10) gettering by beam induced damage. (From Hill,83dd.)

(83aa and bb, 83ee and 83ff), (8) cleaned surface for Schottky contact;   here the adjustment of barrier height (79k) should be added, (10) beam-induced damage to form gettering sites for impurities (Sec.VII).

If our concern is extended to structuring of the devices, attention should be paid to the newly developing ion beam lithography. Here refereance is made to a review by Adeshida (83gg) and a conference 83hh). Furthermore, thinking on SPE-type rapid annealing, results summarized in Sec.III should also be mentioned. Of course (and fortunately), this list is still incomplete, as new ideas emerge in this mature, but still rapidly growing area.

Figure 43, again by Hill (83dd), based on published experimental data, summarizes different choices for full annealing of implanted layers by solid phase processes. Two lines belong to 40 keV low dose boron and to a 40 keV high dose arsenic. For a full anneal, the layer should be subjected to a time-temperature procedure in the heavier shaded area. This involves dopant redistribution. Tolerable values of

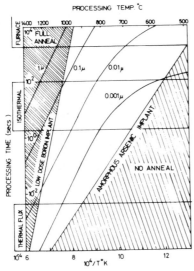

Figure 43. Time-temperature diagram for anneal and redistribution of dopant by solid state mechanisms. Example covers 40 keV low dose boron and 40 keV amorphizing arsenic implants. Tolerable redistributions of $10^{16}/cm^2$ arsenic are drawn by a family of curves using $3(DT)^{\frac{1}{2}}$ approximation. (From Hill,83dd.)

redistribution (lines for arsenic are marked with diffusion distances) bring in further limitations in choosing low energy pulse processing ("Thermal flux"), rapid annealing ("Isothermal") or furnace.

Further examples of applications of novel annealing techniques should be mentioned from the review of Hess et al. (83ii).

X. CONCLUSIONS

When in 1983 I was writing the first edition of this chapter, it was accepted that, as hottest point, the field of pulsed annealing (laser, electrons) should be covered. Within four yerars, we have been witnessing enormous changes of directions. The BF_2^+ story looks a lot cleaner now, and many things have happened in the field of FIB, high energy implantation not to forget fabrication of SOI by implantation. Most of these fields were dormant, or in the shade, just a few years ago.

If one plunges now into a risky forecast, probably will not be seriously mistaken, that the coming years will draw the proper image of today's fashionable teachniques, like research during previous years did with LA.

A possible hot point in near future might be combined techniques, where ion beams are used sequentially or simultaneously with other layer deposition techniques allowing atom-by-atom layer fabrication - maybe with focused "Direct writing".

The author is tempted to make another forecast, as a promising area to those, who build an ion deceleration optics into their high-current machine with a target in (local) UHV ambient (87v). With a system like this and, maybe adding multiple ion sources, a unique layer-by-layer fabrication tool is at hand working with extreme low, but adjustable energies and a high-fluence sequence of ions.

XI. ACKNOWLEDGEMENTS

Thanks are due to J.F. Ziegler for his editorial help of different kind "bridging" over distances. Technical co-operation of Dr. G. Battistig and Mrs. L. Imre is kindly acknowledged making the time schedule running.

XII. REFERENCES

66a M.Watanabe and A.Tooi, Jpn. J. Appl. Phys. 5, 737 (1966)
67a P.V.Pavlov and E.Shitova, Sov.Phys. 12, 11 (1967)
67b B.E. Deal, E.H. Snow and A.S. Grove, "Microelectronic
    Technology, Boston Technical Publ., Cambridge Mass.,
    1967. p.171.
69a D.J.Mazey and R.S.Nelson, Radiat. Eff. 1, 229 (1969)
69b F.L.Vook and H.J.Stein, Radiat. Eff. 2, 23 (1969)
69c R.S. Nelson, Radiat.Eff. 2, 47 (1969)
69d P. Sigmund, Phys.Rev. 184, 383 (1969)
70a J.W. Mayer, L. Eriksson and J.A. Davies "Ion Implantation
    in Semiconductors", Academic Press, New York, 1970.
70b V.M. Gusev and M.I. Guseva, Personal communication,
    published later in "Priroda" No.12, p.26, 1979, in Russian
70c O. Meyer and J.W. Mayer, Radiat.Eff. 3, 139 (1970)
71a F.H. Eisen and L.T. Chadderton, eds. "Ion Implantation"
    Gordon Breach, New York, 1971.
71b I. Ruge and J. Graul, eds. "Ion Implantation in
    Semiconductors". Verlag, Berlin and New York, 1971.
71c S.T. Picraux, W.H. Weisenberger and F.L. Vook,
    Radiat.Eff. 7, 101 (1971)
71d H.Muller, H.Ryssel and I.Ruge, in 71b, p.85)
71e K.Gamo, M.Iwaki, K.Masuda, S.Namba, S.Ishiwara and
    I.Kimura, in 71a, p. 459)
71f J.Gyulai, O.Meyer, R.D.Pashley and J.W.Mayer,
    Rad. Effects 7, 17 (1971)
72a S.Namba, K.Masuda, K.Gamo, M.Iwaki in S.Namba ed. "Proc.
    US-Japan Seminar Ion Implantation in Semiconductors"
    Kyoto, Jpn. Soc. for the Promotion of Science, 1972, p.1
72b T.M. Buck, K.A. Pickar, J.M. Poate and C.M. Hsieh,
    Appl.Phys.Lett. 211, 485 (1972)
73a G. Dearnaley, J.H. Freeman, R.S. Nelson and J. Stephen
    "Ion Implantation". North-Holland, Amsterdam, 1973.
73b B.L. Crowder, ed. "Ion Implantation in Semiconductors
    and Other Materials". Plenum, New York, 1973.
73c H.Ryssel Thesis, Technical Univ. Munich, 1973
73d N.N.Gerasimenko, A.V.Dvurechenskiy, G.A.Kachurin,
    N.B.Pridachin and L.S.Smirnov, Sov. Phys. Semicond. 6,
    1588 (1973)
73e C.R. Fritzsche and W. Rothemund, J.Electrochem.Soc.
    120, 1603 (1973)
73f T.E. Seidel and R.L. Meek, In "Ion Implantation in
    Semiconductors and Other Materials" (B.L. Crowder, ed.),
    Plenum, New York, 1973, p.305.
74a J. Gyulai, L. Csepregi, T. Nagy, J.W. Mayer and H. Muller
    (1974). Le Vide, p.416.
74b P.Blood, G.Dearnaley and M.A.Wilkins, J.Appl.Phys. 45, 5123
    (1974)
75a S. Namba and K. Masuda, Adv.Elecron Phys. 37, 263-330 (1975).
75b S. Namba, ed. "Ion Implantation in Semiconductors", Plenum,
    New York, 1975.

75c H. Muller, W.K. Chu, J. Gyulai, J.W. Mayer, T.W. Sigmon and
    T.R. Cass, Appl.Phys.Lett. 26, 292 (1975).
75d G.Holmen, A.Buren and P.Hogberg, Radiat. Eff. 24, 51 (1975)
75e H. Muller, J. Gyulai, W.K. Chu, J.W. Mayer and T.W. Sigmon,
    J.Electrochem.Soc. 122, 1234 (1975)
75f K.Nomura, Y.Hirose, Y.Akasaka, K.Horie and S.Kawazu,
    In 75b, p. 681)
75g T.E. Seidel, R.L. Meek and A.G. Cullis, In "Lattice
    Defects in Semiconductors 1974" (F.A. Huntley, ed.) Inst.
    of Physics, London, 1975, p.494.
75h E.I.Shtyrkov, I.B.Khaibullin, M.M.Zaripov, M.F.Galyautdinov
    and R.M.Bayazitov, Fiz. Tekh. Polupr. 9, 2000 (1975)
76a P.D. Townsend, J.C. Kelly and N.E.W. Hartley, "Ion
    Implantation, Sputtering and Their Applications".
    Academic Press, New York, 1976.
76b G. Carter and W.A. Grant, "Ion Implantation of
    Semiconductors". Edward Arnold, London, 1976.
76c F. Eisen and J.W. Mayer, In "Treatise on Solid State
    Chemistry" (N.B. Hannay ed.) Vol.6B, Plenum, New York,
    1976, pp.125-201.
76d M. Miyao, N. Natsuaki, N. Yoshihiro, M. Tamura and T.
    Tokuyama, In "Proc.7th Conf.Solid State Devices, Tokyo".
    Jpn.J.Appl.Phys. 15.Suppl. 15-1, 57 (1976)
76e L. Csepregi, J.W. Mayer and T.W. Sigmon,
    Appl.Phys.Lett. 29, 92 (1976)
76f J.W. Mayer, L. Csepregi, J. Gyulai, T. Nagy, G. Mezey,
    P. Revesz and E. Kotai, Thin Solid Films 32, 303 (1976).
76g J.B. Benjamin and G. Dearnaley, In "Applications of
    Ion Beams to Materials, 1975" (G. Carter, Colligon and
    C. Grant, eds.), Inst.Phys.Conf.Series No.28,
    Inst.Phys., Bristol, 1976, p.141.
76h L. Csepregi, W.K. Chu, H. Muller, J.W. Mayer and T.W.
    Sigmon, Radiat. Eff. 28, 227 (1976)
76i T.W. Sigmon, L. Csepregi and J.W. Mayer,
    J. Electrochem.Soc. 123, 1116 (1976).
77a F. Chernow, J.A. Borders and D.K. Brice, eds. "Ion
    Implantation in Semiconductors", Plenum, New York, 1977.
77b T. Mitsuishi, Y. Sasaki and H. Asami,
    Jpn.J.Appl.Phys. 16, 367 (1977).
77c N. Yoshihiro, M. Miyao and T. Tokuyama, In "8th Symp.
    "Ion Implantation in Semiconducors". Inst. Phys. Chem.
    Res.1977, p.31 (in Japanese), through 78b
77d L. Csepregi, E.F. Kennedy, T.J. Gallagher, J.W. Mayer
    and T.W. Sigmon, J.Appl.Phys. 48, 4234 (1977).
77e E.F. Kennedy, L. Csepregi, J.W. Mayer and T.W.
    Sigmon, J.Appl.Phys, 48, 4241 (1977).
77f R.B.Fair in "Semiconductor Silicon" H.R.Huff ed.,
    Electrochem. Soc.,1977, p.968
77g L. Csepregi, R.P. Küllen, J.W. Mayer and T.W. Sigmon,
    Solid State Comm. 21, 1019 (1977).
77h G. Mezey, T. Nagy, J. Gyulai, E. Kotai, A. Manuaba,
    T. Lohner and J.W. Mayer, In 77a, p.49.
77i R.A. Moline, In 77a, pp.319-331.

77j E.F. Kennedy, L. Csepregi, J.W. Mayer and T.W. Sigmon,
    In 77a, p.511.
77k N. Natsuaki, M. Tamura, M. Miyao and T. Tokuyama,
    Jpn.J.Appl.Phys. 16-1, 47 (1977).
78a H. Ryssel and I. Ruge "Ion Implantation", Teubner,
    Stuttgart, 1978.
78b T. Tokuyama, M. Miyao and N. Yoshihiro,
    Jpn.J.Appl.Phys. 17, 1301-1315 (1978).
78c W.-K. Chu, J.W. Mayer and M.-A. Nicolet,
    "Backscattering Spectrometry". Academic Press,
    New York, 1978.
78d J. Suski, I. Krynicki, H. Rzewuski, J. Gyulai and
    J.J. Loferski, Radiat.Eff. 35, 13 (1978).
78e W.F. Tseng, J. Gyulai, T. Koi, S..S. Lau, J. Roth and
    J.W. Mayer, Nucl.Inst.Meth. 149, 615 (1978).
78f L. Csepregi, E.F. Kennedy, J.W. Mayer and T.W. Sigmon,
    J.Appl.Phys. 49, 3906 (1978).
78g P. Revesz, M. Wittmer, J. Roth and J.W. Mayer,
    J.Appl.Phys. 49, 5199 (1978) and
    M. Wittmer, J. Roth, P. Revesz and J.W. Mayer,
    J.Appl.Phys. 49, 5207 (1978).
78h K.Izumi, M.Doken and H.Ariyoshi, Electron. Lett. 14, 593
    (1978)
78i W.H.Kool, H.E.Rosendaal, L.W.Wiggers and F.W.Saris,
    Radiat. Eff. 36, 41 (1978)
78j M. Wittmer and T.E. Seidel, J.Appl.Phys. 49, 5827 (1978).
78k K.N. Tu and J.W. Mayer, In "Thin Films-
    Interdiffusion and Reactions" (J.W. Poate, K.N. Tu and
    J.W. Mayer, eds.), Chapter 10, Wiley, New York, 1978.
79a G.K. Wolf, Top.Curr.Chem. 85, 1-88 (1979).
79b J. Gyulai, T. Lohner and E. Pasztor, eds. "Ion Beam
    Modification of Materials" Vols. I,II and III, KFKI Publ.,
    Budapest, 1979, partly also in Radiat.Eff. 47-49 (1980).
79c G. Foti, In 79b, p.611-628, also in Radiat.Eff. 48, 161
    (1980).
79d J. Gyulai and P. Revesz, In "Defects and Radiation
    Effects in Semiconductors, 1978" (J.H. Albany, ed.),
    Conf.Series No.46, Inst.Phys., London, 1979, pp.128-147.
79e L.W. Wiggers and F. Saris, In 79b, p.583. also in
    Radiat. Eff. 49. 177 (1980).
79f S.S. Lau, S. Matteson, J.W. Mayer, P. Revesz, J. Gyulai,
    J. Roth, T.W. Sigmon and T. Cass, Appl.Phys.Lett.
    34, 76 (1979).
79g M.E. Roulet, P. Schwob, I. Golecki and M.-A. Nicolet,
    Electronics Lett. 15, 529 (1979).
79h E. Kotai, T. Nagy, O. Meyer, J. Gyulai, P. Revesz, G.
    Mezey, T. Lohner and A. Manuaba, In 79b, p.573, also in
    Radiat.Eff. 47, 27 (1990).
79i P. Revesz and J.W. Mayer, phys.stat.sol.(a) 54, 513 (1979).
79j J.F. Götzlich, K. Haberger, H. Ryssel, H. Kranz and
    E. Traumuller, In. 79b, p. 1419, also in Radiat.Eff. 47,
    203 (1980).
79k W.-K. Chu, M.J. Sullivan, S.M. Ku and M. Shatzkes,
    In 79d, p.193, also in Radiat.Eff. 47, 7 (1980).

79l H. Ryssel, H. Kranz, P. Bayerl and B. Schmiedt,
    In 79d, p.461. also in Radiat.Eff. 48, 125 (1980).
79m K.B. Winterbon, In 79b, pp.35-43, also in Radiat.Eff. 48, 97
    (1980).
80a J.A. Davies and L.M. Howe, In "Site Characterization
    and Aggregation of Impanted Atoms in Materials" (A. Perez
    and R. Coussement, eds.) Plenum, New York, 1980, pp.7-32.
80b C.L. Anderson, G.K. Celler and G.A. Rozgonyi, eds.
    "Laser and Electron Beam processing of Electronic
    Materials", Electrochem.Soc., Inc., Princeton, 1980.
80c L.M. Howe, M.H. Rainville, H. Haugen and D.A.
    Thompson, Nucl.Instr.Meth. 170, 419 (1980).
80d D.A. Thompson, A. Golanski, H.K. Haugen, L.M. Howe
    and J.A. Davies, Radiat.Eff.Lett. 50, 125 (1980).
80e K. Tsukamoto, Y. Akasaka and K. Kijima,
    Jpn.J.Appl.Phys. 19, 87 (1980).
80f U. Littmark and W.O. Hofer, Nucl.Instr.Meth. 168, 329 (1980).
81a D.A. Thompson, Radiat.Eff. 56, 105-150 (1981).
81b R.E.Benenson, E.N.Kaufmann, G.L.Miller and W.W.Scholz, eds.,
    "Ion Beam Modification of Materials, I-II",
    North-Holland, Amsterdam, 1981;
    Nucl. Instr. Meth. 182/183 (1981).
81c C.M.McKenna, P.J.Scanlon and J.R.Winnard, eds.
    "Ion Implantation Equipment and Techniques",
    North-Holland, Amsterdam, 1981;
    Nucl. Instr. Meth. 189 (1981)
81d J.F. Gibbons, L.D. Hess and T.W. Sigmon, eds. "Laser and
    Electron-Beam Solid Interactions and Material Processing"
    MRS Symp.Proc. Vol.1. North-Holland, New York, 1981.
81e J. Narayan and T.Y. Tan, eds. "Defects in
Semiconductors" North-Holland, New York, 1981.
81f B.H. Kear, J.W. Mayer, J.M. Poate and P.R. Strutt, In
    "Metallurgical Treatises" (J.K. Tien and J.F. Elliott,
    eds.), Metallurg.Soc.of AIME Publ., 1981, pp.321-342.
81g H. Krautle, In 81b Nucl.Instr.Meth. 182/183, 625 (1981).
81h L.M. Howe and M.H. Rainville, In 81b, Nucl.Instr.Meth.
    182/183, 143 (1981).
81i T.W. Sigmon, IEEE Trans.Nucl.Sci. NS-28, 1767 (1981).
81j I. Golecki, G. Kinoshita and B.M. Paine, In 81b,
    Nucl.Instr.Meth. 182/183, 675 (1981).
81k T. Inoue and T. Yashii, In 81b, Nucl.Instr.Meth. 182/183, 683
    (1981).
81l G.F.Cembali, G.P.Merli and F.Zignani, Appl. Phys. Lett. 38,
    808 (1981)
81m J.S. Williams, C.E. Christodoulides, In 81b,
    Nucl.Instr.Meth. 182/183, 667 (1981).
81n I. Ohdomari, K.N. Tu, K. Suguro, M. Akiyama, I. Kimura
    and K. Yoneda, Appl.Phys.Lett. 38, 1015 (1981).
81o D.M. Scott and M.-A. Nicolet, In 81b
    Nucl.Instr.Meth. 182/183, 655 (1981).
81p D.M. Scott, L. Wielunski, H. von Seefeld and M.-A. Nicolet,
    In 81b, Nucl.Instr.Meth. 182/183, 661 (1981).
81q D. Lecrosnier, J. Paugam, G. Pelous, F. Richou and
    M. Salvi, J.Appl.Phys. 50, 5090 (1981).

81r L.A. Christel, J.F. Gibbons and S. Mylroie, In 81b,
     Nucl.Instr.Meth. 182/183, 187 (1981).
81s T.J. Magee, C. Leung, H. Kawayoshi, B. Furman, C.A.
     Evans,Jr. and D.S. Day, Appl.Phys.Lett. 39, 260 (1981).
81t T.J. Magee, C. Leung, H. Kawayoshi, L.J. Palkuti,
     B.K. Furman, C.A. Evans,Jr., L.A. Christel, J.F. Gibbons
     and D.S. Day, Appl.Phys.Lett. 39, 564 (1981).
81u J.W. Mayer, B.-Y. Tsaur, S.S. Lau and L.-S. Hung, In 81b,
     Nucl.Instr.Meth. 182/183, 1 (1981).
82a G.L. Miller, M. Soni, M. McDonald, E.N. Kaufmann and
     R.L. Fenstermacher, In 82g, p.363.
82b J.M. Poate and J.W. Mayer, eds. "Laser Annealing of
     Semiconductors", Academic Press, New York, 1982.
82c L.C. Feldman, J.W. Mayer and S.T. Picraux "Materials
     Analysis by Ion Channeling", Academic Press, New York, 1982.
82d J.M. Poate, in "Treatise on Materials Science and
     Technology, Vol.24. Preparation and Properties of Thin
     Films" (K.N. Tu and R. Rosenberg, eds.), Academic Press,
     New York, 1982. pp.213-236.
82e H.Ryssel and H.Glawischnig, eds. "Ion Implantation
     Techniques", Springer Series on Electronphysics,
     Vols.10 and 11, Springer, Berlin, Heidelberg, New York, 1982
82f B.R. Appleton and G.K. Celler, eds. "Laser and Electron-
     Beam Interactions with Solids", MRS Symp.Proc.Vol.4,
     North-Holland, New York, 1982.
82g S.T. Picraux and W.J. Choyke, eds. "Metastable Materials
     Formation by Ion Implantation", MRS Symp.Proc. Vol.7.,
     North-Holland, New York, 1982.
82h B.H. Kear, B.C. Giessen and M. Cohen, eds. "Rapidly
     Solidified Amorphous and Crystalline Alloys", MRS Symp.
     Proc. Vol.8., North-Holland, New York, 1982.
82i S.S. Lau and J.W. Mayer, In "Treatise on Materials Science
     and Technology, Vol.24. Preparation and Properties of Thin
     Films", (K.N. Tu and R. Rosenberg, eds.), Academic Press,
     New York, 1982, pp.67-111.
82j I. Suni,, G. Göltz, M.G. Grimaldi, M.-A. Nicolet and
     S.S. Lau, Appl.Phys.Lett. 40, 269 (1982).
82k J. Narayan and O.W. Holland, In 82g, p.117.
82l G.Cembali, M.Finetti, P.G.Merli and F.Zignani,
     Appl. Phys. Lett. 40, 62 (1982)
82m G.A.Kachurin and N.P.Stenina, Sov. Phys. Semicond. 16, 738
     (1982)
82n P.F.Byrne, N.W.Cheung and D.K.Sadana, Appl. Phys. Lett. 41,
     537 (1982)
82o A.G. Cullins, H.C.. Webber, N.G. Chen, J.M. Poate and
     P. Baeri, Phys.Rev.Lett. 49, 219 (1982).
82p C. Hill, In "Laser Annealing of Semiconductor" (J.M. Poate
     and J.W. Mayer, eds.), Academic Press, New York, 1982,
     pp.479-558.
82q L.J. Chen, L.S. Hung, J.W. Mayer and J.E.E. Baglin, In 82g,
     p.319.
82r A. Lietoila, A. Wakita, T.W. Sigmon and J.F. Gibbons,
     J.Appl.Phys. 53, 4399 (1982).

83a J.W. Mayer and J. Gyulai, In "Applied Atomic Collision
     Physics", Vol.4. (S. Datz, ed.), Academic Press, New York,
     1983, pp.545-575.
83b B. Biasse, G. Destefanis and J.P. Gaillard, eds.
     "Ion Beam Modification of Materials", Nucl.Instr.Meth.
     209/210 (1983).
83c J. Narayan, W.L. Brown and R.A. Lemons, eds.
     "Laser-Solid Interactions and Transient Thermal Processing
     of Materials", MRS Symp.Proc. Vol.13, North-Holland,
     New York, 1983.
83d S. Mahajan and J.W. Corbett, eds. "Defects in
     Semiconductors", MRS Symp.Proc. Vol.14, North-Holland,
     New York, 1983.
83e R. Ludeke and K. Rose, eds. "Interfaces and Contacts",
     MRS Symp.Proc. Vol.18, North-Holland, New York, 1983.
83f L.S. Smirnov, ed. "A Survey of Semiconductor Radiation
     Techniques", MIR Publ. Moscow, 1983.
83g G.W.. Cullen, ed. "Single-Crystal Silicon on
     Non-Single-Crystal Insulators", J. Crystal Growth 63
     (No.3), 429-582 (1983).
83h J.S.. Williams, In 83c, p.621-632.
83i G.F. Cerofolini, G. Ferla, G.U. Pignatel, F. Riva and
     G. Ottaviani, Thin Solid Films 101, 263 (1983).
83j G.Peto, T.Lohner and J.Kanski, Nucl. Instr. Meth.B 209/210,
     447 (1983)
83k H. Ishiwara, K. Naruke and S. Furukawa, In 83b,
     Nucl.Instr.Meth. 209/210, 689 (1983).
83l B.-Y Tsaur and C.H. Anderson, Jr., J.Appl.Phys. 54,
     6336 (1983).
83m D.K. Sadana, J. Washburn and C.W. Magee, J.Appl.Phys,
     54, 3479 (1983).
83n E. Elliman and G. Carter, In 83b, Nucl.Instr.Meth. 209/210,
     663 (1983).
83o J.S. Williams and K.T. Short, In 83b,
     Nucl.Instr.Meth. 209/210, 767 (1983).
83p G.Zimmer and H.Vogt, IEEE Trans. Electr. Dev. ED-30, 1515
     (1983)
83r H.W.Lam and R.F.Pinizzotto, J. Cryst. Growth 63, 554 (1983)
83s R.F. Pinizzotto, In "Single-Crystal Silicon on
     Non-Single-Crystal Insulators" (G.W. Cullen, ed.)
     J.Crystal Growth 63, 559-582 (1983).
83t G.A.Kachurin and V.A.Mayer, Sov. Phys. Semicond. 17, 278
     (1983)
83u M. Wittmer, C.-Y. Ting and K.N. Tu, J.Appl.Phys. 54,
     699 (1983).
83v D.M. Scott and M.-A. Nicolet, In 83b,
     Nucl.Instr.Meth. 209/210, 297 (1983).
83w P. Revesz, J. Gyimesi and E. Zsoldos, J.Appl.Phys. 54,
     1860 (1983).
83x D. Lecrosnier, In 83b, Nucl.Instr.Meth. 209/210, 325 (1983).
83y D. Hagmann. In 83b, Nucl.Instr.Meth. 209/210, 683 (1983).
83z T. Izumi, T. Matsumori, T. Hirao, Y. Yaegashi, G. Fuse and
     K. Inoue, In 83b, Nucl.Instr.Meth. 209/210, 695 (1983).

83aa S.S. Lau, B.-X. Liu and M.-A. Nicolet, In 83b,
     Nucl.Instr.Meth. 209/210, 97 (1983).
83bb L.S. Hung, M. Nastasi, J. Gyulai and J.W. Mayer,
     Appl.Phys.Lett. 42, 672 (1983).
83cc N.N.Gerasimenko, A.E.Gershinskiy, A.Yu.Surtaev, S.A.Sokolov,
     E.I.Cherepov, B.I.Fomin and G.V.Timofeeva,
     phys. stat. sol.(a) 78, K151 (1983)
83dd C. Hill, In 83b, p.381.
83ee B.-X. Liu, W.L. Johnson, M.-A. Nicolet and S.S. Lau,
     In 83b, Nucl.Instr.Meth. 209/210, 229 (1983).
83ff C.J. Palmström and R. Fastow, In 83b, p.715.
83gg I. Adeshida, In 83b, Nucl.Insr.Meth. 209/210, 79 (1983).
83hh H. Ahmed, J.R.A. Cleaver and G.A.C. Jones, eds.
     "Microcircuit Engineering 83", Academic Press, New York,
     1983, pp.135-171.
83ii L.D. Hess, G. Eckhard, G.A. Kokorowski, G.L. Olson,
     A. Gupta, Y.M. Chi, J.B. Valdez, C. Bito and E.M. Nakaji,
     In 83c, p.337.
84a J.F. Ziegler, ed. "Ion Implantation: Science and Technology",
     Acad.Press, New York, 1984.
84b R.A.Powell, J. Appl. Phys. 56, 2837 (1984)
84c R.Ghez, A.S.Oehrlein, T.O.Sedgwick, F.F.Morehead and
     Y.H. Lee, Appl. Phys. Lett. 45, 881 (1984)
84d R.B.Fair, J.J.Wortman and J.Liu, J.Electrochem. Soc. 131,
     2387 (1984)
84e S.J.Pennycook, J.Narayan and O.W.Holland, J. Appl. Phys.
     55, 837 (1984)
84f P.L.F.Hemment, E.A.Maydell-Ondrusz, K.G.Stephens,
     J.A.Kilner and J.B.Butcher, Vacuum 34, 203 (1984)
84h M.Tamura, S.Shukuri, T.Ishitani, M.Ichikawa and T.Doi,
     Jpn. J. Appl. Phys. 23, L417 (1984)
84i S.Shukuri, Y.Wada, T.Tamura, H.Masuda and T.Ishitani,
     Ext. Abstracts, 1984 Int. Conf. Solid State Dev. Materials,
     Kobe, p.91
85a B.M.Ullrich, ed. "Ion Beam Modification of Materials, I-II"
     North-Holland, Amsterdam, 1985;
     Nucl. Instr. Meth. B7/8 (1985)
85b J.F.Ziegler and R.L.Brown, eds. "Ion Implantation
     Equipment and Techniques", North-Holland,Amsterdam;
     Nucl. Instr. Meth. B6 Nos.1-2 (1985)
85c T.E.Seidel, D.J.Lischner, C.S.Pai, R.V.Knoell, D.M.Maher and
     D.C.Jacobson, Nucl. Instr. Meth. B7/8, 251 (1985))
85d T.E.Seidel, R.Knoell, G.Poli, B.Schwartz, F.A.Stevie and
     P.Chu, J.Appl. Phys. 58, 683 (1985)
85e J.Narayan, O.W.Holland, W.h.Christie and J.J.Wortman,
     J. Appl. Phys. 57, 2709 (1985)
85f M.Delfino and M.E.Lunnon, J. Electrochem. Soc. 132, 435
     (1985)
85g N.W.Cheung, in Proc. SPIE, Vol. 540, 1985, p.2
85h P.L.F.Hemment, R.F.Peart, M.F.Yao, K.G.Stephens,
     R.J.Chater, J.A.Kilner, D.Meekison, G.R.Booker and
     R.P.Arrowsmith, Appl. Phys. Lett. 46, 952 (1985)
85i D.E.Davies, IEEE Electron. Dev. Lett. EDL-6, 397 (1985)

85j N.Natsuaki, K.Ohyu, T.Suzuki, N.Kobayashi, N.Hashimoto and
    Y.Wada, Ext. Abstr. 17th Conf . Solid State Dev. and
    Materials, Tokyo,1985, p. 325
85k M.Delfino, E.K.Broadbent, A.E.Morgan,B.J.Burrow and
    M.H.Norcott, IEEE Electron. Dev. Lett. EDL-6, 591 (1985)
85l T.P.Chow, W.Katz and G.Smith, Appl. Phys. Lett 46, 41
    (1985)
86a H.Ryssel and I.Ruge, "Ion Implantation" John Wiley,
    Chichester 1986
86b A.V. Dvurechenski, R. Groetzchel and V.P. Popov, Phys.Lett. A
    116, 399 (1986)
86c R. Angelucci, P. Negrini and S. Solmi, Appl. Phys. Lett.
    49, 1468 (1986)
86e I-W.Wu, R.T.Fulks and J.C.Mikkelsen, Jr., J.Appl. Phys. 60,
    2422 (1986)
86f S.D.Brotherton, J.P.Gowers, N.D.Young, J.B.Clegg and
    J.R.Ayres, J. Appl. Phys. 60, 3567 (1986)
86g D.A.Hansen, D.B.Poker and B.R.Appleton, Nucl. Instr. Meth.
    B16, 373 (1986)
86h M.Delfino, D.K.Sadana, A.E.Morgan and P.K.Chu,
    J. Electrochem. Soc. 133, 1900 (1986)
86i G.K.Celler, P.L.F.Hemment, K.W.West and J.M.Gibson,
    Appl. Phys. Lett. 48, 532 (1986)
86j B.-Y.Mao, P.-H.Chang, H.W.Lam, B.W.Chen and J.A.Keenan,
    Appl. Phys. Lett. 48, 794 (1986)
86k A.Mogro-Campero, R.P.Love, N.Lewis, E.L.Hall and
    M.D.McConnell, J. Appl. Phys. 60, 2103 (1986)
86l J.Stoemenos, J.Margail, C.Jaussaud, M.Dupuy and M.Bruel,
    Appl. Phys. Lett. 48, 1470 (1986)
86m D.E.Davies, J.A.Adamski and E.F.Kennedy, Appl. Phys. Lett.
    48, 347 (1986)
86n B.A.MacIver and K.C.Jain, IEEE Trans. Electr. Dev. ED-33,
    1953(1986)
86o S.Nakashima, Y.Maeda and M.Akiya, IEEE Trans. Electr. Dev.
    ED-33, 126 (1986)
86p D.P.Vu and J.C.Pfister, Appl. Phys. Lett. 48, 50 (1986)
86q S.Cristoloveanu, J.H.Lee, J.Pumfrey, J.R.Davis,
    R.P.Arrowsmith and P.L.F.Hemment, J. Appl. Phys. 60, 3199
    (1986)
86r T.Tokuyama, personal communication
86s J.Linnros and G.Holmen, J. Appl. Phys. 59, 1513 (1986)
86t T.Banwell, M-A.Nicolet, R.S.Averback and L.J.Thompson,
    Appl. Phys. Lett. 48, 1519 (1986)
86u M.Tamura, N.Natsuaki, Y.Wada and E.Mitani, J. Appl. Phys.
    59, 3417 (1986)
86v A. Stolmeijer, IEEE Trans. Electr. Dev. ED-33, 450 (1986)
86w F.-S.J.Lai, L.K.Wang, Y.Taur, J.Y.-C.Sun, K.E.Petrillo,
    S.K.Chicotka, E.J.Petrillo, M.R.Polcari, T.J.Bucelot and
    D.S.Zicherman, IEEE Trans. Electron. Dev. ED-33, 1308 (1986)
86x M.Tamura and N.Natsuaki, Jpn. J. Appl. Phys. 25, 1474 (1986)
86y F.G.Kuper, J.Th.M.De Hosson and J.F.Verwey, J. Appl. Phys.
    60, 985 (1986)
86z S.S.Todorov, S.L.Shillinger and E.R.Fossum, IEEE ELectron.
    Dev. Lett. EDL-7, 468 (1986)

86aa A.H.van Ommen, H.J.W.van Houtum and A.M.L.Theunissen,
     J. Appl. Phys. 60, 627 (1986)
86bb D.L.Kwong and N.S.Alvi, J. Appl. Phys. 60, 688 (1986)
86cc I.Ohdomari, T.Chokyow, H.Kawarada, K.Konuma, M.Kakumu,
     K.Hashimoto, I.Kimura and K.Yoneda, J.Appl. Phys. 59, 3073
     (1986)
86dd D.Hagmann, D.Steiner and T.Schellinger, J.Electrochem.Soc.
     133, 2597 (1986)
86ee J.T.Borenstein, J.T.Jones, J.W.Corbett, G.S.Oehrlein and
     R.L.Kleinhenz, Appl. Phys. Lett. 49, 199 (1986)
86ff A.Nylandsted Larsen, S.Yu.Shiryaev, E.Schwartz Sorensen
     and P.Tidemand-Petersson, Appl. Phys. Lett. 48, 1805 (1986)
87a S.U.Campisano, G.Foti, P.Mazzoldi and E.Rimini, eds.
    "Ion Beam Modification of Materials, I-II",
    North-Holland, Amsterdam, 1987;
    Nucl. Instr. Meth. B19/20 (1987)
87b M.I.Current, N.W.Cheung, W.Weisenberger and B.Kirby, eds.,
    "Ion Implantation Technology", North-Holland,
    Amsterdam,1987; Nucl. Instr. Meth. B21 Nos.2-4(1987)
87c P.L.F. Hemment, K.J. Reason, J.A. Kilner, R.J. Chater,
    C. Marsh, G.R. Booker, J.R. Davis and D.K. Celler,
    In 87b, p. 129
87d A.Grob, J.J.Grob and A.Golanski, Nucl. Instr. Meth. B19/20,
    55 (1987)
87q J.F. Marchiando and J. Albers, J. Appl. Phys. 61, 1380
    (1987)
87e M.Servidori, R.Angelucci, F.Cembali, P.Negrini, S.Solmi,
    P.Zaumseil and U.Winter, J.Appl. Phys. 61, 1834 (1987)
87f A.E.White, K.T.Short, J.L.Batstone, D.C.Jacobson, J.M.Poate
    and K.W.West, Appl. Phys. Lett. 50, 19 (1987)
87g K.J.Reeson, P.L.F.Hemment, C.D.Meekison, G.R.Booker,
    J.A.Kilner, R.J.Chater, R.J.Davis and G.K.Celler,
    Appl. Phys. Lett. 50 1882 (1987)
87h E.Sobeslavsky, H.U.Jaeger, U.Kreissig, W.Skorupa and
    K.Wollschlaeger, phys. stat. sol.(1987)
87i J.Narayan, S.Y.Kim, K.Vedam and R.Manukonda, Appl. Phys.
    Lett. 51, 343 (1987)
87j R.G.Elliman, D.C.Jacobson, J.Linnros and J.M.Poate,
    Appl. Phys. Lett. 51, 314 (1987)
87k R.G.Elliman, J.S.Williams, W.L.Brown, A.Leiberrich, D.M.Maher
    and R.V.Knoell, Nucl. Instr. Meth. B19/20, 435 (1987)
87l G.Alestig, G.Holmen and J.Linnros, J. Appl. Phys. 62, 409
    (1987)
87m P.Spinelli, A.M.Cartier and M.Bruel, In 87b,
    Nucl. Instr. Meth. B21, 452 (1987)
87n E.Deng, H.Wong and N.W.Cheung, In 87b,
    Nucl. Instr. Meth. B21, 134 (1987)
87o A.K.Rai, J.A.Baker, D.C.Ingram, A.W.McCormick and
    D.A.Walsh, In 87b, Nucl. Instr. Meth. B21, 466 (1987)
87p A.K.Rai, J.Baker and D.C.Ingram, Appl. Phys. Lett. 51, 172
    (1987)
87r N.Yu, Z.Zhou, W.Zhou, S.Tsou and D.Zhu, Nucl. Instr. Meth.
    B19/20, 427 (1987)

87s A.A.Pasa, J.P.de Souza, I.J.R.Baumvol and F.L.Freire, Jr.,
     J. Appl. Phys. 61, 1228 (1987)
87t D.H.Weiner, S.S.Wong and C.I.Drowley, Appl. Phys. Lett. 50,
     986 (1987)
87u A.E.Michel, W.Rausch, P.A.Ronsheim and R.H.Kastl,
     Appl. Phys. Lett. 50, 416 (1987)
87v B.R. Appleton, S.J. Pennycook, R.A. Zuhr, N. Herbots and
     T.S. Noggle, In 87a, p. 975
87w D.S.Moroi and P.M.Hemenger, Appl. Phys. Lett. 50, 155(1987)
87x H.Wong, E.Deng, N.W.Cheung, P.K.Chu, E.M.Strathman and
     M.D.Strathman, in 87b, p.447

MEASUREMENT OF ION IMPLANTATION

P.L.F. Hemment
Department of Electronic and Electrical Engineering
University of Surrey
Guildford, Surrey GU2 5XH, England

1. INTRODUCTION

The many surface analysis techniques used to evaluate ion implanted layers will be described in this chapter. The emphasis will be directed specifically, but not entirely, towards the electrical evaluation of semiconductor wafers implanted under conditions typically used in device fabrication. In these implanted semiconductor systems (eg. As/Si) it is necessary to determine the damage and dopant depth profiles as well as quantifying the electrical activity. The fundamental aspects of doping by ion implantation, including damage creation and annealing have been covered in Chapters 2 and 3.

In this chapter the measurement techniques are looked upon as tools which provide data to support:-

(i)   Materials science experiments
(ii)   Quality control of wafer processing
(iii)   Diagnosis of equipment failures
(iv)   Machine calibration/standards exercises.

Practical applications are discussed in by Current, this volume.

The detailed evaluation of implanted layers in terms of their composition, microstructure and electrical properties, can only be achieved by applying many complementary analysis methods to this task. Such an evaluation procedure can be long, taking weeks and even months, to complete. Thus it is important to select the most appropriate techniques for a given system and to follow a logical sequence of analyses, which avoids needless duplication of data and the imposition of unnecessary time penalties.

In this chapter the techniques are discussed in the sequence in which they may typically be used, namely (i) non destructive testing, (ii) electrical and (iii) surface analysis. However, in many cases branching will be required to resolve unexpected issues which may come to light as a result of previous analyses. To assist the reader, firstly some background topics will be summarised in Section 2, whilst Section 3 will cover some basic aspects of electrical activity in semiconductors. Specific reference will be made only to semiconductors, mainly silicon, although most of the techniques have much wider applications.

ION IMPLANTATION:
SCIENCE AND TECHNOLOGY

165

2.   BACKGROUND

Ion implantation is inherently a highly controllable method
of doping semiconductors, because it involves the use of ion
beams whose generation, transport and beam purity can be tightly
controlled. However, users, such as materials scientists and
integrated circuit engineers, wish to control the dopant depth
profile, dose and dose uniformity, which the machine operator is
unable to directly set. Instead the operator must rely upon
monitoring and adjustment of many different machine parameters
(voltages and currents). Additionally, there is a need to
minimise contamination caused by surface particulates and
deposited or implanted foreign atoms and ions. For these
reasons, routine evaluation of the ion implanted wafers is
always necessary in order to maintain, over time scales of days,
weeks and months, the optimum performance from the machines.
Within a commercial implantation facility, these checks lead to
improved yields of working devices and enable realistic
performance norms to be defined.

The science and technology required to achieve the above
mentioned goals are discussed in detail in other Chapters. Here
the principal causes of implantation errors are briefly
discussed.

Figure 1: Implanted silicon wafer with dose
variations due to (a) scanning "lock up" and (b)
shadow masking caused by the mounting clips.

2.1   MATERIALS RELATED
Absolute/incident/retained dose: It is customary to determine
the dose (fluence)(ions cm^{-2}) by integrating the beam current
incident upon the sample,

$$\text{Dose } (\phi) = \frac{\int I dt}{meA} \text{ atoms } cm^{-2} \tag{1}$$

where I is the beam current, t the time, m the charge state of
the ion, e the electronic charge and A the area over which the
current is monitored.

Errors may arise principally due to (i) the presence of high energy neutral particles(78a)(80d) and (ii) incomplete suppression of secondary and tertiary electrons(79i)(77f). In a well designed system these errors will be less than 1%.

Equation 1 permits the "incident dose" to be estimated, however, the calculated areal density will always be greater than the actual "retained dose", as there is a finite probability of both ion reflection and target sputtering occurring. The discrepancy will probably be greater than 1% and may be as large as 5%. With increasing incident dose the discrepancy will become larger, as the retained dose will eventually saturate and become independent of the incident dose(72d)(86d).

The losses by reflection and sputtering are additive. Calculations by Smith(75a) of the percentage error for some typical implantation systems are shown in Figure 2. Simple Gaussian depth profiles have been assumed.

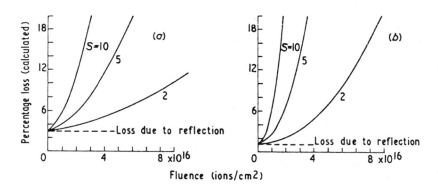

Figure 2: Calculated dose loss due to ion reflection and sputtering (a) 50keV P^+ ions into silicon and (b) 50keV As^+ ions into silicon, from Freeman(76j).

Channelling: Ion channelling occurs when a high energy ion suffers correlated small angle scattering which steers it between the tightly packed atomic rows (axial channelling) or planes (planar channelling) of a crystal lattice. Whilst in these channels the ion suffers almost no nuclear collisions and, thus, is able to penetrate to greater depths beneath the surface, than in an amorphous target. An incident ion will be channelled provided that it makes an angle less than the critical angle with a low index crystallographic direction(70c). Typical values of the critical angle lie between 2° to 5°.

In the context of ion implantation, when reproducible dopant depth profiles are essential, it is necessary to minimise the number of ions which are channelled and this is frequently

attempted by rotating and inclining the wafer by 7° to 10° to
the direction of the incident beam(85g). However, Zeigler(85b)
and others(87f) have shown that a significant fraction of the
ions will still be channelled in higher index axial and planar
channels. In general, during implantation, particularly of
large diameter wafers, the angle of incidence of the ions to the
surface will vary as the beam sweeps across the wafer. Thus,
not only will partial channelling occur but, also, there is
liable to be a lateral variation in the channelled fraction.
This can lead to systematic variations in the dopant depth
profile, which may cause large variations in the electrical
characteristics of the implanted layer after annealing. These
variations will be particularly deleterious to the performance
of small geometry devices and circuits. A solution to the
problem is to preamorphise the surface layer, prior to
implantation of the dopant, when channelling will be completely
suppressed(87d).

Charging/gas release: Charge build up will occur in insulating
samples and in deposited dielectric layers on semiconductor
wafers(83b). Upon breakdown, in high field regions, local
catastrophic surface damage will occur. Breakdown and beam
heating of photoresist will lead to gas release causing
additional charge exchange, between the fast ions and thermal
gas modecules, leading to dose inhomogeneities. These
issues(76g) are discussed by Smith, (this volume).

Crystal orientation: The regrowth rate of amorphous layers and
the thermal annealing of defects in silicon is dependent upon
the crystal orientation (100) (110) or (111). This may lead to
differences in the measured damage and electrical depth
profiles(76i) in partially annealed samples.

Lateral spread: In coming to rest the implanted ions will
suffer lateral scattering, which will give rise to lateral
broadening of the damage and atomic profiles. The magnitude of

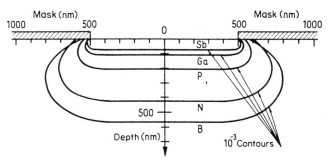

Figure 3: Lateral spread beneath a mask edge.
Ion energy 70keV, from Furukawa, Matsumura and
Ishiwara(75f).

the straggle is similar to the range straggle ($\overline{\Delta R}_p$)(75f). This effect becomes evident at shadow mask edges and, also, has the effect of reducing the dopant volume concentration at the geometric edge of the mask, to a value of 50% (Figure 3). This dose inhomogeneity will effect the performance of small geometry devices, in which as the lateral dimensions are comparable to the straggle.

Recoil implantation: Implantation through a surface layer such as a passivating oxide or bevelled mask edge, can be a cause of unwanted contamination. This occurs as the high energy ions transfer, by collisions, some of their energy to the constituent atoms of the surface layer. Some of these recoiling (knock-ons) particles will have adequate energy to penetrate into the semiconductor, typically to a maximum depth of, say, 100Å when they will create additional damage, degrading the surface layer(75c)(75d). Surface contaminants will be similarly recoil implanted.

Profile modification: Range theories (Chapter 1) are applicable to low dose implantations when the implanted species is merely a trace element in the target matrix. In high dose situations, say >10^{16} ions cm^{-2}, the implanted profile will be modified due to (i) sputter erosion of the surface, which recedes into the target and swelling due to the additional atoms, (ii) loss of the implanted species due to sputtering, (iii) changes in the stopping power due to the implanted impurity atoms and (iv) segregation, precipitate growth and formation of other phases. Some of these aspects have been considered by Jager(87t) for the synthesis of SiO_2 by the implantation of high doses of O^+ ions into silicon. Other workers have modelled metal ion implantations(86f).

2.2 MACHINE RELATED PROBLEMS

Uniformity: Beam scanning methods, which are necessary to achieve lateral dose uniformity are discussed in Chapter X. Here we wish to add that dose inhomogeneities can be caused by excessive tilting of the wafer with respect to the beam. The effect becomes more pronounced with increasing wafer diameter, examples are quoted by Freeman(76j).

Beam purity: Good control of the dose may be compromised by charge exchange processes(87q)(78a) and disassociation of molecular species(76j)(77e). If poor vacuum conditions exist (>10^{-6} torr) in the beam line then neutralization of the high velocity ions, by electron capture from the residual gas molecules, will occur. Since this neutral beam will not be deflected by an electrostatic field, it will form a 'hot' spot on the sample. Ryding et al.(75h) have determined the inhomogeneity, $\Delta N_\square / N_\square$, to be:

$$\frac{\Delta N_\square}{N_\square} = 2.48 \times 10^{14} \; PL \; \sigma_{10} \; \frac{A}{a} \tag{2}$$

where P is the pressure (Pa), L the path length of the particles (cm), σ_{10} the effective cross section (cm^2), A the area over which the beam is scanned and a the area of the non deflected beam.

Dose non uniformity caused by the presence of neutral particles can be reduced by the inclusion of a neutral trap or by improved pumping. In systems using post analysis acceleration there is the additional problem of charge exchange occuring within the accelerator tube when particles of significantly different energy will reach the target(81a)(87q).

Sputter contamination: Exposure to the ion beam of apertures and retaining clips (see Figure 1) will lead to sputter erosion of these components and, inevitably, this results in contaminants being transported to the wafer(79h). Contamination levels can be as high as a few percent of the incident dose. This problem is minimised by the use of low sputtering coefficient materials or by the use of a benign material such as silicon.

Heavy metals and hydrocarbon contamination on the surface of a wafer will suffer recoil implantation (see above). However, this may be reduced by the use of a thin sacrificial screen oxide.

Wafer temperature: Heating of the sample due to the kinetic energy of the beam becomes a problem in high current and, also, high energy machines (Chapter 7,8). For example, a power loading of 1 watt cm^{-2} (10μA at 100keV) on a freely radiating silicon wafer, with no conduction cooling, will give an equilibrium temperature of about 300°C(86d), the actual value being very sensitive to the emissivity. Control of the wafer temperature is important in order to avoid unintentional partial annealing of radiation damage and to avoid degradation of photoresist (88b).

It is customary to provide wafer cooling on high current machines (Chapter 7) to limit temperature excusions to about 100°C. An exception is SIMOX technology(86c) where it is essential to avoid amorphisation of the surface layer by implanting at elevated temperatures of, typically, 500°C to 600°C.

Dose rate (ions cm^{-2} sec^{-1}): The dose rate (instantaneous current density) is one of the more difficult implantation parameters to control as it is dependent upon the beam spot size (and aberrations), which itself is sensitive to the plasma conditions in the ion source (Chapter 5) and ion optics of the system. In practice, implantations are carried out over a wide range of current density from about 1mA cm^{-2}, in many commercial

production machines, to less than $1\mu A$ cm^{-2} in research machines.
The use of polyatomic molecular beams provides a further
variable(71a).

Implantation at high dose rates causes two competing
effects(i) sample heating (see above), which will lead to
partial annealing of radiation damage and (ii) higher defect
densities(69e). Using an epr technique, Crowder(72b) finds a
dependence on the dose rate for 5×10^{15} B^+ cm^{-2} at 200keV at a
beam current density in the range $0.1\mu A$ cm^{-2} to $1 \cdot 0\mu A$ cm^{-2}.
Other systems have been reported by Davis(75e) and Tinsley(74d).

3. ELECTRICAL ACTIVITY IN SEMICONDUCTORS

Substitutional doping of a semiconductor by ion implantation is a two stage process:
        (i)   Implantation
        (ii)   High temperature anneal
The anneal itself serves two purposes, firstly to annihilate
lattice defects and/or achieve epitaxial regrowth of the
amorphous surface layer formed during high dose implantation
and, secondly, to accommodate the dopant atoms on substitutional
lattice sites. The annealing conditions must be tightly
controlled to minimise dopant and impurity redistribution, which
could lead to unintentional profile broadening. Additionally,
there could be loss of electrical activity through dopant
precipitation and segregation(86d). In the present context,
implantation is used to dope both uniformly across whole wafers
and locally within defined areas (windows).

The complete annihilation of lattice defects in implanted
material is seldom achieved and those defects and defect-
complexes which remain will degrade the electrical properties of
the implanted layer. The minority carrier lifetime(τ) is most
sensitive to the presence of damage and the carrier mobility(μ)
is also rapidly reduced by the presence of addditional
scattering centres. The monitoring of these parameters and also
the use of spectroscopic techniques (DLTS)(74f)(FRS)(88e) to
determine the number and positions of electronic (defect)
levels, are very sensitive methods for the evaluation of the
quality of an implanted layer. However, it is the electrical
conductivity (resistivity) which is most widely used to assess
implanted semiconductors, in part because it provides a direct
determination of the electrical activity over a wide dose range
(say, 10^{10} to 10^{16} cm^{-2}) and, also, as the measurement
techniques are relatively quick, easy to apply and adaptable to
many sample geometries.

The electrical conductivity (σ) of a single crystal
semiconductor may be expressed in terms of the density and
transport properties of the free charge carriers. (Moll, 64)
(Smith, 79c) (Kittel, 76c). These parameters are related as
follows:

$$\sigma = \frac{1}{\rho} = ne\mu \quad \text{or} \quad \sigma = pe\mu \tag{3}$$

Where n, p are the number of free charge carriers in n-type and p-type material,respectively, e the electronic charge and μ the mobility (velocity per unit electric field, $(^\nu/_E)$) of the carriers in the particular material. The resistitivity (ρ) is the reciprocal of the conductivity.

Doping: The conductivity of a semiconductor may be increased by as many as nine orders of magnitude through the addition of impurity atoms. This process is known as doping and requires that the impurities (N_D or N_A, cm^{-3}) be accommodated into the crystal structure on substitutional lattice sites and, also, have a chemical valency which differs from that of the host.

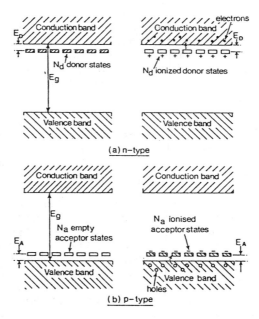

Figure 4: Energy level scheme for a semiconductor containing (a) donor or (b) acceptor impurities.

The band diagrams for both n-type and p-type semiconductors are shown in Figure 4, and it is evident that the extrinsic carriers associated with the localized donor and acceptor states require thermal energy, of magnitude E_D and E_A in order to be excited into the conduction and valence bands, respectively. The conditions $|E_D| \ll Eg$ and $|E_A| \ll Eg$ must be satisfied to ensure that extrinsic conduction dominates at room temperature. In the case of silicon these conditions are satisfied by selecting impurity species from adjacent groups in the Periodic Table, namely, acceptors from Group III (B, Al and Ga) and donors from Group V (P, As and Sb), these having activation energies in the range 0.04 eV to 0.07 eV. The band gap (E_g) of silicon is 1.1 eV.

The number of extrinsic carriers has a strong temperature dependence and for n-type material may be expressed as:

$$n = \frac{(N_D - N_A)}{2N_A} \, N_C \, \exp\,(-E_D/kT) \tag{4}$$

where N_D, N_A are the number of donor and acceptor impurities, respectively and N_C the effective density of states in the conduction band.

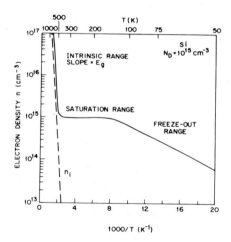

Figure 5: Electron density as a function of temperature for a silicon sample with a donor concentration of 10^{15} cm^{-3}(Sze, 81d).

Figure 5 shows the temperature dependence of the free electron density (majority carriers) in lightly doped (N_D = 1 x 10^{15} cm^{-3}) n-type silicon. At very low temperatures only a few of the donor impurities will be ionized (n≈o) but with increasing temperature more carriers will be excited into the conduction band. This increase in the carrier density will continue until all of the donors are ionized, when n = N_A and the conductivity is in the saturation region. At still higher temperatures direct excitation across the band gap (E_g) will

occur and the population will be dominated by intrinsic
carriers. The mass-action law applies at all temperatures and
the condition $np = n_i^2$ is independent of the added impurities.
    In most laboratory situations involving an electrical
measurement of a semiconductor it is assumed that the conditions
$n = N_D$ and $p = N_A$ are satisfied. As can now be seen, this is
only an approximation, which in silicon is accurate only at and
above room temperature for doping levels not exceeding 10^{18}
cm^{-3}. At higher doping levels measurable deviations will exist
due to incomplete ionization of the dopant atoms. In general,
however, larger discrepancies between the expected and observed
electrical activity, will arise due to other causes such as
compensation by impurities of the opposite type, deep traps due
to residual damage and de-activation by the trapping of dopant
atoms on non-substitutional sites. A low level of activity of
implanted dopants is one of the characteristics of compound
semiconductors, such as GaAs and InP and is due to deactivation
of dopants by the formation of defect complexes.
    In all cases the effective electrical activity may be
defined as:-

$$\frac{n}{N_D} \times 100\% \quad or \quad \frac{P}{N_A} \times 100\%$$

for n and p-type material, respectively.

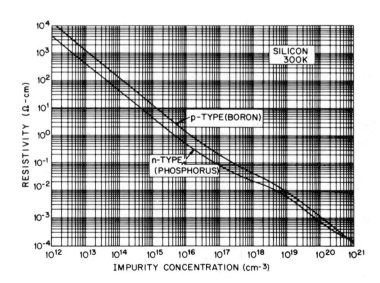

Figure 6: Bulk resistivity versus dopant
concentration for silicon at 300K(Sze, 81d).

Figure 6 shows the dependence of the bulk resistivity of silicon upon the impurity concentration at 300k, assuming 100% activity, after Beadle, Plummer and Tsui(81c).

The movement of charge carriers through a semiconductor at a finite temperature is impeded by scattering processes due, for example, to lattice vibrations (phonons), lattice defects and ionised impurities (Blatt, 68a). The latter are due to the added donors and accepters, which assume a net charge upon release of their extrinsic carriers. This is shown schematically in Figure 4.

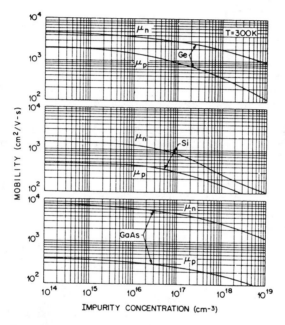

Figure 7: Drift mobility of carriers in silicon, germanium and GaAs at 300K versus dopant concentration (Sze, 81d).

Figure 7 shows the measured carrier mobilities of Si and GaAs plotted against impurity concentration. In each case the electron mobility is higher than the hole mobility, due to the difference in effective mass (m^*).(Blatt, 68a).

4.   EVALUATION TECHNIQUES

4.1   NON-DESTRUCTIVE ASSESSMENT

4.1.1 Visual
An immediate and surprisingly sensitive method of cosmetic
inspection of the surface of a polished wafer is the use of the
unaided eye, with the wafer under bright illumination. Defects
may be seen as bright spots of light, scratches, haze or stains.
Particulates, which may be produced during mechanical handling
of the wafers, are especially troublesome as they mask the wafer
surface from the beam, causing highly locallised dose
variations. Large particulates can easily be detected by the
eye.
        Under favourable conditions changes in the optical
properties of the implanted layers may be sufficient to enable
dose non-uniformity to be observed. Shadow masking by clips,
inadvertantly obscuring part of the wafer from the beam, is
generally easily detected due to the change in the reflectivity
of the exposed damaged/amorphous surface, see Figure 1. Sputter
erosion of clips is a prolific source of metal contamination
(79h).

4.1.2 Light scattering
        A number of fully instrumented optical scattering systems
are available for automatic monitoring of surface defects(82g).
These are based upon the specular reflection of He-Ne laser

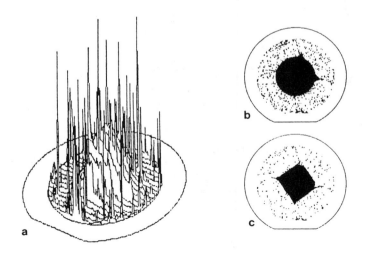

Figure 8: Surface characterisation by elastic light
scattering (a) map of the scattered light intensity,
(b) and (c) sections through the intensity plot(85a).

light and take as a signal the randomly scattered light from surface defects and particulates of sub-micron size. Steigmeier and Auderset(85a) estimate a senstivity limit of 0.4µm for practical systems. Small scale surface features, which are much smaller (vertically) than the wavelength of the illumination can be detected provided the irregularities are correlated and extend over a large area of the surface. Such features give rise to additional scattering (haze).

Recent work has shown that the "haze like" background signal can arise from scattering associated with imperfections in the sub-surface layer(84d). By processing this signal, Steigmeier and Auderset(85a) have demonstrated that elastic light scattering may be used to map the lateral distribution of defects in silicon within the subsurface layer of thickness 0.25µm. The system has good sensitivity and a wide dynamic range, however, the physical nature of the defects and their number density and distribution are unknown quantities. Figure 8 shows experimental data from an O^+ implanted SOI/SIMOX wafer. Sections A and B are cuts through the raw data and clearly show different symmetries which can be traced back to non-uniformities in the dose and implantation temperature across the wafer, respectively.

Inelastic light scattering (Raman scattering) may also be used to determine the crystalline perfection of implanted layers. In crystalline silicon the lattice vibrations give a characteristic sharp line which, in the presence of defects and thus a loss of lattice periodicity, suffers a shift in frequency and an increase in width. These deviations may be used as a measure of the lattice strain or distortion resulting from the implantation and annealing processes. Recently, Harbeke et al.(87m) have used Raman scattering to monitor the properties of high dose O^+ ion implanted silicon. The measurements were carried out using standard equipment with a 100mW laser (5145Å) focussed to a line image of 50µm x 2mm and with a penetration depth of about 6000Å. The ability to determine, retro-spectively, the wafer temperature during implantation (in the range 250°C to 580°C), from measurements of the Raman frequency and linewidth, was demonstrated, and a sensitivity of about 20°C was estimated.

4.1.3 Thermal-wave

A relatively new (1984) laser based system for the measurement of near surface lattice defects and, by inference, retained dose, uses thermal wave techniques(86e)(87s). This method provides a very fast, entirely contactless and non-destructive evaluation of implanted layers. Currently, it is being applied to silicon and GaAs(87r) for local area and wafer mapping of samples implanted over the dose range 10^{10} to

10^{16} ions cm^{-2}. This is a very wide dose range and, uniquely, measurements can be made through oxide or nitride overlayers whose thickness, also, is measured.

Conceptually, the method involves the detection of changes in the thermoelastic and optical properties of the implanted layer, caused by the presence of lattice defects. In the commercial Therma-wave system(86e), the change in the reflectance of the surface is monitored. Thermal waves are

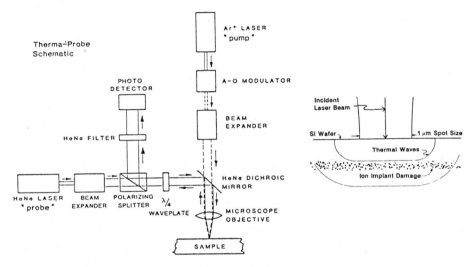

Figure 9: Generation of thermal waves, using a focussed laser beam incident on a silicon wafer containing subsurface damage (from Smith et al. 86e). Figure 10: Schematic of the Therma-probe 150 systems (from Smith et al. 86e).

generated, as illustrated in Figure 9, using a focussed beam ($\approx$1µm) from an Ar laser and the surface reflectivity is monitored with a second "probe" beam from a HeNe laser. The system is shown schematically in Figure 10. The Ar laser is modulated (typically at 1MHz to give a probing depth of $\approx$3µm in silicon) and the change in reflectance is deduced from the output of a phase sensitive synchronous detector. Experiments have shown that the magnitude of the signal, measured in arbitrary thermal wave (TW) units, is dependent upon both the presence of lattice defects and impurity atoms in the implanted layer. However, the signal does not have a simple dependence upon the substrate material nor implantation conditions and, thus, absolute measurements cannot be made and resort to relative measurements and the use of calibration samples is necessary. Figure 11 shows the dependence of the arbitrary thermal wave signal upon dose for 50keV B^+ ions in silicon.

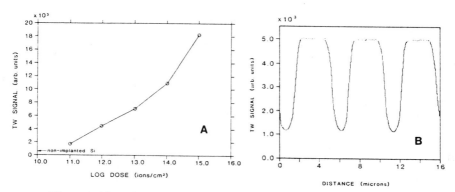

<u>Figure 11</u>: Measured Thermal wave (TW) signals as a
function of (a) dose for 50keV B$^+$ implants into
silicon and (b) position across a fine (2.5μm) pattern
implanted with a low dose (Smith et al. 85).

Wendman and Smith(87c) have reported the use of the
technique for process control of NMOS product wafers, with
implants of 4.0×10^{11} to 8.0×10^{11} B$^+$ cm^{-2} at 50keV and 1.5×10^{12}
to 2.5×10^{12} As$^+$ cm^{-2} at 25keV. A strong correlation between the
thermal wave data and transistor electrical characteristics was
reported.

4.1.4 Optical dosimetry

For purposes of process control, a measurement of the
consistency with which a fixed dose and a particular level of
dose uniformity can be achieved, is required. A novel method
that provides this data is based upon a measurement of the
optical transmission of a photoresist coated glass plate, which
is implanted at the same time as the semiconductor wafers(87a).
Ion damage creates defects in the photoresist and the optical
density is found to vary monotonically with the ion dose over
the range $10^{11}-10^{16}$ ions cm^{-2} for all ion spices over a wide
range of energy (>5keV).

While not an absolute method, the technique enables
uniformity maps to be rapidly generated (80 sec) with a spatial
resolution of 3mm and a dose resolution, typically, 0.1%.
However, as measurements are not made on the actual wafers,
spurious effects due to wafer charging and channelling
(see Section 2) will pass undetected.

4.1.5 Other methods

In addition to the above methods, many other techniques
have been used to evaluate implanted layers, although these are
generally experimental in nature. At the present stage of
development they may not be suitable for a production
environment.

Carrier lifetime: Rehwald et al.(88a) have reported the
measurement of carrier lifetimes in annealed implanted layers
using contactless probing by microwaves(84c). The sample is
located close to a 1.2GHz microwave stripline resonator and the
loading of the circuit by photogenerated (UV lamp) excess
carriers is monitored. Lifetimes are determined from the
temporal decay of the reflected microwave signal. Lifetimes
greater than about 1μs can be measured with a spatial resolution
of about 4mm.

Optical: Optical techniques are used also to analyse high dose
implanted samples, for example, buried dielectric layers where
optical reflectance(87i)(86b) and spectroscopic ellipso-
metry(87p) have been applied. The latter technique enables the
composition of multiple layers to be deduced. Automatic laser
ellipsometry systems, in which the extinction coefficient is
monitored, have been developed for dose uniformity mapping of
silicon and GaAs over the dose range 10^{11}-10^{15} cm^{-2}(87k). The
use of photoluminescence to investigate implanted semiconductors
has been discussed by Palik and Holm(79g).

Nomarski interference contrast microscopy (Sze 83c) is an
alternative and very powerful technique for revealing subtle
changes in morphology, having a sensitivity in the vertical
direction of 100Å - 200Å.

4.2   ELECTRICAL METHODS

It is necessary to anneal the implanted wafers prior to
making electrical measurements in order to activate the dopant.
Exceptions to this generalization occur when the highest dose
sensitivity is required (see below) or when locallised defect
levels are the cause of the desired electrical activity. The
annealing procedures are discussed in Chapter 3. The measurement
techniques are used either to directly determine the average
electrical activity of the implanted dopant (areal density,
cm^{-2}) or, when combined with layer removal or bevelling, to
enable depth profiles (volume concentration, cm^{-3}) to be
determined. Alternatively, the net dopant profile (N_D-N_A) may
be determined from C-V measurements.

The principal measurement techniques fall into two
categories (i) those involving measurements of the resistance of
passive layers and (ii) measurements of the characteristics of
test structures, which may be simple p-n junctions, Schottky
diodes or fully fabricated active devices.

4.2.1 Thermal-probe/junction staining

The thermal-probe, which is based upon the Seebeck effect
(dT/dx), gives rise to a very simple procedure to determine
whether the implanted layer is n- or p-type. The semiconductor
is locally heated to induce a temperature gradient which
produces a potential difference due to the drift of the carriers
with the gradient. If the sample is n-type a positive voltage

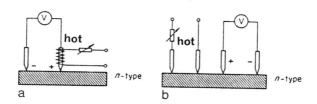

<u>Figure 12</u>: Schematic of the thermal probe (a) two
contacts with one heated and (b) modification of the
four point probe using Joule heating of the
semiconductor.

develops at the hot contact. Figure 12 shows schematically (a)
the simplest two terminal configuration and (b) an adaptation of
the four point probe (see below).
    Alternatively the conductivity type may be determined using
the Hall effect or by plotting the I-V characteristics of a p-n
junction. These methods require contacts to be formed and are
discussed below also.
    A qualitative method to delineate a p-n junction is that of
staining or decorating. The sample is ususally bevelled and an
etchant is selected which preferentially etches or stains one
side of the junction(67a). Historically this technique has
played an important role but is little used today as it is
rather inprecise and unpredictable.

4.2.2 <u>Spreading resistance</u>
    This technique is extensively used to determine depth
profiles, in which case samples are bevelled. The technique
combines high spatial resolution, a dynamic range of 6 (or more)
orders of magnitude with an ability to profile across a p-n
junction. Disadvantages are that the technique is not absolute,
that it calls for considerable artisan skills and that the
equipment is expensive.
    The basis of spreading resistance is that the constriction
and subsequent "spreading" of the current flowing from a small
(point) contact on the surface of a conducting sample, will give
rise to a component of circuit resistance. This spreading
resistance (R_s) may be directly measured when a dc current flows
through a small radius probe (radius 2-10μm) of a hardened
conducting material, such as tungsten carbide or tungsten-osmium
alloy, which is placed on the surface of the sample with a probe
force, typically, in the range 5-20 gms. Under these
conditions, the measured spreading resistance on a semiconductor
(silicon) wafer may be a factor of 10^3 larger than the numerical
value of the resistivity (Figure 13).

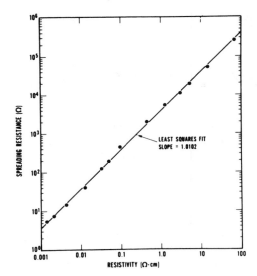

Figure 13: Values of the spreading resistance for (111) silicon plotted against the resistivity (from Zemel, 79e).

Two probe geometries may be considered (Figure 14a & b) and and Holm(67c) has shown that for a single "ideal" ohmic contact forming a hemispherical indentation of radius 'a' on a semi-infinite sample, the spreading resistance is given by:

$$R_S = \frac{\rho}{2\pi a} \tag{5}$$

where ρ is the bulk resistivity of the semi infinite sample.

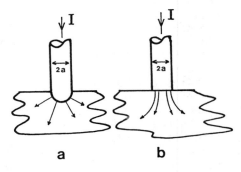

Figure 14: Contact geometries for the spreading resistance probes (a) hemispherical contact and (b) planar contact.

For a single non-indenting contact Holm suggests

$$R_s = \frac{\rho}{4a} \tag{6}$$

In both cases the spreading resistance is proportional to the bulk resistivity and is spatially localized within a distance of about 5 times the contact radius. This enables local resistivity variations in bulk material to be determined with a lateral resolution of, typically, 10μm.

In addition to variations in the geometry of the probe, it is found that the "condition" (or microscopic shape of the probe tip) has a major effect upon the measured values of resistance. This arises, in part, because of the nature of the semiconductor surface, which will have a thin layer of native oxide of thickness 20-50Å, depending upon previous chemical processing. This oxide has high mechanical strength and, if not removed, may cause the probe to form an undesirable metal/insulator/ semiconductor contact. This problem may be overcome by "conditioning" the probe so that the surface has many protrusions which, on a microscale, will each stress the oxide and achieve penetration. The number and distribution of these asperities will depend upon the method of preparing the tip. The preferred technique is that developed by Gorey and Schneider (see Mazur and Gruber, 81b) who used a mechanical device to grind the probe top in a controlled manner. Their work shows that the use of a probe grinder, in combination with a low probe loading, allows precise control of the tip penetration and, thus, good reproducibility of resistance measurements is possible, even on thin layered structures.

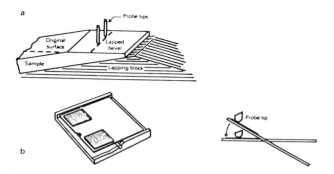

Figure 15: (a) Schematic of the geometry for depth profiling using spreading resistance and (b) the Gorey-Schneider probe grinder and principle of operation(82e).

In practice, it is customary to "condition" the probe at
regular intervals, say, before commencing a series of
measurements. The grinding device is compact and may be placed
on the sample stage and the grinding action is achieved by
moving the probe in a vertical plane so that the tip slides
over the grinding surface, as shown in Figure 15. The grinding
surface is a sapphire plate with a fine (1/4 µm) diamond
abrasive. The grinding action may involve many tens of strokes
(vertical movements), with periodic rotation of the probe about
its major axis. After conditioning, the tip is cleaned with a

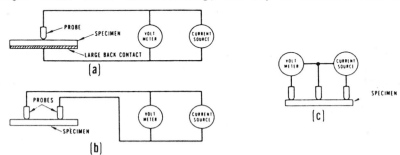

Figure 16: Circuit configurations for spreading
resistance measurements using (a) one, (b) two or
(c) three probes.

swab soaked in an organic solvent. Whilst this method of
conditioning enables self consistent data to be collected, it
still must be remembered that the spreading resistance method is
a comparative technique and frequent checks against known
standard samples are necessary.
Circuit Configuration: Spreading resistance measurements may
be made using one, two or three probes, as shown in Figure 16.
(a) Single probe. This configuration is suitable for
assessment of bulk samples or uniformly implanted layers in
substrates of the same type. However, the circuit arrangement
is little used as a large area (low resistance) ohmic contact
must be formed on the back or side of the sample. In addition,
the presence of buried junctions or planar device structures
cannot be tolerated as they will introduce circuit resistance.
(b) Two probes. In the case of two probes, both are assumed to
contribute equally to the measured spreading resistance,
provided the separation 'ℓ' is such that the bulk resistance
between the probes is negligible compared to the spreading
resistance. In this case the measured resistance is the sum of
the individual values and so $R_s = \rho/2a$. In practice this
configuration is used most frequently and is suitable for
profiling across a p-n junction.

(c) Three probes. In this case only the central probe contributes to the measured resistance and so $R_s = \rho/4a$. However, despite the advantages of using separate current and voltage contacts, this configuration is seldom used because of the difficulties of maintaining the correct alignment of all three probes, particularly on bevelled surfaces used for depth profiling.

Calibration: Spreading resistance values are related to, but are not a direct measure of, specimen resistivity. The principal error is associated with the contact parameter, a, so it is customary to use,

$$R_s = \frac{\rho}{2a(\rho)} \cdot (CF) \tag{7}$$

where $a(\rho)$ is the effective contact radius, which is determined from measurements of standard samples. This parameter also incorporates any non-linearities of the spreading resistance with the bulk resistivity. Parameter (CF) is the "sampling volume correction factor" which accounts for structure variations, layer boundaries and resistivity gradients. (Ehrstein, 79b).

Depth profiling using spreading resistance is a destructive technique, as the profile must be exposed by bevelling the sample at a shallow angle of, say, <0.25°. Under these conditions a wide concentration range ($10^{14} - 10^{20}$ cm^{-3}) may be profiled with a spatial resolution of 200A°-300A°. However, with care and in favourable systems (Si) depth resolutions of a few 10's Å may be achieved. Arbitrary combinations of layer thickness and type may be measured(77b).

Attention must be paid to the preparation of the surface of the bevelled sample and various polishing compounds are available, some relying on mechanical polishing and others being mainly chemical. Mazur and Gruber(81a) have suggested the use of a fine diamond abrasive in an oil-based slurry on a lapped glass plate. The improved reproduciblity and stability of the measured spreading resistance is believed to be due to the absence of water. This procedure is published as an ASTM Standard Practice(74a).

The technique is suitable for relative or absolute measurements of the electrical activity. For relative measurements, which are often used for process control purposes, direct use is made of the spreading resistance data, without conversion to resistivity or dopant concentration. An example of a spreading resistance depth profile is shown in Figure 17. Process control may be achieved by defining upper and lower limits for the value of R_s, using a 'standard' sample.

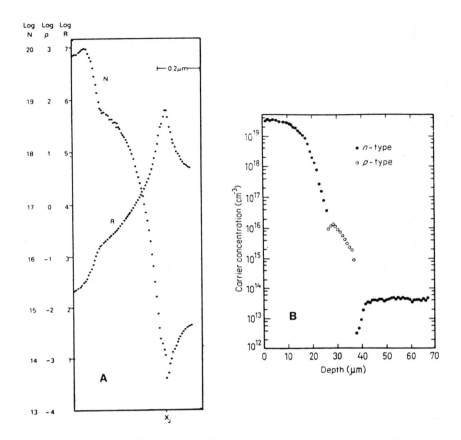

Figure 17: Examples of spreading resistance depth profiles (a) carrier concentration profiles of a BF_2^+ implant into silicon. The junction depth (x_j) is 0.43μm(82e) and (b) carrier concentration profile through a transistor structure (from Ryssel and Ruge, 86d).

Determination of the carrier depth profile is made by reference to equation 7 for which a new value of the sampling volume correction factor must be calculated for each data point. Several models have been developed to handle this computation and these include a unilayer model by Dicky and Ehrstein [79d], which is suitable for thick, slowly graded layers and junction isolated layers. A multilayer model by Schumann and Gardner(69c) (69d) will handle highly graded and multilayer structures. Details of these sometimes cumbersome computations can be found in the literature (Zemel, 79e: D'Avonzo et al, 77b).

   The complete determination of the doping (carrier) profile
is in two parts. Firtly, the application of equation 7 to
determine the resistivity profile, as described above, and,
secondly, the use of mobility data (Figure 7) to make the
conversion from resistivity to dopant concentration.

4.2.3 Sheet Resistance

   The bulk resistivity of a semiconductor has been defined in
equation 3 and the value may be calculated from a measurement of
the bulk resistance (R_B) using:

$$\rho = \frac{wt}{\ell} \cdot R_B \tag{8}$$

where ℓ, w and t are the length, width and thickness,
respectively.

   For thin, ion implanted and diffused conducting layers,
with highly non-uniform doping depth profiles, it is appropriate
to consider a square lamina (w = ℓ) when equation 5 may be
written in terms of a sheet resistance (ρ_S) where

$$\rho_S = \frac{\rho}{t} = R_B \ \Omega/\square \tag{9}$$

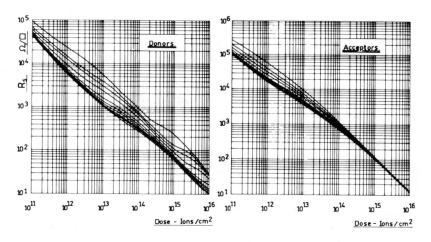

Figure 18: Sheet resistance of ion implanted silicon,
assuming complete electrical activity for (a) donors
and (b) acceptors. The family of curves arises from
mobility variations (see Figure 7) due to different
standard deviations of 0.0025µm, 0.0075µm, 0.0125µm,
0.025µm, 0.05µm, 0.075µm, 0.10µm, 0.125µm, 0.15µm and
0.175µm (Data from Smith and Stephens, 70d).

188

The numerical value of the sheet resistance is independent of the size of the square.

Theoretical values of the sheet resistance of n- and p-type implanted layers in silicon have been published by Ryssel and Ruge(86d) and by Smith and Stephen(70b), Figure 18. Their results have been calculated from the relationship:

$$\rho_s = \frac{dx}{n(x)\mu(n(x))e} \tag{10}$$

where $\mu(n(x))$ is the carrier mobility.

Resistance is determined experimentally by applying Ohm's Law ($R = V/I$) to the measured values of current and voltage across a sample of known geometry. Errors may be minimised by making four terminal measurements in which a constant current is driven through two contacts and the voltage is monitored across another pair of contacts. Equation 11 generally applies, where k is the geometric correction factor and V and I are the measured voltage (PD) and current, respectively.

$$\rho_s = k.\frac{V}{I} \tag{11}$$

The electrical contacts should be ohmic and of low specific resistance. This condition is easily satisfied on medium to heavily doped material when it is adequate to use pressure contacts. Such probe arrays have the major advantage of placing few constraints on the sample size and geometry and require the minimum of specimen preparation. For lightly doped material it is necessary to use metallized or doped contacts, entailing the use of masking procedures. In this case, if photolithography is used also, small areas of micron dimensions may be defined. Small test structures, for quality control purposes, are often included on device wafers.

Four point probe - linear array: The linear four point probe is used extensively to measure the sheet resistance of bulk silicon, and epitaxial, diffused and ion implanted layers. It may be used over the implanted dose range 10^{11} - 10^{16} ions cm^{-3}, corresponding to a range in sheet resistance of 10^1 to 10^4 $\Omega/\square$, see Figure 18. With care an accuracy of $\pm$ 0.1% may be achieved (Market 84h) although values closer to $\pm$ 1% over the dose range 10^{14} - 10^{16} ions cm^{-2} are more typical.

Figure 19 shows a schematic representation of the linear array, where s represents the probe spacing and t the thickness of the conducting layer. Typically the probes will be equi-spaced with a spacing of, say, 250μm. To minimise errors the diameter of the probe tip should be small compared to s, and is ideally a point contact.

In the most general case of a semi infinite sample (t ≫ s), the linear probe may be used to determine the bulk resistivity (ρ), which is given by

$$\rho = 2\pi \; s \; \frac{V}{I} \; \Omega \; cm \qquad (12)$$

where V is the potential difference between probes 2 and 3 due to a current (I) passing through probes 1 and 4.

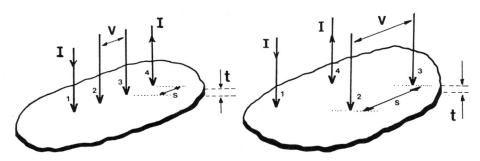

<u>Figure 19</u>: Schematic of the linear four point probe.
<u>Figure 20</u>: Schematic of the square four point probe.

A simplification is possible when measuring ion implanted and diffused layers as the thickness is, typically, less than 1 μm. Thus t ≪ s and the current flow may be considered to be two dimensional. In this case the solution to the integral equation leads to,

$$\rho = \frac{\pi t}{\ell n 2} \frac{V}{I} \qquad (13)$$

and the sheet resistance (ρ_s) may be written as:

$$\rho_s = \frac{\pi}{\ell n 2} \frac{V}{I} = 4.532 \frac{V}{I} = k \frac{V}{I} \qquad (14)$$

where k, the geometric correction factor, is 4.532.
Rymaszewski(69b) has calculated the geometric correction factor for the six possible combinations of potential and current contacts of the linear array (Table 1).

Configuration	Current Contacts	Potential Contacts	k
a	1,4	2,3	4.532
b	1,2	3,4	21.84
c	1,3	2,4	15.50
d	2,4	1,3	15.50
e	3,4	1,2	21.84
f	2,3	1,4	4.532

Table 1: Six configurations of the current and potential probes
for a linear four-point probe, from Rymaszewski(69b).

It is customary to use configuration (a) in Table 1, with
the outer contacts carrying the current, as this leads to the
largest voltage difference for a given current and equi-spaced
probes.
The tabulated values will only give better than 1% accuracy
in ρ_s, when these measurements are made at the centre of finite
samples of linear dimensions greater than about 40 times the
probe spacing. As an example for a probe width spacing of 1mm,
the error will be less than 2% over the central 50% area of an
implanted 50mm wafer, but then increases rapidly for measurements nearer the edge. Values of k are listed in Table 2 for
the linear array for different values of d/s, where d is the
diameter of the sample and s is the probe spacing (Bullis, 74a).

d/s	3.0	4.0	5.0	7.5	10	15	20	40	∞
k	2.226	2.929	3.363	3.927	4.172	4.362	4.436	4.508	4.532

Table 2: Geometric correction factors (k) for a linear four
point probe on circular samples of diameter d (Bullis, 74a).

For small samples or when the probe is near the edge of an
implanted layer, as occurs during uniformity measurements
(Current, 84e), it is necessary to calculate the value of k for
each of the measurement positions. The value is a maximum at
the centre of the implanted area as V is a minimum.
Severin and Bulle(75b) have reported the use of a linear
four point probe using liquid mercury contacts. These have the
advantage of causing no mechanical damage to the surface of the
semiconductor.
Four point probe - square array: The square four point probe
array is shown schematically in Figure 20. It has many features
in common with the linear array plus the advantages, firstly,

that smaller samples may be measured due to the compact layout
of the contacts and, secondly, it has a suitable geometry for
Hall effect measurements.

In the general case of a semi infinite sample ($t \gg s$) the
square array gives for the bulk resistivity:

$$\rho = \frac{2\pi}{(2-\sqrt{2})} \, s. \, \frac{V}{I} = 10.726 \, s.\frac{V}{I} \tag{15}$$

For thin implanted layers, where $t \ll s$, the sheet
resistance is:

$$\rho_s = \frac{2\pi}{\ell n 2} \frac{V}{I} = 9.0648 \, \frac{V}{I} \quad = k\frac{V}{I} \tag{16}$$

Uhlir (1955) has calculated values of k for finite circular
samples of diameter d, as listed in Table 3

d/s	2.0	3.0	4.0	5.0	10.0	20.0	α
k	4.904	6.005	6.880	8.455	8.575	8.933	9.064

Table 3: Geometric correction factors (k) for a square four
point probe on circular samples of diameter d (Uhlir, 55).

Mircea(67b) has determined the values of the correction
factor for this probe array on a square sample of side d. His
results are shown in Figure 21 where it may be seen that k tends
asymptotically to a value of 4.532 as d tends to s. This
limiting situation has the same geometry, and hence the same k
value, as Van der Pauw samples, which are discussed later. These
calculations have been extended by Stephens et al.(71b) who have
considered the causes of systematic errors and one of their
conclusions is that, for symmetric geometries, the apparent
(measured) sheet resistance will always by greater, or in the
limit equal to, the actual value.

The simplicity of the Four-Point Probe technique, namely
the use of pressure probes to make ohmic contact to the
implanted layer, introduces its own problems. If the force on
the probe is too large, the probe may damage the surface and
pierce through the implanted layer, making a short circuit to
the substrate. Smith(77d) recommends that the force should be
restricted to 20 gms wt. when it is possible to contact layers
of 400Å thickness (say, 40 keV As$^+$). Care, also, should be
taken to clamp rigidly both the probe assembly and wafer, to
avoid relative movement. The tips of the probes must be

maintained in a clean state. Smith(77d) reports that reliable
measurements are only possible after driving a current through
all of the probes, to form good ohmic contacts.

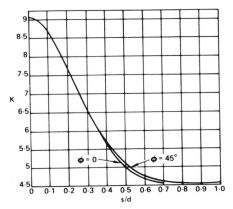

Figure 21: Dependence of the correction factor (k) on
the ratio s/d for the square four point probe of
spacing s on a square sample with sides of length d
(Zemel, 79e).

Conventional four point probe sheet resistance techniques
become unreliable for low doses in the range 10^{11}-10^{13} ions
cm^{-2}, due to surface effects and leakage currents to the
substrate. A development, which overcomes this limitation,
involving the use of a high resistivity substrate in conjunction
with a special cleaning procedure for passivating the silicon
surface, has been reported by Smith et al.(85c)(also see below
and Section 5.4). By this means they are able to measure sheet
resistance values as high as 100kΩ/$\square$ (2 x 10^{11} B^+ cm^{-2})
Six point probe: The six point probe has been developed to
enable measurements of lightly doped (high resistivity) layers
to be made. The instrument consists of a linear four point
probe, used in the conventional manner, plus an additional pair
of probes which contact the back of the wafer at locations
opposite the current probes (Figure 22). These fifth and sixth
contacts enable the potential difference (V'), due to the
leakage current (I_ℓ), to be monitored. This current is given
by:

$$I_\ell = \pi \, w \, V' \left(\rho \left(\ell n \, \frac{3s}{w} + 0.5772 \right) \right)^{-1} \qquad (17)$$

where w is the substrate thickness, ρ the substrate resistivity
and s the spacing of the four point probe.

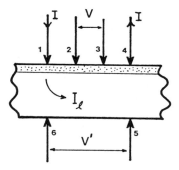

<u>Figure 22</u>: Schematic of the six point probe.

Knowledge of the leakage current enables the four point
probe data to be corrected, facilitating measurements of layers
lightly doped with $10^{11} - 10^{12}$ ion cm^{-2}. In addition, a
measurement of the potential difference between the front and
back surfaces of the wafer makes possible a correction to the
layer thickness, to allow for penetration of the depletion
width.

<u>Test Patterns</u>: The reliability of measurements of doping and
doping uniformity using the four-point probe, depends critically
upon a knowledge of both the wafer and probe geometry.
Significant errors may be introduced if the implanted layer has
an irregular shape. These contraints can be overcome if
photolithography is used to define a matrix of small samples of
fixed dimensions and adequate size to accept the probe. In this
case k is constant over the whole wafer and may be determined
from expressions due to Smits(58a), which have been confirmed by

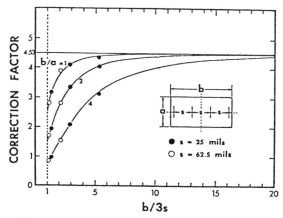

<u>Figure 23</u>: Geometric correction factor (k) versus
$b/_{3s}$ for a linear four point probe symmetrically
positioned within a rectangular test pattern of
length b and width a (Perloff et al.77c).

Perloff(77c) for a linear probe positioned symmetrically within
rectangular test patterns, Figure 23. Unfortunately errors are
still present due to variations in the probe spacing, and a new
source of random error is caused by dimensional variations in
the test patterns.

Photolithography has been used by Buckler and Thurber(76b)
to fabricate planar test structures for the measurement of bulk
resistivity in processed wafers. The structures are analogous
to the mechanical square array and were formed using bipolar
processes and the test patterns included in the National Bureau
of Standards mask set NBS-3 (Buehler, 76d). These structures
have good spatial resolution (57 μm x 57 μm) and may be used for
detailed mapping of the resistivity to monitor device processing
steps. Measurement of high resistivity material is possible by
the use of an emitter doping schedule to form low resistance
contacts.

Linear Resistors: Some of the errors associated with the
four-point probe can be avoided if 2 or 4 terminal linear
resistors are fabricated, using photolithography to define the
structures.

A disadvantage is that many more processing steps are
required, including photolithography, a second implantation to
form the ohmic contact pads, deposition of metal contacts,
thermal processing and mesa etching. Each of these processes
will introduce random errors, but there are advantages:- (i)
the data is more easily interpreted, (ii) errors associated with
the contacting probes will be minimal and (iii) reliable
measurements can be made on wafers implanted with low doses or
at high energies, when the conducting layer will be buried.

Smith and Stephen(70b) have concluded, however, that in
most situations four-point probe measurements will be as
reliable and reproducible as the linear resistor and that there
is little to be gained by undertaking the additional processing
steps necessary to fabricate the resistors.

Van der Pauw: An alternative four-terminal resistor structure,
which is less sensitive than the linear resistor to dimensional
variations in the photolithographic masks and is particularly
suitable for Hall effect measurements (to determine μ_H), is that
defined by Van der Pauw(58c and 58d), who stipulates that the
following conditions must be satisfied:
(i) the sample must be a lamina which is uniform and
     continuous.
(ii) the ohmic point contacts must be at the periphery.

The four contacts are applied in any position on the
periphery of the sample, which may have any desired shape,
Figure 24(a). A symmetric structure, which is suitable for
profile measurements using a layer removal technique is shown in
Figure 24(b).

The geometry of the structure used in the NBS Process
Evaluation Mask Set (NBS-3) is shown in Figure 25.

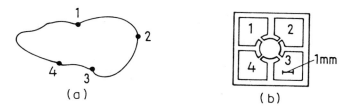

Fig.24: Examples of Van der Pauw samples (a) general case and (b) a symmetric sample which is also suitable for depth profiling.

In all cases:-

$$\rho_s = \frac{\pi}{\ell n2} \frac{(R_a + R_c)}{2} f(R_a/R_c)$$ (18)

where, with reference to Fig. X: $R_a = V_{23}/I_{14}$ and $R_c = V_{43}/I_{12}$. The function $f(R_a/R_c)$ satisfies a transcendental expression and has the values shown in Table 4. When the structure is symmetric $R_a = R_c$, and $f(R_a/R_c) = 1$ when,

$$\rho_s = \frac{\pi}{\ell n2} \frac{V}{I} = 4.532 \frac{V}{I}$$ (19)

R_a/R_c	$f(R_a/R_c)$
1	1.0000
1.3	0.9941
1.8	0.9711
2.0	0.9603
3.0	0.9067
10.0	0.6993

Table 4 Tabulated values of the Van der Pauw function $f(R_a/R_c)$(Van der Pauw, 58c).

The free-carrier concentration and mobility may both be obtained, provided the thickness, t, of the conducting lamina is known by using equation 18 and

$$R_H = \frac{t}{B} \frac{\Delta V_{24}}{I_{13}}$$ (21)

where R_H is the Hall Coefficient, and ΔV_{24} is the Hall voltage due to the applied magnetic field, B. (see 4.2.4)

Measurements by David(76f), using the symmetric structure shown in Figure 25, have given better than 0.1% agreement with equation 8 as long as the condition $d_1/a \geq d_2/a$ is fulfilled. In all cases it is desirable to make measurements for two current-voltage configurations,(equation 18), to eliminate geometric sources of error. By doing so, Perloff(77c) has reported measurement reproducibilities of better than 0.2% for 10 μm x 10 μm resistors.

Failure to locate the contacts as the periphery of the sample introduces errors which Van der Pauw has estimated (Figure 26). Exact calculations for geometries is which the contacts are not located on the periphery have been reported by Buehler and Pearson(66).

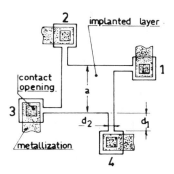

Figure 25: A Van der Pauw structure used for process evaluation (Buehler, 76d).

Undoubtedly, the fabrication of resistors with the Van der Pauw geometry involves many processing steps to produce the required planar structures, but the gains are considerable, these being excellent reproducibility, freedom to choose the size and shape of the resistor arrays, and relative independence from errors due to photolithographic mask distortions.

4.2.4 Hall Effect

The carrier concentration may be directly calculated from measurements of the Hall voltage, which arises due to the displacement of charge carriers in a magnetic field.

Figure 27 shows a schematic of a Hall sample where an electric field is applied along the x axis and a magnetic field along the z axis. For a p-type sample, with holes as the majority carrier, the Lorentz force acting on the moving carriers exerts a downwards force, causing the charge to accumulate at the lower surface of the sample. Thus an electric field develops between the top and bottom surfaces, which

	$\Delta\rho/\rho$	$\Delta R_H/R_H$
(a) 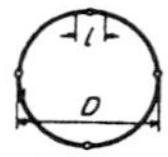	$\approx \dfrac{-\ell^2}{16D^2 \ln 2}$	$\approx \dfrac{-2\ell}{\pi^2 D}$
(b) 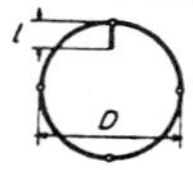	$\approx \dfrac{-\ell^2}{4D^2 \ln 2}$	$\approx \dfrac{-4\ell}{\pi^2 D}$
(c) 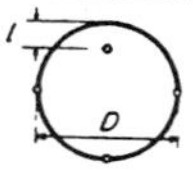	$\approx \dfrac{\ell^2}{2D^2 \ln 2}$	$\approx \dfrac{-2\ell}{\pi^2 D}$

Figure 26: Errors associated with Van der Pauw samples with a non ideal geometry(58c).

reaches an equilibrium value when it exactly balances the Lorentz force. The resulting Hall voltage (V_H) may be measured with a high input impedence voltmeter and is given as,

$$V_H = R_H J_x B_z W \tag{22}$$

where W is the width of the sample and R_H is the Hall coefficient, which may be given approximately as (Blatt, 68a)

$$R_H = \frac{-1}{en} \quad \text{or} \quad R_H = \frac{+1}{ep}$$

for n and p-type material, respectively. Thus, the carrier concentration and the identity of carrier type (electron or hole) can be directly obtained from the Hall measurement provided one type of carrier dominates.

The Hall mobility is defined as:

$$\mu_H = \frac{R_H}{\rho} \tag{23}$$

The Hall coefficient may be determined from measurements of samples with the Van der Pauw geometry, shown in Figure 25 using equation 21.

Carrier depth profiles may be determined by a layer removal method developed by Petritz(58b) and Johansson et al.(70a), which yields:-

$$\mu_i = \frac{(\dfrac{R_H}{\rho_s^2})_i - (\dfrac{R_H}{\rho_s^2})_{i+1}}{(\dfrac{1}{\rho_s^2})_i - (\dfrac{1}{\rho_s^2})_{i+1}} \tag{24}$$

and

$$n_i = \frac{(\frac{1}{\rho_s})_i - (\frac{1}{\rho_s})_{i+1}}{e t_i \mu_i} \tag{25}$$

where R_H and ρ_s are the Hall coeffient and sheet resistivity, respectively. These parameters are measured before (i) and after (i+1) removal of the i-th layer of thickness t_i. The areal density (cm^{-2}) may be obtained from $\Sigma n_i t_i$, assuming 100% electrical activity.

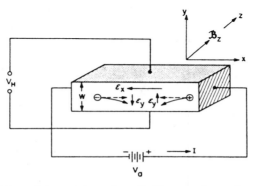

Figure 27: Schematic of a Hall sample, showing the circuit configuration.

The main source of error in the profile will be uncertainty in the depth scale. For silicon this error may be minimised by using one of a number of highly controllable anodic oxidation methods in which an oxide of known thickness is formed and then chemically dissolved. Barber et al.(76a) has characterised a system using ethylene glycol with 0.04N KNO_3, 2.5% water and 1-2 g/litre of $Al(NO_3)_3.9H_2O$. It is reported that the system is relatively insensitive to current density and temperature and that at room temperature silicon is consumed at a rate of 2.20 A/V, up to 280 volts. Alternatively a cyclic anodization and etch procedure, using amyl phosphate and 15:1 buffered HF to remove 200Å thick layers, may be used. For removal of thin layers of compound semiconductors it is customary to use either anodic oxidation or dilute etchants (Yamamoto and Memura, 82c; Smith, 68b). Other layer removal techniques have been described by Ryssel and Ruge(86d).

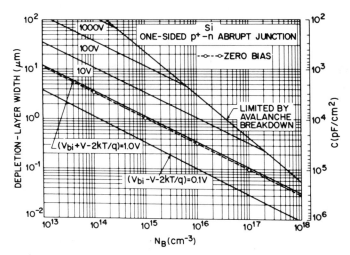

Figure 28: Depletion layer width and capacitance per unit area as a function of the doping for an asymmetric, abrupt junction in silicon. The dashed line is for zero bias (from Sze, 81d).

4.2.5 Capacitance-Voltage

The techniques discussed so far have depended upon d.c. measurements of current flow within a semiconductor. A second family of techniques will now be considered in which a.c. measurements of depletion layer capacitance are made to determine the total charge and/or charge distribution.

When the doping species in a semiconductor changes abruptly from donor to acceptor, a p-n junction will exist. Two important consequences result which are the formation of a depletion region and the existence of a diffusion potential or built in potential (V_{bi}). The electric field far removed from the junction must be zero and therefore it follows that the total negative charge in the p-type material must exactly balance the total positive charge in the n-type semiconductor:

$$N_A \, x_p = N_D \, x_n \qquad\qquad (26)$$

where x_p and x_n are the widths in the p-type and n-type material, respectively. It will be seen that when $N_D = N_A$ the depletion regions will extend an equal distance into both material types. When the doping is asymmetric with $N_A \gg N_D$, the depletion region will, essentially, extend only into the n-type semiconductor.

It may be shown (Sze, 81d) that the total depletion width $(x_p + x_n)$ is:

$$W = \left(\frac{2\varepsilon}{e} \frac{N_A + N_D}{N_A N_D} V_{bi}\right)^{1/2} \tag{27}$$

With asymmetric doping (p^+n) and after including a correction term for the majority carriers, the junction width at thermal equilibrium becomes:

$$W = L_D\left(\frac{2eV_{bi}}{kT} - 4\right)^{1/2} \tag{28}$$

where L_D is the Debye length defined as:

$$L_D = \left(\frac{\varepsilon kT}{e2N_D}\right)^{1/2} \tag{29}$$

Figure 28 shows the dependence of the depletion width upon doping for an asymmetric p^+n junction in silicon. Figure 29 shows the Debye length as a function of doping. It should be noted that when capacitance- voltage data is used to determine depth profiles, the depth resolution is limited to the order of the Debye length. From the figures it can be seen that, at room temperature, in silicon doped to 10^{15} cm^{-3}, the depletion width is about 1 μm and also that $W = 9L_D$.

A consequence of having a depletion layer is the existence of a junction capacitance which is defined as:

$$C = \frac{dQ_c}{dV} \tag{30}$$

where dQ_c is the incremental increase in charge due to an incremental change in applied voltage, dV.

For the asymmetric p^+n junction the capacitance per unit area is given by (Sze, 81d):

$$C + \left(\frac{e\varepsilon N_D}{2}\right)^{1/2}[V_{bi} - V - 2kT/e]^{1/2} \tag{31}$$

This may be rewritten as:

$$\frac{d\left(^1/c^2\right)}{dV} = \frac{2}{e\varepsilon N_D} \tag{32}$$

Thus by plotting $1/c^2$ versus V for the asymmetric junction, a straight line will result where the gradient gives the substrate doping concentration (N_D), and the intercept gives ($V_{bi} - 2kT/q$).

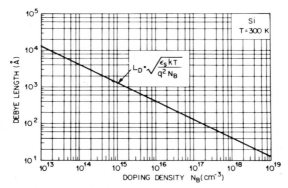

Figure 29: Debye length in silicon as a function of doping concentration (from Sze, 81d).

For reasons of experimental convenience these measurements are frequently carried out on Schottky diodes, either formed by the evaporation of a metal contact or using a liquid mercury probe(75b). The analysis of the metal - semiconductor barrier is similar to that for the asymmetric junction and equation 32 may be used.

C-V depth profiling: The concept of depth profiling by the measurement of capacitance was first proposed by Schottky in 1942. During the last two decades the method has been extensively used, with variations of the basic method being reported by Copeland(69a), Miller(72c) and Nakhamanson(76e).

The presence of a p-n junction or Schottky barrier is required and depth profiling is possible because, with increasing negative bias, the depletion region penetrates further into the semiconductor and in so doing exposes an additional number of donors or acceptors. The presence of these additional charged centres modifies the capacitance of the depletion layer and, if this change in capacitance is monitored as a fuction of depletion width, the doping profile may be determined using a differential form of equation 31 which gives:

$$N_D(w) = \frac{c^3(V)}{e\varepsilon \dfrac{dC(V)}{dV}} \qquad (33)$$

The depletion width is given by:

$$W = \frac{\varepsilon}{C(V)} \qquad (34)$$

202 P. L. F. HEMMENT

The range of depths which may be profiled are controlled by the depletion depth.

The experimental procedure involves measurements of the differential capacitance as a function of bias. In general this will require the superposition of a high frequency bias voltage upon a fixed or slowly varying voltage applied across the depletion layer. Various profiling systems have been described in the literature (Copeland 69a,: Miller 72c) and these may be grouped according to the frequency of the applied bias. These systems are generally automated and provide direct plots of concentration against depth.

Care should be taken to establish the validity of capacitance-voltage profiles in any given analysis, as discrepancies may occur due to the following limitations:
(i) Capacitance-voltage data is insensitive to abrupt changes in the doping profiles which occur within a distance less than a Debye length. Typically, this restricts the depth resolution to about 0.1 μm.
(ii) The capture and emission of carriers from deep trapping levels, within the energy gap, will change with bias and frequency and this can be the cause of gross anomalies in the depth profiles.
(iii) In compensated semiconductors the capacitance measurements give the net doping profile due to (N_D-N_A) or (N_A-N_D) in n and p-type material, respectively.

The applications and limitations of this profiling technique have been discussed by Wu(75g), Zemel(79e) and Ryssel and Ruge(86d).

4.2.6 Device Parameters

Lightly doped samples cannot easily be measured by sheet resistance nor capacitance - voltage methods and, thus, for these samples, it is necessary to estimate the electrical activity from measurements of device parameters. A disadvantage of this approach is the remoteness of such measurements from the physical feature of interest, namely, the electrical activity. This can lead to the accumulation of large errors.
Threshold Voltage: For the MOSFET there is a particular gate voltage which will enable a conducting channel to form between the source and drain. The voltage at which this occurs is the threshold voltage (V_T) which Sze(81d) has given for the narrow-channel MOSFET as:

$$V_T = V_{FB} + 2\phi_B + \frac{[2\varepsilon e N_A(2\phi_B + V_{BS})]^{1/2}}{C_i}\left(1 + \frac{\pi}{2}\frac{W}{z}\right) \qquad (35)$$

where Φ_B is the potential difference between the Fermi level and intrinsic level, C_i is the capacitance per unit area, ε the semiconductor permittivity, V_{FB} is the flat band voltage, V_{BS} the substrate reverse bias, Z the channel length and W the depletion width. The variation of the threshold voltage with substrate doping for n-channel and p-channel $A\ell$-SiO_2-Si devices is shown in Figure 30, due to Streetman(80a).

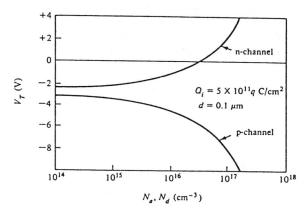

Figure 30: Variation of threshold voltage (V_T) with substrate doping for n-channel and p-channel $A\ell$-SiO_2-Si devices (from Streetman, 80a).

The threshold voltage may be determined experimentally from a linear extrapolation to zero current, of the curve of source-drain current (I_{DS}) against gate-source voltage (V_{GS}), for a given small drain-source bias (V_{DS}), as:

$$I_{DS} = \frac{Z}{L} \mu_n C_i \left(V_{GS} - V_T\right) V_{DS} \qquad (36)$$

where L is the channel length and μ_n the electron mobility (Sze, 81d). Small changes in the doping level ($\Delta N_A = 10^{15} - 10^{16}$ cm^{-3}), due to the implantation of 10^{10}-10^{11} ions cm^{-2} may be determined by monitoring the change in the threshold voltage (ΔV_T), using equations 35 and 36. The overall uncertainty in the estimate of the dopant concentration may be large, as the results are sensitive to variabilities in the device geometry and uncertainties in the electronic properties of the device materials.

The ability to detect small changes in the doping level depends upon an instrumental capability to measure small changes in I_{DS} and V_{GS}. For n-channel silicon devices on substrates with $N_A = 10^{14} - 10^{15}$ cm^{-2} and an oxide thickness of 500Å it is possible to detect changes of about 5×10^{10} cm^{-2} for a ΔV_T of

about 100 mV (Glaccum, 82b). The uncertainty in the threshold
voltage will be about 30 mV, whilst measurements in different
laboratories show discrepancies which are typically 50 - 100 mV.
Achieving these levels of reproducibility calls for careful
control of all processing steps and, in particular, the
inclusion of control devices on the same wafer.
Saturation Current and Pinch-off Voltage: The performance of a
field effect transistor depends upon the doping concentrations
in the active volume of the device, which in a high frequency
transistor will have sub-micron dimensions. The doping level in
these small volumes may be determined by monitoring the
saturation current (I_{DS}) and pinch-off voltage (V_P). Livingstone
et al.(80b) have reported measurements on GaAs MESFET structures
fabricated in semi insulating substrates doped by ion
implantation with 10^{12}-10^{13} Se^+ cm^{-2} at 225 keV.
     The saturation current may be expressed as:

$$I_{DS} = NeV_{sat}Wd \tag{37}$$

where N_D is the dopant concentration, V_{sat} the saturated drift
velocity, W the device width and d the thickness of the active
layer.
     The pinch-off voltage has been defined by Sze(81d) as:

$$V_P = V_{FP} + 2\phi_B + \frac{(4\epsilon e N_d \phi_B)^{1/2}}{C_i} \tag{38}$$

where a is the channel depth and ϵ the permittivity.
     Both parameters have a linear dependence upon N_D but,
because of their additional dependence upon inaccessible
geometric factors, precise determinations of the doping level
cannot be achieved.

4.2.7 Electrochemical Technique

     The constraints on the depth which may be profiled by the
C-V technique have been overcome by Ambridge and Faktor(74b) who
have combined electrochemical dissolution of a semiconductor
sample with Capacitance - Voltage measurements using the
electrolyte as a liquid Schottky contact. By this means
continuous depth profiles may be plotted where the carrier
concentration is determined from a measurement of the
capacitance, whilst the depth is derived from the time integral
of the dissolution current (charge). Profiles may be readily
determined with the carrier concentration spanning four orders
of magnitude. In principle there is no limit to the depth,
however, in practice lateral variations in the dissolution rate
leads to a degradation of the electrolyte/ semiconductor
interface.   This imposes a constraint which is dependent upon

the particular semiconductor system under study. However, by a
suitable choice of dissolution conditions, it is possible to
profile successfully to a depth of 10μm to 50 μm.

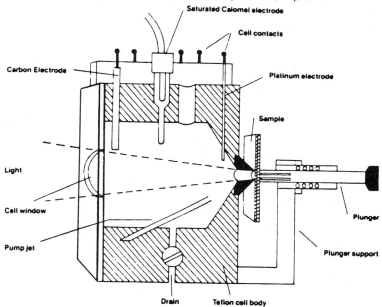

Figure 31: Schematic of the cell used in an
electrochemical profiler (Ambridge and Faktor, 74b
and Polaron Equipment Ltd., 84i).

Figure 31 shows a schematic of the electrochemical cell
developed for a commercial version of this profiler (Polaron
Ltd., 84i). The sample may be of any size greater than 0.5 cm x
0.5 cm and is placed against a small aperture in the wall of the
Teflon cell. The region to be profiled is thus defined by this
aperture, which typically has an area of 0.1 cm^2 or 0.01 cm^2.
The samples require no deposited electrical contacts as the
electrolyte forms the Schottky barrier and small probes, which
are external to the cell, form the ohmic "back" contacts. The
technique has been successfully applied to III-V, II-VI compound
semiconductors and also to silicon. Both n- and p-type material
can be profiled although it is necessary to illuminate n-type
samples as the dissolution process depends upon a supply of
holes.
        The instrument was originally developed to profile GaAs and
other compound semiconductors, for which there are a number of
suitable electrolytes. The choice of electrolyte for silicon is
limited and the technique has only recently been applied to this
semiconductor using H_2SO_4/NaF in aqueous solution. Figure 32
shows a hole depth profile from a p-type silicon sample

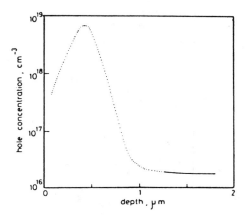

Figure 32: Depth profile of the hole concentration in a p-type silicon sample implanted with 3×10^{14} B^+ cm^{-2} at 190keV and annealed at 900°C, measured using the electrochemical profiler. (Sharpe et al.79f).

implanted with 3×10^{14} B^+ cm^{-2} at 190 keV (Sharpe et al.79f).

4.2.8 Contact Resistance

The spreading resistance method cannot generally be applied to III-V semiconductors due to high values of contact

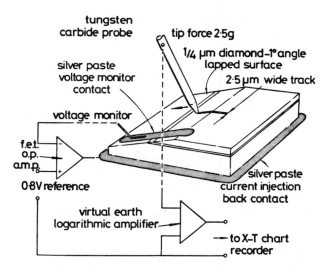

Figure 33: Schematic of a contact potential profiler reported by Goodfellow et al. (77a).

resistance. Goodfellow et al.(77a) have exploited this feature
in a depth profiling technique which utilizes the contact
resistance between a metal probe and an angle lapped surface.
The resistance is found to be related to the doping level and/or
the composition of the compound sample. The probe has a tungsten
carbide tip of 12μm radius, loaded to 2.5gm, which is drawn
slowly along the lapped surface whilst the contact resistance at
a fixed bias voltage (0.8V) is monitored and displayed on a
chart recorder. A schematic of the instrument is shown in
Figure 33. Measurements on samples with a 1° bevel show a depth
resolution of better than 1000Å. The contact area is estimated
as 5 μm and the sampling depth as 400Å. Electrical profiles of
III-V multilayered structures including GaAs, GaAlAs, InP and
GaInAsP are reported for doping levels of 10^{17}-10^{19} carriers
cm^{-3} in n- and p-type materials.

4.3 SURFACE ANALYSIS

Electrical measurements are unable to provide a complete
description of the composition and structure of implanted
layers. It is therefore customary to use surface analysis
techniques to complement the electrical methods, especially to
resolve ambiguities in the interpretation of data due, for
example, to unknown levels of activation and compensation of the
implanted dopants, as discussed in Sections 2 and 3. A recurring
problem is the determination of the absolute value of the
retained dose.

The most useful surface analysis techniques, which may be
used for impurity depth profiling are listed below and their
strengths and weaknesses are highlighted.

4.3.1 Rutherford backscattering and ion channelling

In Rutherford backscattering analysis the energy
distribution of ions, which are backscattered from the sample,
is measured. The incident ions are generally 1MeV to 3MeV He^+
and the sample is usually mounted on a goniometer to permit
alignment in a channelling (low index) direction(85h)(77g).

Strengths
- It is an absolute measurement technique.
- The sample is not destroyed.
- Channelling gives damage depth profiles and atom site location
  of heavy impurities.
- Exceedingly high sensitivity to heavy mass impurities on the
  surface is achieved.
- Volume sensitivity to heavy mass elements is typically
  10^{18}-10^{19} cm^{-3}.
- Depth resolution is 50Å to 300Å, depending upon the system
  geometry.
- Fast analysis is possible.

Weaknesses

- The technique has low sensitivity to low mass impurities.
- A high energy particle accelerator is required.

4.3.2 Secondary Ion Mass Spectrometry (SIMS)

During SIMS analysis a low energy ion beam is used to sputter erode the surface of the sample. The ejected ions are mass analysed and the intensities of the various ionic species are measured in order to determine the composition of the sample(85i)(85h).

Strengths
- Isotopic sensitivity is available.
- Very high sensitivity to most elements (eg. 10^{14} B cm^{-3}) is achieved.
- Small areas may be analysed ($\approx\mu m$).

Weaknesses
- Matrix effects occur.
- The sample is destroyed.
- The technique is not absolute and use of standard samples in necessary.

4.3.3 Auger electron spectroscopy

Auger spectroscopy is a well developed technique which is available in many laboratories(79e). The technique has good spatial resolution, has medium sensitivity ($10^{18}-10^{19}$ cm^{-3}) to light and heavy elements and is also sensitive to the chemical (bonding) state.

4.3.4 Ion induced nuclear reactions/X-rays

These techniques have some experimental similarities to Rutherford Backscattering which they complement as they are suitable for light element analysis(71)(73)(77g)(85h).

5.   NEW PROBLEMS AND NEW SOLUTIONS

The shrinking device geometries that are required to achieve higher performance circuits make increasing demands for improved materials and new and sometimes novel structures. Examples of developments that have come into prominence during recent years are synthesis of compounds by high dose implantation (eg. silicides(88d) and SIMOX technology(86c)), deep implants using MeV ion beams(87g) and very shallow doped layers(86a)(87f). Analysis of these thin film materials presents special problems. In this section examples of diagnostic techniques which have been developed to tackle some of these problems, will be described.

5.1   THERMAL OXIDATION REPLICATION

Hill et al.(85d) report a novel analytical technique for rapid two dimensional surveys of the surface location of implanted arsenic near the edges of mask windows used in the

fabrication of 1μm geometry silicon integrated circuits. A
spatial resolution of 100Å and a sensitivity of about 10^{20} As
atoms cm^{-2} is claimed. The technique is based upon replication
of the arsenic concentration profile, at the window edge, by
exploiting the dependence of the oxidation rate upon arsenic
concentration. By this means thin SiO_2 films are formed whose
thickness is a function of the impurity concentration. The
films are analysed by monitoring the transmitted electron
current in a scanning transmission electron microscope.
Figure 34 shows (a) the processing sequence and (b) a schematic
cross sectional diagram of a thin film, showing the estimated
arsenic concentration contours.

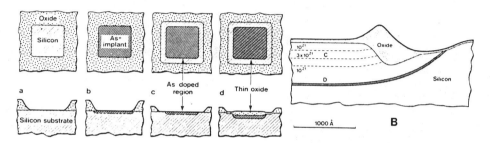

Figure 34: (a) Process sequence to fabricate a
thermal oxide replica of the arsenic distribution
implanted into silicon through an oxide window,
namely, etch, implant, open and grow oxide.
(b) Schematic of a cross section of a replica with the
estimated arsenic concentration contours (from Hill et
al.85d).

5.2 TWO DIMENSIONAL PROFILING
    A technique to determine the distribution of implanted
boron in silicon under an oxide mask edge with a spatial
resolution of 600Å x 600Å and sensitivity of 10^{17} B^+ cm^{-3}, has
been reported by Hill et al.(87n). The principle of the
technique is to fabricate identical planar resistor structures,
electrically isolated from the substrate and each other, and to
monitor resistance changes as thin layers of different
geometries are removed by anodic sectioning. Essential stages
in the procedure are (i) fabrication of test resistors of
accurately known dimensions, (ii) automatic sectioning and
conductance measurements and (iii) analysis to generate a two
dimensional distribution. Figure 35 shows the experimentally
determined boron concentration contours in a section under a
mask edge.

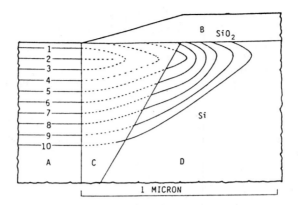

Figure 35: Experimentally determined two dimensional
boron concentration contours in a section through a
mask edge. The concentrations are from 9.5×10^{18}
cm^{-2} to 1.5×10^{17} cm^{-2} and were determined using
resistor structures (from Hill et al.87n).

5.3 Gated MOS devices

The characteristics of both the Van der Pauw and
capacitance-voltage techniques may be exploited by the use of
gated MOS structures. These structures consist of a lateral Van
der Pauw resistor above which an MOS capacitor is fabricated.
The principle of operation is to use the depletion layer,
associated with the MOS capacitor, to control the effective
thickness of the resistor. Thus, by progressively increasing
the bias applied across the capacitor, it is possible to
determine the differential sheet resistance and Hall voltage,
from which the carrier concentration and mobility depth profiles
may be determined. (see Section 4.2.4).

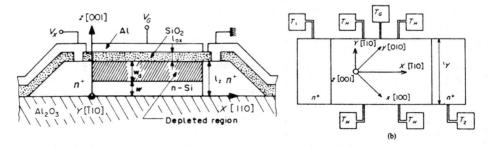

Figure 36: Gated MOS device (a) cross section and (b)
plan view of the resistor structure (Lee et al.82d).

Gated MOS devices have been used by Lee et al.(82d) to study the transport properties of silicon on sapphire structures. Figure 36 gives a cross section of the device geometry. These structures call for full MOS processing but, due to the use of photolithography, they provide a means of determining the electrical activity and carrier transport properties of small volumes in fully processed device wafers.

5.4 DOUBLE IMPLANT (SHEET RESISTANCE) METHOD

Low dose implants in the range 10^{10} to 10^{12} cm^{-2} are commonly used during the fabrication of MOS transistors. Unfortunately the sheet resistance techniques described in Section 4.2 lose sensitivity and become unreliable in this low dose regime. These problems are overcome in an indirect measurement technique described by Smith et al.(85c)(87b). The principle of the method is to implant the unknown low dose into a previously implanted and annealed layer of high conductivity.

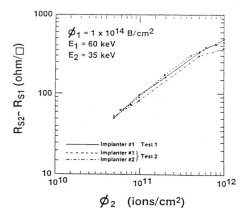

Figure 37: Experimental data showing the sensitivity of the double implant technique to detect doses of 35keV B^+ ions in the range 5×10^{10} ions cm^{-2} to 1×10^{12} ions cm^{-2} (from Smith et al.85c).

The increase in sheet resistance, ΔR, due to defects created during the second implantation, is monitored and by comparison against previously calibrated samples may be used to determine ion doses as low as 1×10^{10} cm^{-2}. Details of experimental procedures are given by the authors. The linear dependence of ΔR upon dose is shown in Figure 37 for samples implanted with 35keV B^+ ions over the dose range 4×10^{10} cm^{-2} to 6×10^{11} cm^{-2}. Advantages of the technique are (i) the sample does not require to be annealed, (ii) doses as low as 1×10^{10} cm^{-2} may be measured, and (iii) test wafers may be reused after a regenerative anneal.

Technique	Dose Range (cm^{-2})	Advantages	Disadvantages
Spreading resistance	$10^{10} - 10^{16}$	. can be used on arbitrary shapes and structures . sample preparation relatively easy . spans widest dose range . resolution good (10 - 20 μm)	. demands a great skill in sample and contact preparation . data interpretation is complex . probe damages the surface . apparatus is expensive
Four point probe -double implant -surface treatment	$10^{13} - 10^{16}$ $10^{10} - 10^{12}$	. technique highly developed and widely used . reproducibility better than ±0.5% . measurements fast and simple . a double implant can extend range to 10^{10} cm^{-2}	. best for high doses . probes damage the surface . lateral resolution poor (1mm)
Six point probe	$10^{11} - 10^{16}$	. has many of the features of the four point probe	. plane samples required (wafers)
Van der Pauw	$10^{12} - 10^{16}$	. quantitative . insensitive to geometric errors . suitable for Hall measurements . reproducibilities of ±0.02% reported for 10 μm x 10 μm samples	. requires lithography
Capacitance - Voltage	$10^{13} - 10^{16}$	. non destructive profiling . accuracy good in working range . measurements fast and easy	. requires a p-n or Schottky contact . depth and concentration ranges are restricted . not suitable for highly doped material . great care required in the interpretation of data
Device Parameters	$10^{10} - 10^{13}$	. suitable for the lowest doses	. extensive processing required . interpretation of data difficult . large, unknown errors may be present
Ion Scan	$10^{11} - 10^{16}$	. no processing required	. indirect
SIMS	$10^{10} - 10^{18}$	. has the highest sensitivity	. interpretation of data difficult
RBS	$10^{14} - 10^{18}$	. quantitative . good depth resolution	. lateral resolution poor (≈ 1mm) . suitable for heavy mass impurities . equipment very expensive

Table 5: Measurement techniques, showing their dose range and strength and weaknesses.

5.5   BRIDGING METHOD - MeV IMPLANTS

High energy (MeV) implantation is being studied for a number of applications(85e). As the implanted layer is buried, typically to a depth of 1μm to 3μm, new problems arise, which frustrate attempts to determine the sheet conductivity and dose uniformity. Keenan(87h) has reported the use of a second shallower ("bridging") implantation, which provides electrical contact to the deep layer so that the four point probe mapping technique (Section 4.2.3) (Chapter 4) may be used.   The measurement procedure involves measurements of the single and double implanted layers. This data is interpreted using the algorithm

$$R_m = \left(\frac{1}{R_2} - \frac{1}{R_1} \right)^{1/2}$$

(39)

where R_2 is the combined sheet resistance of the two layers, R_1 the resistance of the initial (shallow) layer and R_m is the unknown value for the deep layer. Thus, R_2 is a parallel combination of R_1 and R_m. The author has made uniformity maps of MeV implanted layers over the dose range 1×10^{13} to 1×10^{14}.

6.   CONCLUDING REMARKS

A detailed evaluation of implanted layers demands the application of many complementary analytical methods. Some of the most commonly used, and by inference the most useful and rapid, techniques have been discribed. Throughout the chapter emphasis has been given to electrical measurement techniques. The relative merits of the various methods are summarized in Table 5.

In any measurement situation there is a conflict between the simplicity of the measurement procedures and the attainment of a desired accuracy, precision and dynamic range. With the increasing use of computers, what were previously laborious procedures have now been automated to produce high quality detailed analysis. This trend will continue and in the future we may confidently look forward to continual upgrading and refining of the techniques to meet the increasing demands for more detailed analysis of implanted layers.

ACKNOWLEDGEMENTS

    The author wishes to thank colleagues for their help and advice and acknowledges the value of discussions with B.J. Smith, P.A. Leigh, K. Nicholas, P.D. Scovell, A.C. Glaccum, N.G. Emerson, G. Harbeke, E.F. Steigmeier, W. Rehwald, C. Hill, P. Rosser, A. Lane, C. Cristoloveanu and K. Steeples and thanks Mrs B. Dore for her patience and care in typing this manuscript.

REFERENCES

(79) E.H. Hall, Ann. J. Math. 2, 287, 1879.
(42) W.   Schottky, Zeithchrift fur Physik, 118, 539, 1942.
(54) L.B. Valdes, Proc. Inst. Radio Engrs. 42, 420, 1954.
(55) A. Uhlir, Bell Systems Tech. J. 34, 105, 1955.
(58a) F.M. Smits, Bell Systems Tech. J. 37, 711, 1958.
(58b) R.L. Petritz, Phys. Rev., 110, 1254, 1958.
(58c) L.J. Van der Pauw, Philips Res. Repts., 13, 1, 1958.
(58d) L.J. Van der Pauw, Philips Tech. Rev., 20, 220, 1958.
(58e) F.M. Smits, Bell Systems Tech. J., 37, 711, 1958.
(64) J.L. Moll, Physics of Semiconductors, McGraw-Hill, New York, 1964.
(66) M.W. Buehler and G.L. Pearson, Solid State Electronics, 9, 395, 1966.

(67a) J.F. Gibbons, Proc. Inter. Conf. Appl. Ion Beams
      Semiconductor Technology [Ed. P. Glotin], Grenoble, p.561,
      1967.
(67b) A. Mircea, J. Sci. Instr. $\underline{41}$, 679, 1967: see also
      D.S. Perloff Solid State Electronics, $\underline{20}$, 681, 1977.
(67c) R. Holm, Electrical contacts Theory and Application,
      Stringer Verlag, 1967.
(68a) F.J. Blatt, Physics of Electronic Conduction in Solids,
      McGraw-Hill, 1986.
(68b) A. Smith, Electron. Letts., $\underline{4}$, 332, 1968.
(69a) J.A. Copeland, IEEE Trans. Electron. Devices $\underline{ED-16}$, 445,
      1969.
(69b) R. Rymaszewski, J. Phys. E, $\underline{2}$, 2, 170, 1969: and also
      F.E. Wahl and D.J. Perloff, Proc. 8th Int. Conf. on
      Electron. and Ion Beam Science and Technology, Seattle,
      W A, May 1978.
(69c) P.A. Schumann and E.E. Gardner, J. Electrochem. Soc. $\underline{116}$,
      87, 1969.
(69d) P.A. Schumann and E.E. Gardner, Solid State Electronics
      $\underline{12}$, 371, 1969.
(69e) F.L. Vook and H. Stein, Rad.Effects, $\underline{2}$, 23, 1969.
(70a) N.G.E. Johansson, J.W. Mayer and O.J. Marsh, Solid State
      Electronics, $\underline{13}$, 317, 1970.
(70b) B.J. Smith, J. Stephen and G.W. Hinder, Measurement of
      Doping Uniformity in Semiconductor Wafers, AERE-R7085.
(70c) J.W. Mayer, L. Eriksson and J.A. Davis, Ion Implantation
      in Semiconductors, New York, 1970.
(70d) B.J. Smith and J. Stephen, Theoretical Calculations of
      Resistance of n- and p- type implanted silicon, AERE
      Report No. R7097, 1970.
(71a) H. Muller and H. Ryssel, Ion Implantation in Semi-
      conductors (Eds. I. Ruge, J. Graul) Springer-Verlag, p.85,
      1971.
(71b) A.E. Stephens, J.J. Mackey and J.R. Sybert, J. Appl. Phys.
      $\underline{42}$, 2592, 1971.
(71c) G.A. Amsel, J.P. Nadai, E. D'Artemare, D. David, E. Girard
      and J. Moulin, Nucl. Inst. Meths, $\underline{92}$, 481, 1971.
(72a) S. Furukawa, H. Matsumura and H. Ishiwara, Jap. J. Appl.
      Phys. $\underline{11}$, 134, 1972.
(72b) B.C. Crowder, Proc. US - Japanese Seminar Ion Implantation
      in Semiconductors, Ed. S. Namba, Jap. Soc. for the
      Promotion of Science, p.63, 1972.
(72c) G.L. Miller, IEEE Trans. Electron. Devices $\underline{ED-19}$, 1103,
      1972.
(72d) G. Carter, J.N. Baruak and W.A. Grant, Rad.Effects, $\underline{16}$,
      107, 1972, Rad.Effects, $\underline{16}$, 101, 1972.
(73) T.J. Gray, R. Lear, R.J. Dexter, F.N. Schwettmann and
      K.C. Wierner, Thin Solid Films, $\underline{19}$, 103, 1973.

(74a) W.M. Bullis, Standard Measurements of the Resistivity of Silicon by the Four Point Probe Method, NBS1R 74-496, August 1974: see also D.S. Perloff, J. Electrochem. Soc., 123, 1745, 1976.

(74b) T. Ambridge and M.M. Faktor, Electron. Letts. 10, 10, May 1974.

(74c) J.H. Freeman, Applications of Ion Beams to Materials, Warwick 1975, Inst. of Phys. Conf. Ser. II, 28.

(74d) A.W. Tinsley, G.A. Stephens, M.J. Nobes and W.A. Grant, Rad. Effects 23, 165, 1974.

(74e) J.R. Ehrstein (Ed.) Spreading Resistance Symposium. NBS Special Publications 400 - 10, US Dept of Commerce, Washington DC, 1974.

(74f) D.V. Lang, J. Appl. Phys., 45, 7, 3023, 1974.

(75a) J.F. Gibbons, W.S. Johnson and S.W. Mylrou, Projected Range Statistics, Stroudsburg, USA, 1975.

(75b) P.J. Severin and H. Bulle, J. Electrochem. Soc., 122, 134, 1975 Eds. Palik and R.T. Holm, in Zeinel (79).

(75c) T.W. Sigmon, W.K. Chu, H. Muller and J.W. Mayer, Ion Implantation into Semiconductors, Ed. S. Namba, New York, 1975, p.633.

(75d) R.A. Moline, G.W. Reullinger and J.C. North, Atomic Collisions in Solids Vol. 1, Ed. J. Day, 1975.

(75e) J.A. Davis, G. Foti, I. Howe, J.B. Mitchell, K.B. Winterton, Phys. Rev. Lett. 34, 1441, 1975.

(75f) H. Matsumura and S. Furukawa, Jap. J. Appl. Phys., 14, 1783, 1975.

(75g) C.P. Wu, E.C. Douglas and C.W. Mueller, IEEE, ED-22, 6, 319, 1975.

(75h) G. Ryding, A.B. Willkower, IEEE, MFT-4, 21, 1975.

(76a) H.D. Barber, B.H. Lo and J.E. Jones, J. Electrochem. Soc., 123, 1404, 1976.

(76b) M.G. Buehler and W.R. Thurber, IEEE ED-23, 8, 968, 1976.

(76c) C. Kittel, Introduction to Solid State Physics, Wiley, New York, 1976.

(76d) M.G. Buehler, Semiconductor Measurement Technology, NBS Special Publication, 400-22, 1976.

(76e) R.S. Nakhamanson, Solid St. Electron., 19 87, 1976.

(76f) J.M. David, Nat. Bur. Stand. Spec. Pub., 400-19, 44, 1976.

(76g) K. Nicholas, J. Phys. D, 9, 393, 1976.

(76h) L. Cserpegi, J.W. Mayer and T.W. Sigman, Appl. Phys. Letts. 29, 92, 1976.

(76i) L. Csepregi, W.K. Chu, H. Muller, J.W. Mayer, Rad. Effects 28, 227, 1976.

(76j) J.H. Freeman, Applications of Ion Beams to Materials, Ed. G. Carter, Inst. of Phys. Conf. Ser. No. 28, 340, 1976.

(77a) R.C. Goodfellow, A.C. Carter, R. Davis and C. Hill, Electron. Lett. 14, 328, 1977.

(77b) D.C. D'Avanzo, R.D. Rung and R.W. Dutton, Tech. Report
      5013-2, Stanford University, California, 1977.
(77c) D.S. Perloff, F.E. Wall and J. Conragon, J. Electrochem.
      Soc., 124, 582, 1977: and also M.G. Buehler and
      J.M. David, NBS Special Publication 400-29, 64, 1979.
(77d) B.J. Smith and J. Stephen, Revue de Phys. Appl. 12, 493,
      1977.
(77e) J.H. Freeman, D.J. Chivers and G.A. Gard, Nucl. Inst.
      Meths, 143, 99, 1977.
(77f) C.M. McKenna, Ion Implantation Equipment and Techniques,
      Trento, Aug. 1977.
(77g) Ion Beam Handbook for Materials Analysis. Ed. J.W. Mayer
      and E. Rimini, Academic Press 1977.
(78a) P.L.F. Hemment, Inst. Phys. Conf. Ser. No. 38, 117, 1978.
(79a) H.H. Weider, Nardistructive Evaluation of Semiconductor
      Material and Devices, Ed. J.N. Zemel, Plenum Press, 1979.
(79b) J.R. Ehrstein, Nordestructive Evaluation of Semiconductor
      Material and Devices, Ed. J.N. Zemel, Plenum Press, 1979.
(79c) R.A. Smith, Semiconductors, Cambridge University Press,
      London 1979.
(79d) D.H. Dickey and J.R. Ehrstein, Semiconductor Measurement
      Technology NBS Spec. Pub., 400-48, 1979.
(79e) J.N. Zemel, Nondestructive Evaluation of Semiconductor
      Materials and Devices, Plenum Press, 1979.
(79f) C.D. Sharpe, P. Lilley, C.R. Elliot and T. Ambridge,
      Electron. Letts., 15, 20, 623, 1979.
(79g) E.D. Palik and R.T. Holm in Nondestructive Evaluation of
      Semiconductor Materials and Devices, Ed. J.N. Zemel,
      Plenum Press, 1979.
(79h) P.L.F. Hemment, Vacuum, 29, 11/12, 439, 1979.
(79i) S. Matteson and M.A. Nicolet, Nucl. Inst. Meths, 160, 2,
      301, 1979.
(80a) B.G. Streetman, Solid State Electronic Devices,
      Prentice-Hall, New Jersey, 1980.
(80b) A.W. Livingstone, P.A. Leigh, N. McIntyre, I.P. Hall,
      J.A. Bowie and P.J. Smith, Solid State Electronics, 1981.
(80c) P.L.F. Hemment, Inst. Phys. Conf. Ser. No. 54, 1980.
(80d) P.L.F. Hemment, Prac. of Second Inter. Conf. Low Energy
      Ion Beams, Bath 1980, Inst. of Phys. Conf. Ser. No. 54,
      1980.
(81a) D.A. Jamba, Nucl. Inst. Meths. 189, 253, 1981.
(81b) R.G. Mazur and G.A. Gruber, Solid State Technology,
      November 1981.
(81c) W.F. Beadle, R.D. Plummer and J.C.C. Tsui, Quick Reference
      Manual for Semiconductor Energies, 1981.
(81d) S.M. Sze, Physics of Semiconductor Devices, Wiley, 1981.
(82a) P. Eichinger and H. Ryssel, Ion Implantation Techniques,
      Eds. H. Ryssel and H. Glawischnig, Springer-Verlag 1982.
(82b) A.C. Glaccum, private communication.
(82c) A. Yamamoto and C. Memura, Electron. Lett. 18, 63, 1982.

(82d) J.H. Lee, S. Cristoloveanu and A. Chovet, Solid State
      Electron. 25, 9, 947, 1982.
(82e) M. Pawlik and S.M. Davidson, GEC J. Sci. and Tech. 48, 2,
      119, 1982.
(82f) P.L.F. Hemment, Ion Implantation Techniques, Ed. H. Ryssel
      and H. Glawischnig, p.209, 1982.
(82g) R. Iscoff, Semiconductor International, Cahners Publishing
      Co., USA, November 1982.
(83a) Four Dimensions Inc. Hayward, CA94545, USA.
(83b) C.P. Wu et.al, RCA Review 44, 48, 1983.
(83c) VLSI Technology, Ed SM Sze, McGraw Hill, 1983.
(84c) T. Yamazaki, Y.I. Ogita and Y. Ikegani, Jap. J. Appl.
      Phys. 23, 322, (1984).
(84d) E.F. Steigmeier and H. Auderset, J. Electrochem. Soc. 131,
      1693, (1984). US Patent 4, 391, 524, (1983).
(84e) M. Current and M.J. Current, Ion Implantation Science and
      Technology Ed. J.F. Ziegler, Academic Press, 1984.
(84f) J. Gyulai, Ion Implantation Science and Technology
      Ed. J.F. Ziegler, Academic Press, 1984.
(84g) S. Mader, Ion Implantation Science and Technology
      Ed. J.F. Ziegler, Academic Press, 1984.
(84h) M.I. Current, Solid State Technology, 101, November 1983.
(84i) Polaron Equipment Ltd., Watford, Herts, England.
(85a) E.F. Steigmeier and H. Auderset, RCA Review 46, 3, 1985.
(85b) J.F. Ziegler and R.F. Lever, Appl. Phys. Letts. 46, 385,
      1985.
(85c) A.K. Smith, D.S. Perloff, R. Edwards, R. Kleffinger and
      M.D. Rigik, Ion Implantation Equipment and Techniques,
      Eds. J.F. Ziegler and R.L. Brown, North Holland, 1985.
(85d) C. Hill, P.D. Augustus and A. Ward, Inst. Phys. Conf. Ser.
      No. 76, 11, 477, 1985.
(85e) H.W. Cheung, Proc. Inter. Soc. Optical Eng. (SPIE), Vol.
      530, 2, 1985.
(85f) W.L. Smith, M.W. Taylor and J. Schmur, SPIE Conf. Vol.
      530, 201, Los Angeles, Jan. 1985.
(85g) N.L. Turner, M.I. Current, T.C. Smith, D. Crane and
      R. Simonton, SPIE Conf. Vol. 530, 55, Los Angeles, Jan.
      1985.
(85h) See Seventh Int. Conf. on Ion Beam Analysis, Berlin, July
      1985.
(85i) See Int. Conf. SIMS V, Washington, Oct. 1985.
(86a) M. Horiuchi and K. Yamaguchi, IEEE ED-33, 2, 260, 1986.
(86b) T.I. Kamins and J.P. Colinge, Elect. Letts. 22, 1236,
      1986.
(86c) P.L.F. Hemment, Mat. Res. Soc. Sym. Proc. Vol.53, 207,
      1986
(86d) H. Ryssel and I. Ruge, Ion Implantation, John Wiley &
      Sons, p.231, 1986.
(86e) W.L. Smith, A. Rosenawaig, D.L. Willenborg, J. Opsal and
      M.W. Taylor, Solid State Technology, p85, Jan. 1986.

218   P. L. F. HEMMENT

(86f) N.Q. Lam and G.K. Leaf, Mat. Res. Soc. Symp. Proc. Vol. 51, 1986.
(87a) J.R. Golin, N.W. Schell, J.A. Glaze and R.G. Ozarski, Nucl. Inst. Meths. B21, 542, 1987.
(87b) A.K. Smith, W.H. Johnson, W.A. Kelnan, M. Rigik and R. Kleppinger, Nucl. Inst. Meths. B21, 529, 1987.
(87c) M.A. Wendman and W.L. Smith, Nucl. Inst. Meths. B21, 559, 1987.
(87d) S.R. Wilson et.al., Ion Implantation Technology, Ed. M.I. Current, North Holland, 1987, p.433, Nucl. Inst. Meths, B21, 433, 1987.
(87e) R.B. Liebert, D.F. Downey and V.K. Basra, Nucl. Inst. Meths. B21, 391, 1987.
(87f) S.B. Felch and R.A. Powell, Nucl. Inst. Meths. B21, 486, 1987.
(87g) D.C. Ingram, J.A. Baker, D.A. Walsh and E. Strathman, Nucl. Inst. Meths. B21, 460, 1987.
(87h) W.A. Keenan, Nucl. Inst. Meths. B21, 563, 1987.
(87i) S.M. Bunker, P. Siashansi, M.M. Sanfacon and S.P. Tobin, Appl. Phys. Letts. 50 (26), 1900, 1987.
(87j) F. Ferrieu, D.P. Vu, C.D. Anterroches, J.C. Oberlin, S. Maillet and J.J. Grob, J. Appl. Phys. 62, (8), 3458, 1987.
(87k) Microelectronic Manufacturing and Testing, Lake Publishing Corp., USA, Feb. 1987.
(87m) G. Harbeke, E.F. Steigmeier, P.L.F. Hemment, K.J. Reeson and L. Jasbrgebski, Semiconductor Sci. Technol., 2, 10, 687, 1987.
(87n) C. Hill, P.J. Pearson, B. Lewis, A.J. Holden, R.W. Allen, Proc. ESSDERC '87, University of Bologna 1987, p.923.
(87o) S.M. Bunker, P. Siashansi, M.M. Sanfacon and S.P. Tobin, Appl. Phys. Letts. 50 (26), 1900, 1987.
(87p) F. Ferrieu, D.P. Vu, C.D. Anterroches, J.C. Oberlin, S. Maillet and J.J. Grob, J. Appl. Phys. 62, (8), 3458, 1987.
(87q) K. Brack, W. Euen, D.Hagman, Nucl. Inst. Methos, B21, 405, 1987.
(87r) R.T. Blunt and A.R. Lane, GaAs and Related Compounds, Crete 1987 and A. Lane and R.T. Blunt, Chemtronics, 1987.
(87s) W.L. Smith, A. Rosenawaig, D.L.Willenborg, M.W. Taylor, Nucl. Inst. Meths, B21, 537, 1987.
(87t) H.U. Jager, Phys. Stat. Sol. A, 103, 1987.
(88a) W. Rehwald, SIN c/o RCA Laboratory, Zurich, Switzerland. Private communications, P.L.F. Hemment et al., Ion Implantation Techniques, Kyoto, June 1988.
(88b) T.C. Smith, this volume.
(88c) K.P. Homewood, P.G. Wade and D.J. Dunstan, J. Phys. E, Sci. Inst., 1988.
(88d) See Proc. Fall Meeting Materials Research Society, Symposium H, Boston, December 1987.

Ion Implantation Technology

An INTRODUCTION TO ION SOURCES

Dr. Ken G. Stephens

Professor of Electrical Engineering
Head of Department of Electronic and Electrical Engineering
University of Surrey
Surrey, Guildford
GU2-5XH   England

The purpose of this chapter is to lead a non-specialist through the complexity of the fundamental physical processes which occur in the production of positive and negative ion beams. Since many ion sources are gas, or vapour based sources, some introduction to conduction in gases and the behaviour of a gaseous plasma is given. The principles of extracting ion beams from such plasmas are explained followed by brief descriptions of some of the more commonly used ion sources in the semiconductor industry.

CHAPTER SUMMARY

1 -   **Introduction**

2 -   **Physical Processes of Ion Production**

3 -   **Electrical Breakdown of a Gas**

4 -   **The Gas Plasma and Extraction from It**

5 -   **Typical Ion Sources**

6 -   **Beam Quality and Control**

1.   INTRODUCTION AND REVISION

The creation of positive and negative ions in ion sources
is not a modern discovery of this microelectronic age; it has
been a technical challenge for many years. In the development
of the accelerators for nuclear physics research in the 30's and
40's and in the need to investigate how to achieve isotope
separation by the use of charged ions during the second world
war, the ion source was a key factor. However, since it was
realised in the late 1960's that the implantation of ions into
semiconducting substrates was a significant part of the process
of achieving even smaller dimensions, combined with higher
packing densities, the needs of the semiconductor industry have
resulted in the development of ion sources specifically for implantation purposes. It is with these sources that we are
concerned in this chapter.

Although positive ion sources are most commonly used in ion
implanters, during the 1980's there has been much discussion on
the possible usefulness of higher energy (0.5 MeV to 6 MeV) ions
for semiconductor devices. One of the cheaper routes to
providing such ions is to use tandem accelerators requiring the
input of negative ions to the first stage of the tandem. As
intense negative heavy ion beams have been required for use in
recent atomic and nuclear physics research programmes, there is
a considerable research base to draw from. Thus some consideration of negative ion sources is included here.

The purpose of this Chapter is to lead a non-specialist
through the complexity of the fundamental physical processes
which occur in the production of positive and negative ion
beams. Since many ion sources are gas, or vapour based sources,
some introduction to conduction in gases and the behaviour of a
gaseous plasma is given. The principles of extracting ion beams
from such plasmas are explained followed by brief descriptions

of some of the more commonly used ion sources in the semiconductor industry.

In the later sections the quality and control of an ion beam
is discussed insofar as this is dependent on the ion source. It
is necessary to consider the 'optical' quality of a beam,
emittance and brightness, as well as the content and energy
spread of ions emitted from sources.

The chapter is not a technical review of ion sources, as
there are many such reviews published: Freeman (73a), Wilson
and Brewer (73b), Septier (67a), Sidenius (78a), Aitken (82a),
and Alton (81a) to name some.

1.1   Positive Ions

The simplest atom, hydrogen, consists of a single positively
charged nucleus, (a proton), with one electron in orbit around
it. This electron is only allowed to occupy discrete energy
levels according to the rules of quantum mechanics. Its lowest,
or ground, state corresponds to the unexcited atom where the
electron has its maximum binding energy. To produce H^+ at least
this binding energy must be provided, in an endothermic reaction, to remove the electron. This required energy is known as
the ionisation potential (energy) and is usually expressed in
electron volts (eV) where 1 ev is 1.6×10^{-19} joules. Excited
electronic states of the H atom (H^*) can be produced if energies
less than the ionisation potential are involved.

As the atomic number of the nucleus is increased by the
addition of more protons, so more electrons are required for
neutrality, and these electrons are again added according to the
rules of quantum mechanics, in shells, i.e. only allowed
discrete energy levels.

The first stable closed shell has two electrons, the second
8, and the third 8 (and so on in a somewhat more complicated

pattern). The atoms corresponding to closed shells of 2, 2 + 8,
2 + 8 + 8, are the rare gas group VIIIA atoms, helium, neon and
argon respectively.

Group 1A atoms such as lithium, sodium and potassium have
one electron in the outer shell, whereas group VIIA atoms such
as fluorine, chlorine and bromine each lack one electron from an
otherwise closed shell.

The ionisation potentials (binding energies) of a selection
of atoms is given in Table 1.

Note that Group 1A atoms such as lithium, sodium and caesium
have low ionisation potentials, followed closely by Group IIA
atoms such as magnesium, whereas the group VIA (oxygen) and VIIA
(fluorine) have much higher ionisation potentials. Note also
that the energy required to remove successive electrons, in
order to create multiply charged ions, increases rapidly.

1.2   *Negative Ions*

Because negative ions have small electronic binding energies
(called electron affinities), they are more fragile than their
positive counterparts. Whereas the formation of a positive ion
is an endothermic reaction, i.e. energy has to be supplied to
the system, the processes involved in the attachment of electrons to neutral atoms or molecules are exothermic, with energy
given out by the system. The binding energy or electron
affinity, E_A, of a negative ion is a measure of its stability
and its ease of formation. An atom (or molecule) can form a
stable negative ion if it possesses less potential energy when
an electron is attached to it than when it is in its normal
ground state.

If E_0 is the normal atom ground state and E_1 the ground
state energy of the atom with an attached electron, then the

TABLE   1

IONISATION   POTENTIALS   AND   ELECTRON   AFFINITIES

OF   SELECTED   ATOMS   OR   IONS

Atom or Ion	Periodic Table Group No	Ionisation Potential[1] (eV)	Electron Affinity[2] (eV)
Li	IA	5.4	0.6
B	IIIA	8.3	0.3
O	VIA	13.6	1.5
O$^+$		35.1	
O^{++}		54.9	
F		17.4	3.4
Na		5.1	0.5
Mg		7.6	<0
Aℓ	IIIA	6.0	0.5
P	VA	10.5	0.7
Ga	IIIA	6.0	0.3
As	VA	9.8	0.8
Cs	IA	3.9	0.5
Cs$^+$		25.1	
A	VIIIA	15.7	<0
He	VIIIA	24.6	<0
Ne	VIIIA	21.6	<0
Ne$^+$		41.1	

1) From Kaye and Laby, 73c.

2) From Hotop and Lineberger, 75a.

electron affinity $E_A = E_0 - E_1$. E_A must be positive for stable negative ion formation. Negative values of E_A refer to unstable negative ion states.

About 75% of the naturally occurring elements have positive electron affinities and values of E_A for some elements are given in Table 1.

The probability of forming a negative ion depends significantly on E_A as well as the formation method used. We look at some of these methods in a later section.

2.   PHYSICAL PROCESSES OF ION PRODUCTION

The sources used in ion implantation machines are usually those in which ions are created in a gas or vapour by ionisation and not those in which the source is a well defined surface of known shape and position. The creation of the ions is induced primarily by the bombardment of electrons from a hot filament or cold cathode but the rate of ion production is determined by several factors including: the electron current, the pressure and geometry of the gas discharge, and the ionisation cross section of the atoms within the gas or vapour. Secondary electrons are also produced in ionisation processes and these may be energetic enough to contribute to further ionisation. Thus it is important to establish a basic understanding of the electrical conduction in gases, especially the nature of a gas plasma. A gas plasma is an ionised gas which contains no net space charge, and thus it contains equal numbers of positive and negative charges. It can be treated as a conductor with a conductivity which is dependent on the electron density and mobility. Before we look in more detail at the behaviour of plasmas we should explain further the collision processes.

2.1   Collisions

For a gas to conduct, electrons or charged ions must be present and free to move when an electric field is applied. Although ions may contribute to conduction in a gas, it is usually a small contribution since the ion is significantly heavier, and therefore, has more inertia than the electron. However, as we shall see later the presence of ions in a discharge is of great importance for other reasons.

The effect of a gas on the passage of charged particles (electrons), between two electrodes depends on the pressure of gas since this determines the average distance, or mean free path, travelled by the particle between collisions.

We are only interested in the statistical result of many collisions not in each or any one collision. The best way of describing the probability of various types of collision is to define an effective cross section or area for collision, by analogy with spherical particles. The collision cross section, usually denoted by the symbol σ, is the effective area of one atom or particle for a particular type of collision. If the particle density is n cm^{-3}, the total geometric area or total cross section = $n\sigma$ cm^{-1}.

The simplest definition of cross section is for the so—called elastic collision process, similar to billiard ball collisions where kinetic energy is conserved before and after a collision. If particles have a diameter of d any one will collide with another which lies with its centre within a cylindrical volume of diameter 2d, (the axis of the cylinder is the path of the first particle).

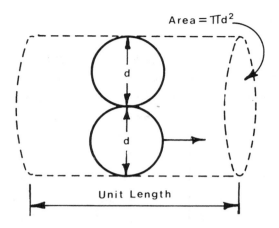

Fig. 1.   Cross Section Concept

This volume is πd^2 per unit length of path, which contains
$n\pi d^2$ particles. The average number of collisions per particle
per unit distance is, therefore, $n\pi d^2$. The mean free path λ,
which is the average distance travelled between collisions is:

$$\lambda = \frac{1}{n\pi d^2}$$

Ignoring electron/electron collisions, then the m.f.p. (λ)
of an electron depends only on the collison rate with atoms. We
assume the electron to be a point of zero diameter.

So then
$$\lambda = \frac{4}{n \pi d^2}$$

The cross section per atom is effectively an area $\frac{\pi d^2}{4}$
in this simple case. This is the probability of an electron
colliding elastically with a single atom. However as the mass
of the smallest atom (hydrogen) is 1840 times that of an

electron, the energy imparted to an atom by an electron in an elastic collision is small. Thus we look to inelastic processes to which the concept of cross section may also be applied, for ionisation processes.

An inelastic collision involves a change in the internal excitation energy of the atom when interacting with an electron. Total energy is conserved but the ratio of potential to kinetic energy is changed.

This type of interaction is all important in producing electrons and ions within a gas which is conducting electricity. The electrodes may produce electrons but large currents will not be able to flow in a gas unless electrons are produced within the gas volume. [In some applications, of course, a gas is used as a dielectric, in which case electron production must be inhibited as much as possible to achieve the highest breakdown voltage.]

We shall look at several types of inelastic electron –atom collisions, and a few atom – atom collisions.

2.2 How to Excite and Ionise

2.2.1 Electron Collisions

The excitation of electrons to higher energy levels, sometimes leading to ionisation, is important in a gas discharge. For simplicity we shall discuss electron collisions with atoms and ions only, even though molecules may be involved in many real situations.

The kinetic energy of the electron must be enough to provide the necessary energy difference between levels, say, E_m and E_n.

If W_1 = kinetic energy of the electron and E_{mn} is the energy required to raise a valency electron from the state m to n. Then if $W_1 > E_{mn}$ we get excitation.

The ionisation energy is the energy needed to take an electron from an orbit to infinity or freedom. It is greatest when the valence electron is in the ground state i.e. unexcited. The outermost, least tightly bound electron has the lowest ionisation energy, usually known as the first ionisation potential. (see Table 1)

If electrons collide with atoms which already have electrons in excited, higher energy states, then ionisation will be achieved more easily, i.e. the cross section will be greater at any given energy.

Note that in these ionisation–excitation processes, the kinetic energy of the incident electron is used to change the potential energy of the bound electrons. The probability of releasing these bound electrons to produce an ion will be a function of the electron energy and the atom involved.

This type of inelastic collision is sometimes known as one "of the first kind".

This implies there is a collision of the "second kind". In this latter case the electron with kinetic energy interacts with an atom in an excited state. This excitation or potential energy is transferred to the incident electron as kinetic energy. You see that this is the reverse of the ionising process of the "first kind".

2.2.2 Collisions of the second kind between atoms or between atoms and ions

A neutral atom A of one kind collides with a neutral atom B of another which is in a metastable state.

If the energy of this state e V_{meta} is $>$ e V_i the ionisation energy of the first atom A, ionisation of A will occur.

This is known as the Penning effect and is important as a means of ionising mixtures of rare gases in discharges since

Neon has a metastable state at 16.5 electron volts with a long
lifetime. Common molecular gases have a V_i of about 15 volts
and so are easily ionised by this method. Note that since the
atoms are 'large' the collision probability or cross section is
large. Thus the process is efficient.

2.2.3 Processes for negative ion formation

Alton (83a) reviews the various mechanisms of forming
negative ions. Basically there are three groups of processes:
electron impact, charge exchange and surface ionisation.

(i) Electron impact

There are three dominant processes worth mentioning here:
three body, or ternary, collisional transfer, dissociative
attachment and polar dissociation. In a dense gas at low
electron energies the following three body collision can occur:

$$e^- + A + B \rightarrow A^- + B$$

where A and B are atoms or molecules, e^- is the electron.
Alternatively electrons may be stably attached to atoms
during their interactions with molecular neutrals producing
reactions:

$$e^- + AB \rightarrow A^- + B \quad \text{or}$$
$$e^- + AB \rightarrow A + B^-$$

This dissociative attachment can be looked upon as a three
body process where the excess energy which is released during
the reaction is transferred to the fragments.
Polar dissociative attachment is characterised by the

equation:

$$e^- + AB \rightarrow A^+ + B^- + e^-$$

In this case the electron is not itself captured but only serves as a catalyst to molecular dissociation and ionisation into positive and negative ions.

(ii) Charge exchange

When ions collide with neutral atoms or molecules, the ions may lose or capture electrons. The cross sections for these capture and loss processes depend on the ion and atomic species, and the relative velocity of the colliding particles. If a thick target is used so that several collisions are probable then the emerging ion beam will have a charge distribution determined by the statistics of the capture and loss processes for the particular interactions.

If an ion beam is passed through a vapour of atoms of group IA or IIA metals, which have low ionisation potentials, there is a good probability, if the "thickness" of the target is such that several collisions are probable, that a substantial fraction of the incident ions will capture two electrons to form negative ions.

O'Connor and Joyce (87a) measured the emerging equilibrium charge state distributions for 45 keV beams of $^{11}B^+$, $^{31}P^+$ and $^{76}As^+$ ions after passing through sodium vapour of thickness up to 6×10^{15} atoms/cm^2. Equilibrium charge state fractions of B^-, P^- and As^- were 10%, 21% and 15% respectively.

The processes of electron transfer are complex but it has been established for the production of He^- in a Cs vapour that the electron transfer takes place in sequential collisions:

$$He^+ + Cs \rightarrow He^{o*} + Cs^+$$
$$He^{o*} + Cs \rightarrow He^{-*} + Cs^+$$

This type of reaction may be important in creating negative ions of elements which have negative electron affinities, since many of these ions can be formed in metastable excited states with reasonable efficiencies.

(iii)   Surface ionisation phenomenon

Surface ionisation producing positive or negative ions can occur when atoms or molecules impinge on a hot metal surface and are absorbed for some average period of time before re-evaporation occurs. Positive ion production in this way has been utilised, not often in ion implantation, but the lack of chemically stable low work function materials inhibited the use of this technique for negative ion production.

The importance of surface adsorbates on the negative ion formation processes has been recognised, since the discovery by Krohn (62a) that negative ion yields are enhanced by sputtering a material in the presence of an alkali metal. Direct correlations bewteen the negative ion yields and surface work function have been established by Yu (78b).

Examples of such sources are given later.

2.3   Effects of Electrodes

Electrodes are normally used to provide the electrons which initiate a discharge. Emission from them may also maintain the discharge in a steady state.

Electrons are emitted from electrodes by:

(i)   Thermionic emission

The current density J is given by the Richardson–Dushman equation:–

$$J = AT^2 \exp(-\phi e/kT)$$ where A is a constant, ϕ is the work function, e is the electronic charge, and T the absolute temperature.

The emission increases as ϕ decreases or as T increases, Oxides and low ϕ metals are used for such electrode surfaces.

(ii)   Photoelectric emission

If the energy of the quantum of light is greater than the work function of the surface then electron emission occurs.

(iii)   Electronic impact

Electrons in the range of 100 eV to 1 keV are the most efficient in creating secondary electrons at an electrode surface.   The number of secondary electrons produced per primary electron is the secondary electron coefficient.   For metals it is about 1.5 but may be much higher in insulating materials. This process is important in high frequency, AC, discharges but not in DC discharges.

(iv)   Positive ion impact

In this case the nature and condition of the surface is important.

(v)   Neutral atom impact, especially for metastable atoms.

(vi) Field emission.

If there is a very high field at a surface, electron bonds are broken producing electron emission directly. It is a tunnelling effect. A field of 10^6 V cm^{-1} gives a reasonable emission.

2.4   Effects of a Magnetic Field

The force $\underline{F}$ on a charged particle in a magnetic field B is: $\underline{F} = e \, (\underline{v} \times \underline{B})$ where e is its charge, and v its velocity. The force gives rise to a circular motion if v has a component perpendicular to the field B. The angular velocity of this motion is $\omega_c = eB/m$ where ω_c = cyclotron frequency for a mass m.

If the perpendicular component to the field B is u the radius of curvature of this component is

$$ r = \frac{u}{\omega_c} = \frac{mu}{eB} $$

i.e. the radius of curvature depends only on this component. Any parallel component to the magnetic field, v_1 will be unaffected by B so the combined motion in a uniform field B is a helix of constant pitch, with the axis of the helix pointing in the direction of B.

If an electric field $\underline{E}$ is applied perpendicular to B the analysis of the motion is much more complicated. In fact, the resulting motion is a helical one along an axis perpendicular to both B and E. (If there is a component of E along that of B this produces drift in that direction).

Fig. 2. Motion of electron in crossed E, B field

Thus it is clear that the gyration of electrons caused by a magnetic field gives an increased path length for an electron moving from cathode to anode, thereby increasing its collision probability. Hence the breakdown voltage is decreased. If the applied electric field is alternating then at frequencies near to the cyclotron frequency ω_c, there is a resonant effect, with electrons gaining energy rapidly from the field and thus lowering the breakdown field considerably.

2.4.1 The Magnetic Mirror

If the value of B increases from P_1 to P_2 as shown in Fig. 3, the helical path of a charged particle in this converging field is of interest.

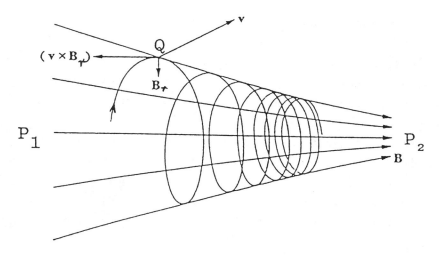

Fig. 3.   The Magnetic Mirror

As B is increasing at point Q, for example, there is a
radial component of magnetic field, B_r, pointing towards the
centre of the helical path. Thus there is a force acting to
decelerate the motion along the axis of the spiral, given by
($\underline{v}$ x $\underline{B_r}$), and this force slows down the particle drift velocity
in the field direction. In fact it can be shown that the
particle motion not only slows but actually goes to zero and
reverses so that this system acts as a magnetic mirror which can
be used to create a magnetic bottle for charged particles.

Since however the magnetic force is always normal to the
velocity of the particle, no work is done on the particle and
its total velocity stays constant; the velocity component along
the axis reduces but that perpendicular to the axis increases.

An example of such capture of charged particles is that of
the Van Allen belt where particles are retained by the
converging lines of magnetic fields from the earth.

3.   ELECTRICAL BREAKDOWN OF A GAS

The nature of gas discharges is heavily dependent on pressure. We shall consider, as an example the current –voltage relationship or a discharge in a gas at a pressure of less than 1 torr.

Electrons to initiate the discharge may be produced by electrodes and/or by ionising events in the gas. Under the conditions we are considering, cosmic rays at sea level and radiation cause between about 1 to 10 ion pairs to be produced per cc per sec. If a field $\underline{E}$ of about 1 V cm^{-1} is applied, a current flows which initially obeys Ohms Law: $J = \sigma \underline{E}$ (where σ = conductivity and J is the current density).

If it is assumed that the electrons contributing to the discharge are produced only by ionisation within the gas (i.e. ignoring electron production by the electrodes), at an ionisation rate per unit volume of $\frac{dn}{dt}$, then the current density J is given by:

$$J = x \ \frac{dn}{dt} \ e$$

where e is the charge on an electron and x is the spacing between electrodes.

This neglects any recombination effects of free electrons with other ionised atoms.

If $\frac{dn}{dt}$ remains constant, as E increases J will saturate as Fig. 4 shows.

The rate of removal of the charges is then equal to the rate of creation, $\frac{dn}{dt}$.

The saturation current density is about 10^{-9} A cm^{-2} but this can be increased by increasing the external radiation or by heating the electrodes to provide an extra source of electrons.

After saturation, further increases of voltage (field) will again result in an increase in J until breakdown occurs. This breakdown voltage depends on several parameters (see 3.1 below) but it is typically several hundreds of volts.

The region immediately prior to breakdown is known as the Townsend discharge region after the first investigator (15a). It is an avalanche region directly comparable to the avalanche effect in a reverse biased p–n junction.

Note that the voltage of this effect is at least a factor 10 greater than that in a solid state diode.

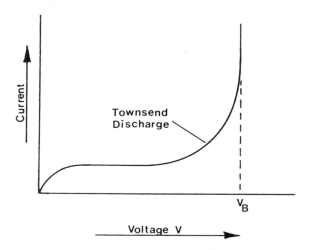

Fig. 4.   Townsend Discharge

Beyond the breakdown voltage the discharge becomes current controlled. See regions C, D, E, F in Fig. 5.

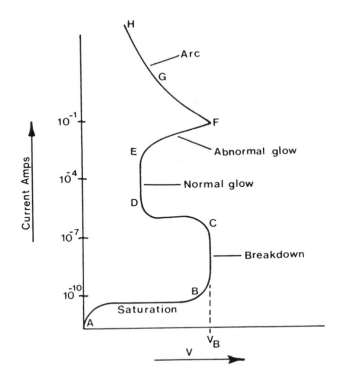

Fig. 5. Characteristics of Gas Discharge

As the behaviour is current dependent after C we should really plot this with current as the independent variable (i.e. x axis).

3.1   Paschens Law

The breakdown voltage of a gas depends on parameters such as: gas purity, gas pressure, type of gas, electrode spacing, type of materials used in electrodes and container, and the shape of the electrodes and container. One important law has been established to cover the combined effect on the breakdown voltage of gas pressure and electrode spacing in an otherwise fixed system. This law known as Paschen's law states that the breakdown voltage depends on the <u>product</u> of gas pressure P and the electrode spacing d for a given system.

i.e. V_B varies as (P x d), where V_B is the breakdown potential.

If this voltage V_B is plotted against the product Pd, there is a minimum in the curve. In other words there is minimum sparking or breakdown potential at a particular value of Pd.

Fig. 6 shows typical curves for 2 gases.

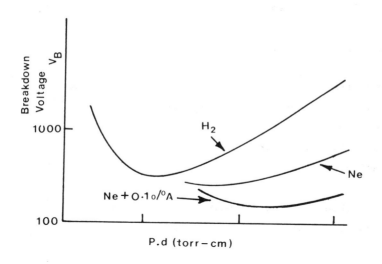

Fig. 6. Breakdown voltage against P.d.

For a given electrode spacing d this implies that breakdown is less likely at <u>both</u> high and low gas pressures. The reasons for this are that for values of Pd greater than that corresponding to the minimum V_B, (Pd)$_{min}$, the number of electron – atom, and atom –atom collisions increases, so that for any given applied voltage or field the energy gained between

collisions is therefore less, reducing the probability of
ionisation in a collision. Thus the breakdown potential
increases as Pd increases.

As Pd is reduced below (Pd)$_{min}$ the collision rate gets
smaller but the energy gained between collisions increases. Thus
the minimum results from competition between collision frequency
and energy gained between collisions.

In air at atmospheric pressure the breakdown field is about
33 kV cm^{-1}.

Note that the Penning effect discussed in section 2.2.2.
causes the reduction in the breakdown voltage observed in Neon
when 0.1% of Argon is added.

3.2   *The DC low pressure (<1 Torr) glow discharge*

The DC glow observed at low pressures in a discharge cavity
tube is one of the most familiar of discharges and is well
described by Howatson 76(a). It is easy to produce and maintain
and has a distinctive appearance depending on the residual gas
in the cavity.

Fig. 7   shows typical variations of voltage, field and space
charge density, ρ, as a function of x.

$$\left[\begin{array}{c} \text{Note that } E = \dfrac{-dV}{dx} \\ \text{and } \epsilon \dfrac{dE}{dx} = -\rho \end{array} \right]$$

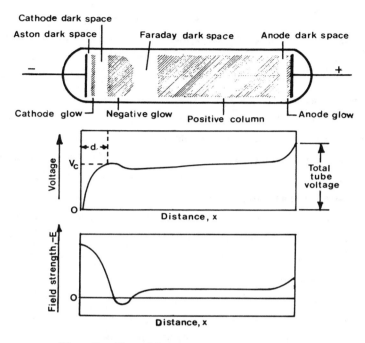

Fig. 7. Glow Discharge at Low Pressure

A qualitative view of the discharge follows:—

In equilibrium the discharge is maintained by electrons leaving the cathode as a result of positive ion bombardment. They are initially accelerated but on meeting a high density of positive ions, some recombination occurs between the slow electrons and positive ions, giving the cathode glow. Most of the voltage is dropped from the cathode to the negative glow region and electrons are accelerated to high velocities such that recombination is improbable but excitation and ionisation probable producing electron multiplication in the negative glow region. In this region the electrons are slowed until the energy available for ionisation and excitation is exhausted. The Faraday dark space follows where some recombination and diffusion occurs and the electrons are accelerated again by the small constant field E. The positive column is a region of

ionised gas where equilibrium exists i.e. ionisation rate =
recombination rate + diffusion losses. The column is a plasma
(see section 4), which is an ionised gas with no net space
charge density. This equilibrium is upset at the anode where
electrons are sucked from the plasma, equilibrium being
maintained by a small negative space charge near the anode. (See
Howatson (76a) for a full description). In a discharge
operating in the normal glow region of Fig. 5, the cathode glow
covers only part of the surface of the cathode. The coverage
increases with total current until the whole surface is covered.
(Point E, Fig. 5). Then as the current is increased further a
subsequent voltage increase is noted. Sometimes this change at E
is obscured by the cathode glow covering additional surfaces
such as metal supports.

The voltage across the cathode region of length d (Fig. 7)
is known as the cathode fall. It is related to, and less than
the minimum breakdown voltage for the gas and electrode
combination.

3.3   Effects of Pressure in the Glow Discharge

The effect of changing pressure in the low pressure
discharge discussed above is quite pronounced, and follows a
simple rule. The length of the cathode dark space (Fig. 7) is
inversely proportional to pressure and the lengths of the other
regions vary in roughly the same way, except for the positive
column, which will occupy as much of the length of the cavity as
is available for it. When the pressure is so low that the
cathode dark space fills the whole cavity the discharge is
simply like an electron beam. At even lower pressures, the
supply of electrons by the bombardment of positive ions is too
low to maintain the current unless a very high voltage is
applied.

On the other hand an increase in pressure reduces the length of the cathode region and gives a significant increase in the current density there and usually an increase in the field in the positive column. The increase in current density J with pressure follows a P^2 law up to a few Torr pressure but at higher pressures J increases more slowly with P.

It is difficult to maintain a glow discharge above 10 Torr and usually above this pressure the glow will degenerate into an arc.

3.4   DC Arc Discharge

This corresponds to regions FGH of Fig. 5. The arc draws a much higher current for a lower voltage and the current density is high enough to make the positive column very luminous. A typical voltage variation along an arc discharge is shown in Fig. 8. It is similar to that of a glow discharge with large voltage drops at the cathode and at the anode, with a region of constant voltage gradient, which corresponds to the positive column of a glow discharge. Typically the cathode fall is 10 volts across about 0.1 mm.

The current density in an arc can be very high especially at the electrodes where, for example, the cathode spot can be very small giving values of J as high as 10^6 A cm^{-2}.

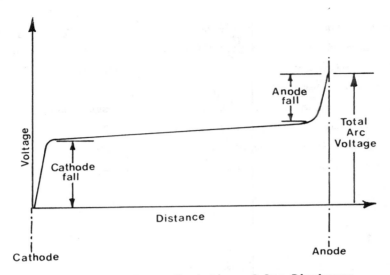

Fig. 8.   Voltage Variation of Arc Discharge

3.5   High Frequency Discharges

3.5.1 Up to 10 MHz

When an AC field is used to break down a gas, instead of a
DC field, there are several regions of interest. For low
frequencies, 1 kHz, the time for 1 cycle is usually greater than
the transit time of the ions and the breakdown voltage is
similar to the DC value. As the frequency increases the value
of V_B may increase and then decrease for the following reasons.
If we consider that the breakdown depends on the balance between
the rate of production and the rate of loss of electrons, then
the saturation region in Fig. 5 is one in which there is a
balance between the electrons provided by the cathode and those
absorbed by the anode. As secondary electrons are created by
gas collisions then the rate of production exceeds the rate of
loss. With an AC field, as the frequency increases, an
increasing number of the charged particles in the gas are
reversed in their drift motion before reaching the electrodes,
or boundaries of the discharge. This reduces the electron loss
and also the rate of arrival of positive ions at the cathode.

The latter effect reduces the rate of production of secondary electrons at the cathode so although the electrons in the gas may make more collisions as they are reversed in direction frequently, it is possible that at some frequencies, the breakdown voltage is increased. At higher frequencies though, the breakdown voltage is reduced as the electron loss is reduced and as the efficiency of the oscillating electrons in producing ionising collisions increases. Thus the electrode effects become unimportant when the electrons oscillate in the field with an amplitude much less than the discharge length and virtually no electrons (or ions) are lost to the electrodes. Electron diffusion is the only mechanism of loss in this case, and one can have a simple picture of breakdown without reference to electrodes; the current flows by the oscillation of a number of electrons in the gas which are self sustaining regardless of whether the field is provided by a voltage between electrodes within the gas.

For given dimensions and gas conditions the breakdown voltage may decrease with frequency above a few MHz. When discharges are created at frequencies above 10 MHz, say, the current is maintained regardless of the electrodes. Such a high frequency field is usually applied without internal electrodes, using either capacitive or inductive coupling. In the former, the AC field is applied to external electrodes near to the outer surface of the discharge bottle. Inductive coupling may be used in which a coil surrounds the discharge region, and the changing magnetic flux within the coil induces a current in the discharge. In effect the discharge forms the secondary circuit of a transformer. This type of discharge has been of importance for the so called radio-frequency (r.f.) ion source referred to later.

3.5.2   Above 1 GHz

If the gas pressure is low, i.e. 10^{-5} to 10^{-4} Torr, and the
applied frequency high, an electron is able to make many
oscillations between consecutive collisions with gas atoms.   At
the frequency corresponding to the cyclotron frequency, (see
section 4.1.3), the electrons gain energy continuously from the
field resulting in a high electron temperature.   This type of
discharge is known as an Electron Cyclotron Resonance (ECR)
source.   Geller et al (83b) used 10 GHz microwaves but the
availability of cheap magnetron oscillators at 2.45 GHz has led
to the development of sources at or around this frequency
(Sakudo et al, 80a and 83c, Ishikawa et al, 83d).

ECR sources can provide larger fractions of multiply charged
ions through multiple ionisation in single impacts of energetic
electrons and through step by step ionisation by multiple
impacts of lower energy electrons.

If the demand is for higher currents of singly charged ions
however it is possible to absorb energy from GHz frequencies,
off resonance, in higher density (pressure) discharges, and this
will be discussed in the next section.

4.   THE GAS PLASMA AND EXTRACTION FROM IT

4.1   Characterisation

We have seen in section 3 the many different ways of
creating a gas discharge or arc and here we are interested in
the positive ion content of any such discharge and how to
extract these positive ions from it.

For an ion source it is important to create a stable region
of positive ions and the basis of such a region is the gas

plasma.

A gas plasma in an ionised gas which contains no net space charge. Such a gas, therefore, contains equal numbers of positive and negative charges, is field free and can be treated as a conductor with a conductivity which is dependent on the electron density (which equals the ion density) and the electron mobility. The positive column region in Fig. 7 is essentially a plasma.

Any plasma may be characterised by reference to several parameters, and those of particular importance are:

1.   The density and type of charged ions n_i, which is equal to the electron density n_e
2.   The density of neutral atoms or molecules (i.e. the gas pressure)
3.   The temperature T
4.   The Debye length λ_D
5.   The cyclotron frequencies of ions and electrons
6.   The plasma (angular) frequency ω_p

The significance of these parameters are explained in the following sections.

4.1.1 Conductivity of a Plasma

The conductivity of a plasma is directly dependent on the number of charge carriers and on the average drift velocity of these carriers when an electric field is applied. More precisely, the current density J flowing in a plasma is given by:

$$J = e(n_i \, v_i + n_e \, v_e)$$

where v_i = average ion drift velocity, n_i = ion density,
v_e = average electron drift velocity, n_e = electron density, and
e is the electronic charge.

In general, $v_i \ll v_e$, so, $J = n_e\,e\,v_e$, and as the
drift velocity, $v_e = \mu_e\,\underline{E}$, (where μ_e = electron mobility and
$\underline{E}$ = electric field), the current density J is:

$$J = n_e\ e\ \mu_e\ \underline{E}$$

Therefore, the plasma conductivity $\sigma = n_e\,e\,\mu_e$, since
$J = \sigma\underline{E}$.

The mobility is dependent on scattering processes and thus
will vary with gas pressure (density of neutrals) and gas
temperature.

The energy gained by electrons from the electric field is
randomised by collisions to give them a nearly Maxwellian
distribution at a temperature T_e, the electron temperature.
Since the energy exchange between electrons and the much heavier
gas atoms is an inefficient process, with the electrons losing
little of their kinetic energy, the electron temperature in the
steady state exceeds the gas temperature T_g. The equivalent ion
temperature, T_i, may also differ from T_g, but usually only by an
insignificant amount.

In the special case of a low pressure, high frequency plasma
where an electron makes many oscillations between consecutive
collisions as referred to in section 3.5, this oscillatory
motion, dissipating no energy to the gas atoms and molecules
constitutes an inductive current lagging the electric field $\underline{E}$ by
90°. If the field applied is $E = E_0 \sin \omega t$, then as the
equation of motion of an electron is:

$$\frac{d}{dt}(m_e\,v_e) = \underline{F} = e\ E_0\ \sin \omega t$$

so
$$v_e = -\frac{E_0 e}{m\omega} \cos \omega t$$

In this case the current density j_e, due to electrons, is
$j_e = n_e e v_e$

$$= -\frac{n_e e^2}{m\omega} E_0 \cos \omega t$$

In the absence of electrons, i.e. a neutral gas of a permittivity approximating to ϵ_0, that of free space, there would be a displacement current density

$$j_D = \epsilon_0 \frac{dE}{dt} = \epsilon_0 \omega E_0 \cos \omega t$$

The total current density in the plasma is therefore:

$$j_e + j_D = \omega E_0 (\epsilon_0 - \frac{n_e e^2}{m\omega^2}) \cos \omega t$$

where $(\epsilon_0 - \frac{n_e e^2}{m\omega^2})$ is the effective permittivity ϵ_{eff}, for the plasma. Since the permeability of the plasma is approximately μ_0, the phase velocity, v_p, in the plasma becomes greater than the free space value and equal to:

$$c / \left[1 - \frac{n_e e^2}{\epsilon_0 m\omega^2} \right]^{1/2}$$

A refractive index k_p may be defined for a plasma as:

$$k_p = \frac{c}{v_p} = \left[\frac{\epsilon_{eff}}{\epsilon_o}\right]^{1/2}$$

$$= \left[1 - \frac{n_e\,e^2}{\epsilon_o\,m\omega^2}\right]^{1/2}$$

i.e. $k_p < 1$ so that radiation entering the plasma from free space is deflected away from the normal.

4.1.2 The Debye Length

The Debye length is a measure of the mean separation of the ions and electrons within the plasma due to random thermal motion. As electrons diffuse faster than ions, they tend to move away from the ions by diffusion until the electric field set up by the separation of the positive and negative charges tends to restore the position to its earlier equilibrium. On average there is a separation of the ions and electrons, λ_D, which depends on the temperature and inversely on the density of charges.

λ_D is given by the equation: $\lambda_D = \left[\frac{2kT\epsilon_o}{n_e\,e^2}\right]^{1/2}$

(where k is the Boltzmann constant, ϵ_o is the permittivity of free space and e is the charge on an electron).

A plasma cannot exist if the dimensions of the vessel are

less than λ_D since the electrons would be lost to the walls. λ_D therefore is a measure of the size requirement to satisfy the condition that macroscopically there is no net space charge in a plasma.

4.1.3 Cyclotron Frequency

The cyclotron frequency occurs when there is a magnetic flux density B associated with a plasma. Any moving charged particle is affected by a magnetic field, and this can have a considerable effect on gas breakdown (as we have seen) since the drift motion of electrons is considerably influenced by B. Any electron moving freely through a field B gyrates about the field lines with an angular frequency ω_C, the cyclotron frequency.

$\omega_C = eB/m$, where e = charge and m_e = mass of an electron

For ions, $\omega_C = eB/m_i$, where m_i = mass of ions

4.1.4 The Plasma Frequency, ω_p

Referring to section 4.1.2 above, the natural frequency of the electrons as they separate from, and return to the positive ions, without collision, is a definition of the plasma frequency ω_p.

A further implication of the effects of such a frequency may be derived from section 4.1.1.

When $n_e e^2 = \epsilon_0 m_e \omega^2$, $k_p = 0$ which means that electric fields (radiation) of frequency ω, where

$$\omega = \left[\frac{n_e e^2}{\epsilon_o m_e}\right]^{1/2}$$

will be reflected by the plasma. This value of ω is ω_p, the
plasma frequency for an electron density n_e. Only radiation at
frequencies greater than ω_p can penetrate the plasma. For
$n_e = 10^{11}$ cm^{-3}, the critical frequency is 2.8 GHz.

Where electron collisions occur in a plasma, imparting
energy to it, the above simple treatment needs modification to
take into account the collision frequency. Essentially however
a wave can only propagate through, and be absorbed by a plasma
if its frequency exceeds ω_p.

This simple model of propagation is modified in the presence
of a magnetic field. There are critical, or cut off frequencies
which depend not only on ω_p but also on the cyclotron frequency
ω_c, according to whether the propagation is parallel, or
transverse, to the magnetic field.

4.2   Stability

In an ion source there may be large variations in the degree
of ionisation and the density of positive charges but it is
usually in the range 10^{10} to 10^{14} ions cm^{-3}. The neutrality of
the plasma is self regulating since the electrons and ions are
able to move to compensate for any local gradient of electric
field. In the steady state, any loss of electrons and/or ions
from the plasma must be equal to their production. Thus the
maximum number of positive ions which can be drawn off the
plasma for a DC ion beam must equal the replacement rate. We
have already seen that the conductivity of the plasma is
dominated by electrons since their mobility is so large, but
this also implies that electrons are more easily lost to the
wall of the container of the plasma by natural diffusion
processes, since the diffusion coefficient increases as the
average particle velocity increases. Thus in the absence of any

other restoring force, electrons would be lost continually from the plasma. However, such a loss is naturally corrected by the plasma assuming a slight (few Volts) positive potential with respect to the container which in most ion sources is the anode. If an electrode is inserted into a plasma, a similar potential difference between the plasma and electrode is set up; the neutrality of the plasma is maintained by the production of a sheath of electrons and ions around the electrode.

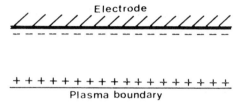

Fig. 9.   Sheath at a Plasma Boundary

If the electrode shown in Fig. 9 is isolated it will acquire a negative potential relative to the plasma. Basically, the fast diffusing electrons initially escape to the probe thereby preventing further escape of electrons from the plasma. If a voltage is now applied to the electrode then the thickness of the sheath will change, to allow this voltage to fall across the sheath. Thus the surface of the plasma remains an equipotential but its position relative to the electrode may vary according to the potential difference which is applied. Extraction of ions from a source is achieved by biasing an electrode negatively, attracting ions from the plasma. However, the shape and position of the emissive surface of the plasma will depend critically on the characteristics of the plasma and on the extraction field applied.

The conditions under which a stable sheath may form have

been studied at some length by Bohm (49a). Ions are produced
within the plasma by primary electrons colliding with neutral
atoms producing secondary electrons with kinetic energies of
2 to 4 eV whereas the ions have little kinetic energy. The
electrons quickly attain some approximate Maxwellian
distribution whereas the ions can only gain kinetic energy from
the small fields created by ion-electron separation within the
plasma, which is considered, macroscopically, to be field free.
A stable sheath only forms when the ion energy gained from the
small electric fields is approximately equal to one half of the
average electron energy. Thus the kinetic energy of the ions
$1/2 m_i v_i^2 = 1/2 k T_e$, where T_e = electron temperature.

So for a stable sheath $v_i = \left[\dfrac{kT_e}{m_i} \right]^{1/2}$

It is important to appreciate that the formation of the
sheath is dependent on the electron temperature (kinetic energy)
within the plasma. Although the ions would normally have a
distribution of energies, they are not necessarily thermalised
and ions are emitted from the plasma surface with a kinetic
energy of 1 or 2 electron volts. If large extraction voltages
are used, this energy can be disregarded.

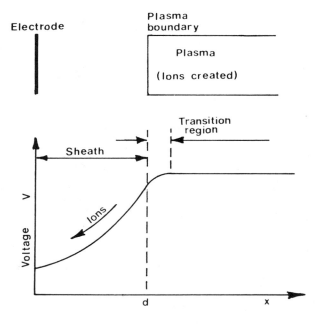

Fig. 10. Voltage across Sheath

The sheath thickness d, can be calculated for simple parallel plate geometries, using the Child–Langmuir formula for space charge limited conditions, (see Bohm 49a):

$$d = K \left[\frac{V^{3/2}}{J_i \left[\frac{m_i}{2e} \right]^{1/2}} \right]$$

where V is the voltage applied, J_i is the ion current density, and m_i is the ion mass.

J_i is given by $n_i e v_i$, where v_i = average ion velocity at the plasma boundary and n_i = ion density at the plasma boundary

$$\text{As } v_i = \left[\frac{kTe}{m_i}\right]^{1/2}$$

$$J_i = n_i \, e \left[\frac{kT_e}{m_i}\right]^{1/2}$$

By substituting some typical values into this equation the order of magnitude of J_i can be determined.   Proper corrections for geometry factors, however, will lower any such values calculated, by at least a factor of 5.

Let us assume a mass $m_i = 40$, a medium ion density $n_i = 10^{12}$ ions cm^{-3} and a high electron temperature of 4 eV ($= kT_e$).   This gives $T_e = 4.6 \times 10^4$ K.

In this case the theoretical maximum value of $J_i = 50m$ A cm^{-2}.

4.3   Ion Beam Extraction

This brief look at the properties of a plasma and its sheath enables us to establish the principles of positive ion extraction and also the problems associated with calculating the precise behaviour of an ion beam during extraction which leads to possible difficulties in focussing and acceleration.

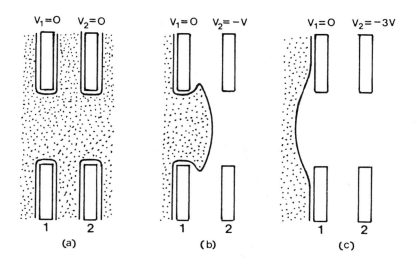

Electrodes 1 – Anodes

Electrodes 2 – Extractors

:::::: Represents Plasma

*Fig. 11. Variations of Plasma Boundary with Extract
Electrode Voltage (after Septier 67a)*

In Fig. 11(a) the anode and extractor electrodes are at the
same potential and the plasma diffuses into the space as shown
forming a sheath at all surfaces; these surfaces will be at a
slightly negative voltage with respect to the plasma. As the
voltage on the extractor electrode is increased negatively (b
and c) the plasma is pushed away from the extractor and in (c)
the surface facing the extractor is concave. Likewise for a
fixed extractor voltage the variation in plasma ion density
causes a change in the emissive equipotential boundary from

concave to convex as the ion density increases (Fig. 12).

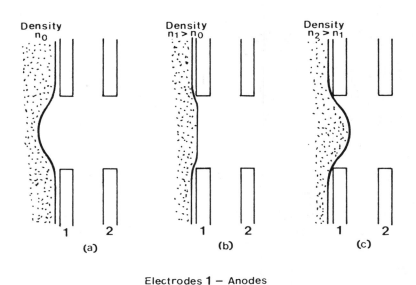

Electrodes 1 — Anodes

Electrodes 2 — Extractors

Represents Plasma

Fig. 12.   Variations of Plasma Boundary with Plasma Density
(after Septier 67a)

If there is a high degree of inhomogeneity in n_i, Septier
(67a) shows that the equipotential surface can be more complex.

It is clear that exact calculations of any 'optical' system
placed at the exit from the source are valid only if the shape
of the plasma boundary is known; i.e. for any particular set of
operating conditions.   Usually an extraction system comprises
the electrode itself followed by a focussing element(s). The

shape of the extraction electrode for ion beams is largely
empirical in spite of much investigation. This is in contrast to
the extraction systems used for electron guns where the emission
surface is fixed and thus calculations for different conditions
can be made. It is useful in this context to define the
perveance of an electron beam: it is the ratio of the beam
current to the three halves power of the extraction voltage. For
electrons the Pierce (40a) design technique is widely used
(Pierce geometry) and the shape of the electrodes ensures that
the electrons in the beam behave as if they were flowing in an
idealised diode system following rectilinear trajectories.

The work of Pierce often forms the basis of studies for
positive ion sources but the varying shape of the plasma
boundary is the controlling factor.

For ions an equivalent perveance (P) can be defined to allow
for the effect of space charge at any given energy (voltage) due
to the lower velocities of the ions. It is:

$$ P = \frac{I}{V_E^{3/2}} \left[\frac{m_i}{m_e} \right]^{1}/_2 $$

where m_i/m_e = ratio of ion mass to electron mass and
V_E = extraction voltage

For high perveances the space charge can cause defocussing
and subsequent divergence of a beam and careful design of the
extract electrode system is required to minimise these effects.

Summary

Any extraction system must yield an intense beam, with the
minimum divergence and as free of aberrations as is possible,
and this usually means using a high voltage and carefully shaped

electrode system.

The extraction of ions can be carried out directly from the source plasma which remains where the ions are created, and this is usually done for low to medium ion densities. In special cases, for very high ion densities, the extraction is carried out from a plasma which has diffused out of the source through a small opening in the anode.

5.   TYPICAL ION SOURCES IN COMMON USE

We have already referred to the many excellent reviews of ion sources which have already been published (Sidenius (78a), Wilson (73b), Septier (67a), Aitken (82a) and Alton (81a) and it is not the purpose of this text to provide a comprehensive review of the many different ion sources in use today. In this section a selection of some sources in use in implanters or accelerators used for ion beam analysis are described briefly.

5.1   The Harwell Freeman Source

One of the most famous designs used in various guises throughout the world is the Harwell Freeman Source (73a).

According to the classification and definition given in a review by Sidenius (78a), this source is of the magnetron type. Magnetic fields feature in many ion source designs to help create stable plasmas, and to increase the ionisation efficiency of the electrons within the source.

Two magnetic fields act upon the plasma in this source. The first of these is about 100 gauss, along the axis of the filament, provided by an external magnet, and the second is that due to the filament current of $\approx$ 150A producing weaker, circular field lines of about 20 gauss, (see Fig. 13). These two

fields produce a stable plasma and the electrons are forced to
move in complicated orbits near to the filament. As the filament
is mounted close to the ion extraction slit, particles emerging
from it will have passed close to the filament surface and the
maximum electron density.

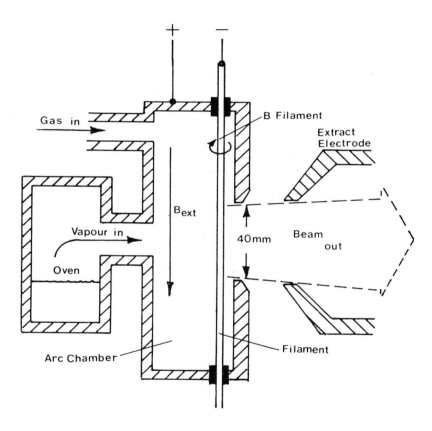

Fig. 13. The Freeman Source

According to Sidenius (78a): "as well as being a very
reliable medium temperature universal source with slit
extraction for milliamperes, the most important feature of this
source is the fixed location of the intense discharge next to

the extraction slit with the formation of a very stable boundary from which an outstanding high quality ion beam may be extracted". The source is noted for the ease with which stable beams can be obtained from most elements in the periodic table. The large rod filament which is about 2 mm in diameter heats the arc chamber and with careful design of heat shielding, temperatures up to 1100°C can be used. The length of the extraction slit is typically 40 mm.

5.2   *Penning ion sources (often called PIG sources, after*
       Penning Ionisation Gauge)

These sources operate in high magnetic fields directed along their axis. The electrons oscillate between a cathode C_1 (which may be hot or cold) and an anti-cathode (or second cathode C_2) which is cold and at the same potential (see Fig. 14). The electrons drift slowly towards the cylindrical anode A which forms the wall of the discharge chamber. If cold cathodes are used, the electron emission is stimulated by ion bombardment. Extraction is either axial, from the centre of the anticathode as shown in the figure, or lateral, through a slit cut in the anode.

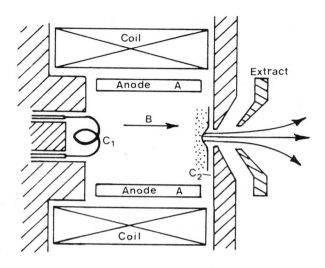

Fig. 14.   Cross Section of High Intensity PIG Source

With a cold cathode the potentials used between C_1, C_2 and A
are 1 or 2 kV and plasma densities are low. With hot filament
sources such as that in Fig. 14 a few tens of volts is
sufficient. Most of the anode voltage is dropped across a layer
close to the filament so electrons are extracted perpendicularly
to the wire surface, and for a typical filament spiral, the
electrons are emitted approximately isotropically. Electrons
emitted close to the axis will be reflected at each of the
cathodes and be trapped by the field lines in a central column,
whilst electrons emitted radially move out to the anode in a
complex spiral. Thus the ionisation probability is increased in
this source, resulting in lower starting and working pressures
and a higher efficiency and ion output.

This hot filament, Penning type source is often referred to
as a Nielsen ion source and many versions exist for low to
medium current applications (10 μA to several 100 μA).

5.3   *High Temperature Hollow Cathode Source*

In the same 10 μA to 500 μA range this source developed by
Sidenius finds much favour as a universal ion source for
research machines.

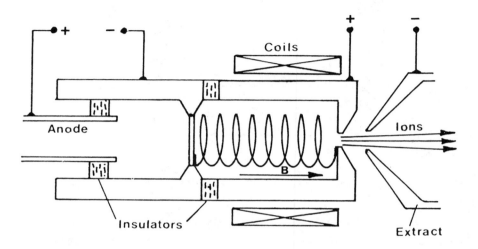

Fig. 15.   Hollow Cathode Source - Cathode Extraction

Fig. 15 shows the version in which the extracted beam is
from the cathode itself. The magnetic field required in this
source is only necessary to compensate for the field due to the
filament since analysis shows that the optimum performance is
for a 'magnetic field free' cathode volume. Thus the field
setting is not critical and the source is easy to use, unlike
the more efficient but sensitive version of the hollow cathode
source with anode extraction. The plasma volume in these sources
is usually less than 0.5 cm^3.

5.4   *Radio-frequency (r.f.) gas ion source*

This source is well established as one of the best sources

for producing light ions from gases and is widely used, for
example, in Van de Graaff accelerators and for low intensity
requirements in implanters. It is mostly useful for natural
gases since other materials might coat the quartz envelope and
degrade its performance.

The source is as shown in Fig. 16. A quartz (or glass)
envelope is surrounded by a coil and the alternating magnetic
field creates the discharge. The r.f. power source is usually a
few hundred volts at 10 to 30 MHz and the plasma pressure is
fairly high at 10^{-3} to 10^{-2} torr.

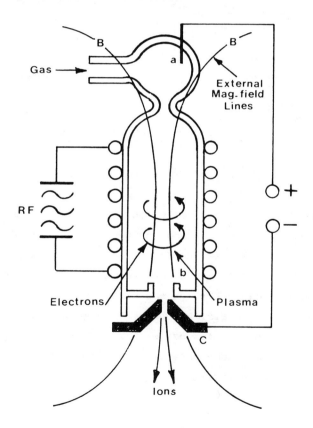

Fig. 16. R.f. Ion Source

5.5   The Duoplasmatron

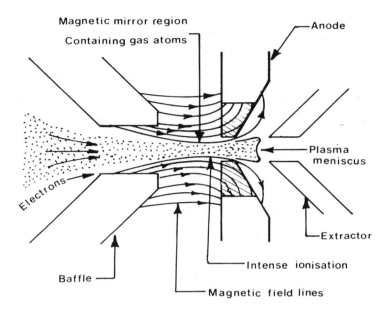

Fig. 17.   Duoplasmatron Source

This source and its many relatives are arc discharge sources but a high density plasma is produced near to the anode aperture by using an intermediate electrode and a strong magnetic lens. Extraction takes place from a so-called expansion cup into which the dense plasma expands through the narrow aperture in the anode. The source is efficient, of high brightness (see section 6.3.2), but useful mainly for gases. Many modifications and adaptations have been made however, in attempts to make this type of source more universal.

5.6   *Microwave Ion Source*

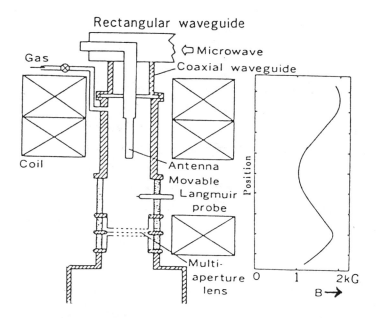

*Fig. 18.   An Off-Resonance Microwave Source for
High Current Singly Charged Ions*

Sakudo and co-workers have made several reports of the
design and operation of their filament free microwave source,
(e.g. 80a and 83c, culminating in a recent review by Sakudo
(87b)). A schematic of an off resonance version of this source
is shown in Fig. 18.

Microwaves at 2.45 GHz are introduced, from an antenna into
the discharge region through a ceramic window, and the gas
required, through a needle valve. The whole of the source is
placed in a solenoid and a magnetic field between 800 – 1800
gauss is applied over the whole discharge region. This field is
a "mirror" field with a ratio of about 2, so that its intensity
varies from 1800 at the two ends of the discharge down to 900 in
the middle. This means that the field value is higher than that

needed for ECR (875G) for the entire discharge region. The aim
of this source is to increase the plasma density to increase the
singly charged ion output. The critical density for which the
plasma oscillation frequency equals 2.45 GHz is only
7.46×10^{10} cm^{-3}. For higher densities only right hand
circularly polarised waves can propagate and efficient energy
absorption can be achieved by oblique propagation in an
inhomogeneous plasma. Several different versions of this source
have been reported. One using a slit shaped beam is shown
schematically in Fig. 19. This source is claimed to be able to
produce 10 mA of P$^+$ and As$^+$ for 100 – 200 hours, which is
several times longer than most conventional hot cathode type ion
sources.

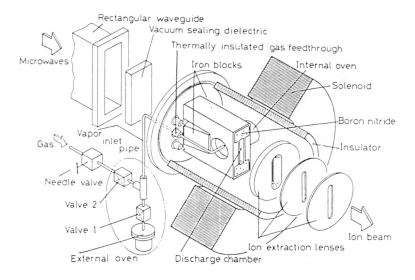

Fig. 19.   Slit Type Microwave Source

An interesting development of the microwave type of source
was reported by Torii et al (87c) for a 50 – 100 mA oxygen ion
implanter (see Fig. 20). Operating as an ECR source at 2.45 GHz

it can achieve a current density of 120 mA/cm^2. The ions are
extracted through grids which have 7 holes, each 5 mm in
diameter, arranged over a circular area with a 20 mm diameter.
Pure oxygen gas is used and the ratio of O$^+$ to total ion output
is more than 80%. This source could have an important
application in producing buried oxide layers in silicon wafers.

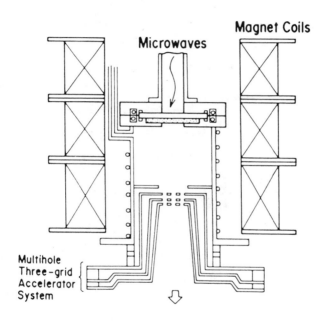

Fig. 20. Microwave Source for O$^+$

5.7   Molten Metal Field Emission Ion Sources

This type of source can be constructed for either liquid or
gaseous feed materials. The liquid metal sources have a
capillary, feeding liquid to the tip of a sharply pointed
electrode. The applied electrostatic field from the so called
Taylor cone, and the extremely high field in the region of the

tip is sufficient to strip electrons directly from the metal
atoms. The voltage required depends on the tip radius and the
needle to electrode distance. However, typically for gallium,
which is a convenient metal, a voltage of 7 kV is needed for a
10 μm radius tip and a needle–electrode distance of 12 mm. The
ions that can be produced by this method are limited to elements
which have a reasonably low vapour pressure at their melting
point and which are chemically compatible with the needles
available. Many different ions have been used such as Cs, Si,
Hg, As, Au, Cu and Ga.

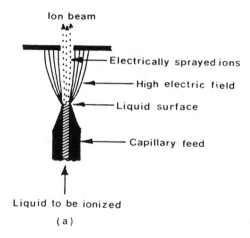

(a)

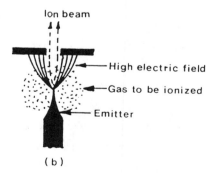

(b)

Fig. 21. Field Emission Sources

These sources have an extremely high brightness, (see section 6.3.2),and are of great interest to workers studying ion beam lithography or direct write implantation for special devices.

5.8   *Negative Ion Sources*

Useful intensities of negative ions may be extracted from a duoplasmatron type source operated with a reversed polarity on the extraction electrode.   Moak et al (59a) were the first to discover this and subsequently many different schemes have been proposed to increase the yield from such an arrangement. Similarly a variety of negative ions have been produced by operating Penning sources in the reversed mode.

The charge exchange method as adapted by O'Connor and Joyce (87a) is shown schematically in Fig. 22.   The charge exchange medium may be sodium, lithium or magnesium.

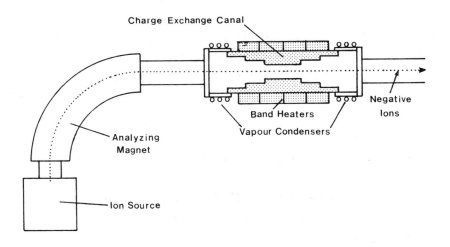

Fig. 22.   A Charge-Exchange Negative Source

Summary

In this section we have covered some of the more important
versatile sources used in both research and production
implantation. As Sidenius (78a) remarks in his excellent
review: "it should be realised, however, that the complexity of
the behaviour of the different chemical elements together with
the various possible kinds of plasma and arc instabilities will
from time to time still be able to create operational problems
even with the most universal ion source system. These problems
will most easily be analysed and solved if the fewest possible
parameters have to be considered. Therefore, the ion source
system which will offer the greatest satisfaction to its user
will almost always be the system with the most simple and
straightforward configuration".

6.   BEAM QUALITY AND CONTROL

For most, but not all, applications in implantation, the
user requires a stable beam of a single atomic species to be
transported from the ion source to the target. Important
considerations are:

  (i)   the content of the beam.
 (ii)   variations in energy or current.
(iii)   the 'optical'quality of the beam which determines
        how well it can be controlled and transported.
 (iv)   loss of content during transport which depends on the
        beam quality.

In this section we shall elaborate on these considerations.

6.1   Beam content

The ions extracted from a source will contain as many different ion species as there are atom species in the discharge. For most applications of implantation, magnetic analysers are used to separate the required ion from the remainder of the output. In some important examples the current of the required ion is only a small fraction of the total ion current extracted. For example boron in elemental form has a very low vapour pressure and temperatures of 2000°C would be necessary to create the necessary partial pressure of boron vapour. Instead boron compounds, such as BF_3, are commonly used. The required boron ion of mass 11, $^{11}B^+$, is diluted first of all by its own isotope boron 10, (which is 20% abundant), but also by fluorine ions such as BF^+ and BF_2^+. In a typical case the $^{11}B^+$ content may only be 20 to 30% of the total output.

In contrast phosphorus element may be used as a feed material for P^+ production and although molecular ions such as P_2^+ and P_3^+ are produced the P^+ content is 70% of the total output.

Thus the current capabilities of a machine varies according to which ion is required.

6.2   Variations in energy and current

Ideally, as we have shown in section 4.2, it is best to extract ions from a stable plasma boundary. This can be realised if the plasma is truly thermalised. However where strong magnetic fields are present and/or reflecting cathodes are used to increase the ionising efficiency of electrons, instabilities can be created in a plasma. Such instabilities may be complex, but in general terms they lead to variations in:

(i)   the position of the plasma boundary.

(ii)   the energy of ions extracted, and

(iii)   the current density which is extracted.

Both (i) and (ii) can cause difficulties in controlling and transporting the beam through the whole system and (iii) makes the job of dosimetry at the target end more difficult.

We shall look at the problems of beam transport from the source in section 6.3 below.

6.3   Beam transport

If a beam, extracted from an ion source, is to be transported over large distances and/or through a series of apertures and lenses, the properties of the beam leaving the source must be known if large losses of ions from the beam are to be avoided. Plasma instabilities can cause variations in the initial optical quality of the beam and we need to understand how we define this quality. The most important definition is that of emittance which essentially is a measure of the divergence of a beam in the transverse plane which is selected for use.

6.3.1   Emittance

Let us look at a beam moving principally in the z direction, which has components of momentum in both the x and y directions such that $p_z > p_x$ and p_y. It is assumed in this treatment that there is no interaction between particles in the beam and that there is no scattering of the particles from residual neutrals in the vacuum system.

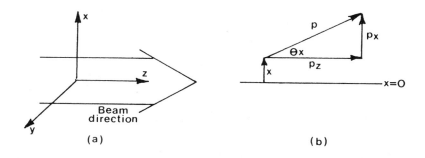

Fig. 23.   Transverse momentum

At any time the state of any ion in the beam can be defined fully by three position co-ordinates, x, y, and z and three momentum coordinates p_x, p_y, p_z. For n particles there are thus n such sets of conditions and this 6 dimensional system is known as 'phase space'.

Let us consider first one dimension (x).   The transverse momentum of an ion is p_x at x.   The slope angle θ_x of this particle at x is given by:

$$\tan \theta_x = p_x/p_z$$

A point representing this particle can be plotted on a p_x -x diagram (which is a transverse momentum –displacement phase diagram).   If similar representative points of all the ions in the same transverse plane are plotted, the area enclosing these points on the p_x -x diagram is called the emittance, in the x phase plane.

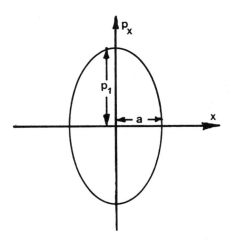

Fig. 24. p_x - x *Phase Space*

This area is also known as the phase space area. This phase space area of particles of the 'x' phase plane (and those of the 'y' plane) remain constant in magnitude, but not in shape as the ions move through the system (Liouvilles theorem (37A)).

Looking in more detail at the emittance in the x phase plane we can see that the divergence θ_x of an ion at x is directly proportional to the transverse momentum p_x, if $p_x < p_z$. (Thus the total momentum is approximately p_z).

The phase space area of a (p_x –x) diagram is shown in Fig. 24.

Furthermore for a beam whose energy does not change, the Liouville theorem will apply equally to a diagram of divergence and position as shown in Fig. 25 since for a small θ_x, $\theta_x \simeq p_x/p_z$ and p_z is assumed constant.

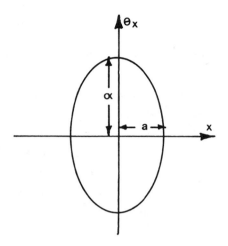

Fig. 25. θ_x - *x Phase Space*

This divergence-position diagram is often quoted as a slope-position diagram (since tan $\theta_x = p_x/p_z$ is the slope) which in this case would be written as x', x.

Clearly, as θ_x is a function of p_z then as the energy of the beam varies so will θ_x. We will consider how to correct for this shortly.

From Fig. 24; above the emittance is assumed to be elliptical and has an area π p_1 and this area is conserved even when the energy of the beam changes. In Fig. 25 the same phase ellipse in the (x, θ_x) plane has an area π a α where $\alpha = p_1/p_z$.

In this case emittance is measured in a mm-mradians and is defined, for an elliptical space, as $1/\pi$ times the area occupied by particles in the (x, θ_x plane).

This emittance is not conserved with changes in energy. Allowance can be made for this however, giving a momentum normalised emittance, ϵ_n, which is invariant with energy for non-relativistic beams such as those used in ion implantation.

ϵ_n is defined as: $\epsilon_n = AE^{1/2}/\pi$

where A = elliptical area (π a α) of the beam in the (x, θ_x)
plane and E = beam energy

If A is in mm-mrad, and E in MeV, then the units of ϵ_n are
mm-mrad MeV$^{1/2}$. (Note: In getting A from the x, p_x diagram, a
factor of $\sqrt{2}$ mE is used (where m = mass of the ion) so that the
value of A allows for different masses of ions). This momentum
normalised emittance is extremely useful since it is directly a
measure of the relative compatibility of beams of different mass
and energy to a given beam transport system which has a given
'acceptance'.

Acceptance here is the complement of emittance and is the
phase space area allowed by the geometry of the particular
system or device. For 100% transmission of a beam, its
emittance must be entirely within the acceptance of the system.

For a cylindrically symmetrical beam the transverse momentum
of an ion is p_r at a radius r.

The slope angle θ_r of this particle at r is given by:

$$\tan \theta_r = Pr/p_z,$$

and the emittance is the area of a $p_r - r$ ellipse.

6.3.2 Brightness

The Brightness of an ion source is also a useful parameter
to define. Brightness is essentially beam intensity per unit
spatial area per unit solid angle, but the various definitions
of emittance affect the definitions of Brightness used.

For a beam which is symmetrical about its axis with
elliptical phase space contours, Brightness (B) is:

$$B = \frac{2I}{A^2}, \qquad \text{where} \qquad A = \frac{\epsilon_n \pi}{E^{0.5}} 1 \qquad \text{and } I = \text{Intensity of beam in } \mu A$$

$$\text{Thus } B = \frac{2I}{\epsilon_n^2 \pi^2} E$$

Typical units of B are therefore $\mu A \ mm^{-2} \ mrad^{-2}$.

Often it is more realistic to determine the phase space area occupied by a core of the beam containing about 90% of the total beam current and then

$$B^{90} = 2(0.9I)/A^2$$

where it is understood that A in this case is that phase space area occupied by a 90% core of the beam.

6.3.3 Why Elliptical Phase Space Contours?

Banford (66a) explains why it is possible to show that elliptical phase space contours are generally to be expected. Any beam of particles are constrained by various means to travel within a defined geometrical space in order to avoid losses to walls. If the restoring force on any particle is proportional to the instantaneous displacement x from the z axis then the motion of that particle is represented by simple harmonic motion about the z axis, (see Fig. 26).

$$\text{Thus } \frac{d^2x}{dz^2} = -\frac{x(2\pi)^2}{\lambda} \qquad \text{...equ.(1)}$$

where λ is a constant and is the wavelength of the particle which will execute a sine wave as it travels in the z direction.

The solution of equation 1 is

$$x = a \sin[(2\pi z/\lambda) + \phi] \qquad ...equ.(2)$$

where ϕ is the phase depending on the initial conditions.

Remembering that emittance is described by the (θ_x, x) or (x', x) diagram, then differentiating equation (2) with respect to z gives

$$x' = a \left[\frac{2\pi}{\lambda}\right] \cos \left[\frac{2\pi z}{\lambda} + \phi\right] \qquad ...equ.(3)$$

Eliminating ($2\pi z/\lambda + \phi$) from equations (2) and (3) yields:

$$\left[\frac{x}{a}\right]^2 + \left[\frac{x'\lambda}{2\pi a}\right]^2 = 1$$

which is the equation of a right ellipse in (x', x) space with major intercepts of (a) and $\left[\frac{2\pi a}{\lambda}\right]$.

In Fig. 26 particles (1) and (2) have the same amplitude a, but different phases whereas particle 4 has the maximum amplitude A allowed by the geometry but the same wavelength. Each point can be represented on the phase space (x', x) diagram and as the beam moves down the z axis, the points move around the ellipse.

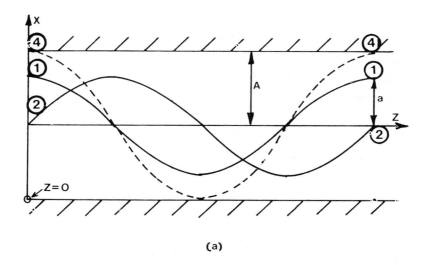

(a)

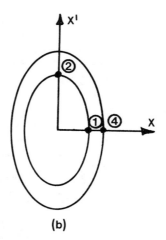

(b)

Fig. 26. (a)   Trajectories of Particles
          (b)   Phase Space Diagram at Z = 0.

Particles such as 4 move around a larger ellipse concentric
with the smaller one shown. Any particles capable of being
transmitted through this hypothetical system of size 2A are
contained within the larger ellipse. The maximum angular
deviation that is accepted is the maximum value of x' which is
$2\pi A/\lambda$.

6.3.4 Examples of the use of emittance

(a)   Extraction of 4 particles from an accelerator.

In Fig. 27 the paths of 4 particles emerging from a hypothetical
accelerator are shown.

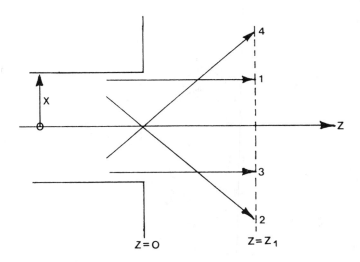

Fig. 27. Paths of 4 particles emerging from hypothetical
accelerator

The phase space diagrams in x' x space at planes
corresponding to z = 0 and z = z_1 are shown in Fig. 28 below.

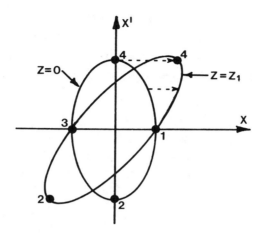

Fig. 28.   Phase Space Diagrams at $Z = 0$ and $Z = Z_1$

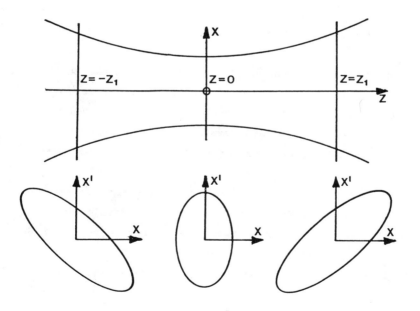

Fig. 29.   Emittance Through a Waist

The tilted ellipse at $z = z_1$ is obtained by shearing the upright ellipse horizontally; it is not a rotation.

Points 1 and 3 representing two particles, remain unaltered since these particles are moving parallel to the z axis.

(b)   Variation of emittance through a waist

We can now see how the emittance diagram changes as a beam converges to a waist and then diverges as shown in Fig. 29.

The shapes of the x', x diagrams for planes $z - 0$ and $z - z_1$ are identical to that of Fig. 28 above. Note that for large values of x (i.e. large distances from the centre line) x^1 is large and positive for the diverging beam at $z - z_1$ but large and negative for the converging beam at $z - - z_1$.

7.   CONCLUDING REMARKS

Having studied this chapter you should appreciate that the science and technology behind even a simple ion source is extensive and has been researched and developed over many tens of years. The many different types of source, some of which have been described, derive from the variety of requirements from users. Even in the relatively short life of ion implantation, big changes have resulted from the requirements for routine production implants. High current densities, a good beam quality and a long time between source servicing are high on the list of priorities, and various microwave sources have been developed recently, largely to fulfil these requirements. Further development of MeV implantation will lead to better negative ion sources for tandem machines and to the availability of sources able to produce a good output of multi-charged ions in order to avoid excessively high terminal voltages.

Ion source science and technology remains therefore an exciting and interesting field of development especially when coupled with the requirement that implanters should be fully controlled by computers.

REFERENCES

15a   J S Townsend, Electricity in Gases, Clarendon Press, Oxford, p.291, 1915

37a   J Liouville, J. Math., Pures. Appl. 2, 16, (1837)

40a   J R Pierce, J. Appl. Phys. 11, 548   A40

49a   D Bohm, Characteristics of Electrical Discharges in Magnetic Fields, A Guthrie and R K Wakerling, McGraw Hill 77, (1949)

59a   C D Moak, H E Banta, J N Thurston, J W Johnson and R F King, Rev. Sci. Instr. 30, 694, 1959

62a   V E Krohn Jr, Appl. Phys. 33, 3523, 1962

66a   A P Banford, The Transport of Charged Particle Beams, Spon Ltd., London, 1966

67a   A Septier, Focusing of Charged Particles, Ed. A Septier, Academic Press, Vol.2, 123, 1967

73a   J H Freeman, from Ion Implantation by Dearnaley G, Freeman J H and Stephen J, N Holland 369, 1973

73b   R G Wilson and G R Brewer, Ion Beams, J Wiley, 11, 1973.

73c   G W C Kaye and T H Laby, Tables of Physical and Chemical
      Constants, Longman, N York, 269, (1973)

75a   H Hotop and W C Lineberger, J. Phys. Chem., Ref. Data 4,
      539, 1975

76a   A M Howatson, An Introduction to Gas Discharges, Pergamon,
      1976

78a   G Sidenius, Low Energy Ion Beams 1977,
      Ed. by K G Stephens, I H Wilson, J L Moruzzi, Inst. of
      Phys., Conf. Series 38, 1, 1978

78b   M L Yu, Phys. Rev. Lett. 40, 574, 1978

80a   N Sakudo, K Tokiguchi, H Koike, I Kanomata, Low Energy Ion
      Beams, 1980, Ed. by I H Wilson and K G Stephens, Inst. of
      Phys., Conf. Series 54, 36, (1980)

81a   G D Alton, Nucl. Inst. & Meth. 189, 15, 1981

82a   D Aitken, Ion Implantation Techniques, Ed. H Ryssel and H
      Glawischnig Springer Verlag, 23, 1982

83a   G D Alton, Proc. Int'l. Ion Eng. Cong. ISIAT '83 and IPAT
      '83, Kyoto, 1, 85, 1983

83b   R Geller, B Jacquot and C Jacquot, Proc. Int'l. Ion Eng.
      Cong. ISIAT '83 and IPAT '83, Kyoto, 1, 187, 1983

83c   N Sakudo, K Tokiguchi, H Koike, Proc. Int'l. Ion Eng. Cong. ISIAT '83 and IPAT '83 Kyoto 373, 1983

83d   J Ishikawa, Y Takeiri and T Takagi, Proc. Int'l. Ion Eng. Cong. ISIAT '83 and IPAT '83, Kyoto, 1, 379, 1983

87a   J P O'Connor and L F Joyce, Nucl. Inst. & Meth. in Phys. Res. B21, 334, 1987

87b   N Sakudo, Nucl. Inst. & Meth. in Phys. Res. B21, 168, 1987

87c   Y Torii, M Shimada and I Watanabe, Nucl. Inst. & Meth. in Phys. Res. B21, 178, 1987

ION OPTICS AND FOCUSSING IN IMPLANTER DESIGN

by

Kenneth H. Purser and Nicholas R. White
General Ionex Corporation, Newburyport, MA. 01950.

INTRODUCTION

FOCUSSING

    Focussing Conditions and Boundary Conditions
    Computer Solutions

ELECTROSTATIC ELEMENTS

    Electric Fields Associated with Acceleration Tubes
    Focal Properties of Einzel Lenses

MAGNETIC ELEMENTS

    Magnetic Analyzers and their Focal Properties
    Introduction of Non-Median Plane Focussing
    The Effects of Pole Edge Rotations on Focussing
    Non-Uniform Field Magnetic Analyzers and Quadrupole Lenses
    Strong Focussing Devices

EXAMPLES

    The Ionex 90° Inflection Magnet
    The Eaton-Nova NV10 Inflection Magnet
    The Focussing used in the IONEX IX1500 Implanter

ION IMPLANTATION:
SCIENCE AND TECHNOLOGY

ION OPTICS AND FOCUSSING IN IMPLANTER DESIGN

by

Kenneth H. Purser and Nicholas R. White
General Ionex Corporation, Newburyport, MA. 01950.

In the present chapter we shall discuss some of the ways in which electric and magnetic fields can be used to interact with charged particles and cause them to be accelerated, mass analyzed and focussed to the end station of an implanter. The examples will include the ion optics of some specialized ion source extraction geometries, the focussing produced by d.c. accelerating tubes, einzel lenses, quadrupole lenses and magnetic fields, both uniform and non-uniform. In most of these examples the focussing properties are critical to efficient operation.

Fundamentally, the formalism for analyzing the optical performance of an implanter has much in common with the procedures used for the design of mass spectrometers, isotope separators and nuclear accelerators. Over the years the workers in these fields have developed elaborate mathematical and computer procedures to precisely predict ion optical trajectories through complicated structures. Improved and extended versions of these techniques[1] are available to the ion implant designer and it is now possible for an engineer to make very sophisticated optical analyses using inexpensive personal computers.

FOCUSSING

The requirement for focussing stems from the fact that the envelope of the ions leaving a source is naturally divergent. One reason for this is that the equilibrium shape of the plasma from which the ions are extracted is unlikely to be a plane so that even at the moment of birth the beam envelope is not cylindrical.

In addition the particles are extracted from a plasma which consists of equal number of ions and electrons with the ions having random velocities corresponding to energies of several tens of electron volts. After extraction and formation into a beam, the individual ions still possess this random velocity which is superimposed to produce angular and energy spreads into the ion beam. In practice energy spreads of a few tens of eV may be present and the envelope of the ion beam may diverge by a few degrees.

A third reason for the divergence is that the space charge of the beam causes repulsion so that an initially parallel beam of particles can become divergent.

Focussing Conditions and Boundary Calculations

If an implanter included no converging elements the envelope of the ion beam would continuously grow larger as it travels from ion source to end-station. Large increases in beam diameter rapidly become unworkable, so in practical implanter designs focussing elements are always present along the beam transport system to constrain the particles and transmit them efficiently to the end station.

In classical light optics focussing is accomplished by the use of glass lenses which produce the effects shown schematically in figure 1. It can be seen by inspection that if a group of parallel rays is to be focussed to a point on the downstream side of the lens individual rays must be deflected by an angle whose tangent is linearly proportional to the distance of the ray from the optic axis of the lens. In the case of light optics, glass lenses having spherical surfaces produce the linear deflections required for focussing, provided the deflection angles are modest.

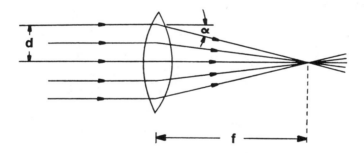

Figure 1.   Glass Lens and parallel light focussing

In the field of ion optics the radial forces required for linear focussing deflections can be produced by passing the particles through both transverse and longitudinal electromagnetic fields or through suitably shaped field boundaries. It can be shown that the focussing effects of such fields can closely parallel the manner in which light rays are deflected and focussed by glass prisms and lenses. Because of this, and also because light optics has a well developed formalism, it is customary in ion optics presentations to make use of the same terminology and formulae. Terms such as focal length, linear magnification, aberrations and principal planes have a one-to-one correspondence between the two fields. Comprehensive accounts of ion focussing can be found in references (2-6).

The focussing effects which occur at the boundary regions of dipole magnets are caused by the fact that electromagnetic fields cannot be terminated abruptly but must make the transition in a smooth, well defined manner. In figure 2, which is a sketch of the magnetic field near the poles of a deflection magnet, it can be seen that the magnetic field bulges out from the gap so that horizontal components of the field are present making it conceptually possible to produce deflections in the vertical direction as well as in the horizontal plane. Clearly, this type of ion focussing depends upon the detailed shape of the boundaries of electromagnetic fields, and over the years much effort has been expended in the development of techniques for calculating such fringing fields.

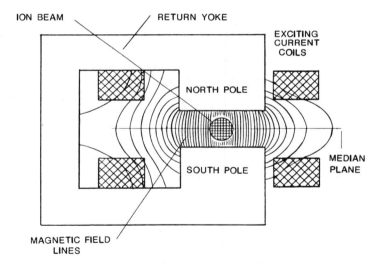

Figure 2 .   A cross section of the magnetic field and scalar potentials near the poles of a deflection magnet

One of the ways that such calculations can be made without the help of a computer depends upon a remarkable fact: The equations which describe electrostatic and magnetostatic fields have exactly the same form as those equations which describe certain other physical phenomena. For example the flow of heat across the width of a thin plate can shed light on the shape of the bulging field which extends from the boundary of a magnet. Figure 3 shows the analogy between heat flow and magnetic field distributions. Although the symbols are different and the quantities that they represent are different, the solutions are identical because the mathematical equations have the same form.

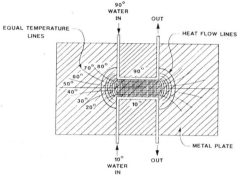

Figure 3   Analogy between heat flow and magnetic field distributions

Computer Solutions

It is also possible today to calculate the shape of electrostatic and magnetic fields using a computer. The calculations are simple but time consuming and depend upon the fact that the distribution of an electromagnetic field throughout all space is completely defined by the value of the magnetic and electrostatic field at the boundaries and by the ion current passing through the region. The boundaries are the magnet poles or acceleration electrodes and the boundary values are set by voltages on electrodes or by currents through coils. The task of the computer is to satisfy the field equations throughout the whole volume of interest, namely Poisson's equation :

$$\nabla^2 V = - \rho/\varepsilon_0.$$

Here, ρ is the density of the charge and ε_0 is the permittivity of free space.

A number of programs have been written which solve this problem. One of us (NRW) has developed a program called SORCERY which runs on a Personal Computer and solves Poisson's equation by a relaxation method[1]. This technique involves overlaying on the wanted geometry a rectangular array of points with fixed values for the boundary electrodes. After the physical boundaries and their potentials have been fed to the computer the program solves Poisson's equation repetitively for each point of the array. When a solution has been found, ion trajectories are traced through the fields. A new solution is calculated, allowing for the space charge of the ion beam, generating a modification to the originally assumed field distribution. After many iterations the corrections at each point become vanishingly small and the field distribution converges to an accurate solution throughout all of the wanted space. Quite complicated systems can be calculated in times of the order of an hour or so.

Figure 4 shows an example of an application of the program SORCERY to a complex calculation of the trajectories of ions produced from the Ionex Model 860 source. The equipotentials have been calculated using the procedures of the previous paragraph. These potentials are then used to make a step by step calculation of ion optical trajectories of the cesium ions which in turn introduces a space charge density of their own which must be used to correct the first order potential solution. This procedure is iterated several times. The final geometry of the source was chosen so that the Cs ions leaving the filament all strike a sputter target at the left of the diagram.

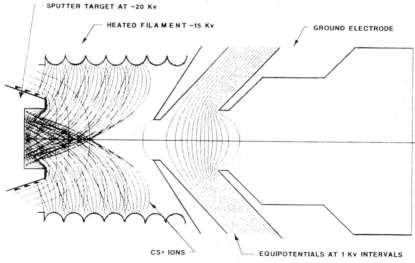

Figure 4.   *Applications of the program SORCERY to calculations of the 860.*

When the Cs ions strike the sputter target negative ions are produced from the surface and these ions are accelerated by the same electric fields to produce a well-defined beam of negative ions from the source. As a second part of the calculation, the program uses the derived equipotentials to calculate the trajectories of the emitted negative ions taking into account the initial energy distribution of sputtered particles. The results of this calculation is shown in figure 5.

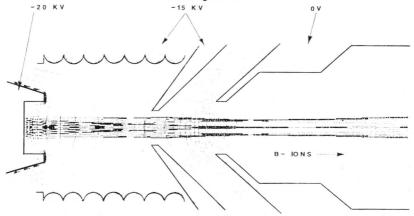

Figure 5.   Calculations of Negative ion trajectories in 860.

The Electric Fields Associate with an Acceleration Tube.

The geometry of a typical acceleration tube is shown in figure 6. The calculated equipotentials are shown in figure 7. At every point the accelerating fields are at right angles to the equipotentials.

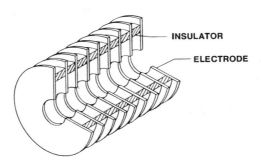

Figure 6.   Typical Acceleration Tube.

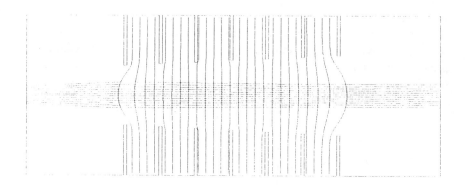

Figure 7.   Calculated Equipotentials.

It can be seen that the field structure can be divided into three sections:

1   A region at the entrance where the surfaces of constant voltage bulge from the end of the tube and the accelerating fields have a converging radial component. <u>In practice there is always a positive lens action at the entrance to a tube and this lens action can have very strong if the injection energy is low.</u>

2   A region in the center of the tube,well away from the ends, where the equipotentials are planes parallel to the tube electrodes. Here, the particles travel along parabola similar to the trajectories of particles in a uniform gravitational field. There is no focussing in this region.

3   An exit region where the surfaces of constant potential again bulge from the end of the tube. The accelerating fields have an outward component so that the particles are defocussed.

In practice the lens strength at the entrance of an acceleration tube often dominates the optical properties of the device and care must be taken to ensure that this lens action does not overwhelm the whole instrument design. A comparatively simple analysis indicates that the focal length of the lens at the tube entrance can be approximated by the following formula:

$$f = 4V_o/E_t$$

Here, V_o is the voltage through which the ion has passed before arriving at the tube entrance and E_t is the electric field in the tube.

For a typical case V_0 = 32kV and E_t = 8kV/cm. Thus, the approximate focal length of the entrance lens is f = 16cm. This focal length is short compared to the characteristic dimensions of an implanter (about 1 meter). The difficulty of such a lens is that with such short focal lengths the designer must be careful that excessively large angles are not introduced into the beam envelope which could lead to later aberrations and beam losses.

In high current implanters, a complication occurs. The mutual repulsion of the ions in the beam tends to strongly defocus it, but a phenomenon call space charge neutralization occurs, whereby electrons created from the residual gas by the beam itself get attracted to the beam and trapped within it, and these electrons counteract the repulsion of the ions to the extent of 98% or so. However, in an acceleration tube these electrons get rapidly accelerated away from the ions. A high current beam will tend to experience a strong de-focussing force in an acceleration tube, and the designer must balance this effect against the strong focussing effect of the tube entrance. Some designs use an adjustable electrode gap, allowing the operator to adjust the focussing power according to the current and voltage levels.

The Focal Properties of Einzel Lenses

The einzel lens is a useful optical element for focussing low energy ion beams of all masses. The focussing fields are derived from voltages applied between three adjacent electrodes, as shown in figure 8. Usually, the outside electrodes are connected to ground potential and the central electrode is connected to a high voltage supply which provides a bias that accelerates or decelerates the incident beam. An important feature of the einzel lens is that the beam enters and leaves the lens with the same energy.

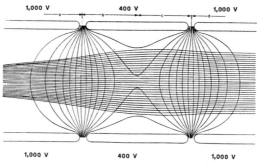

Figure 8.   Einzel Lens.

The operation of the lens in its retarding mode can be understood by considering what happens when a parallel beam of ions enters the lens from the left. When the ions reach the region "a" they experience a radial electrostatic force which accelerates them and pushes them radially away from the axis. In the central region "b" and "c" the reverse is true and the electric fields focus the ions towards the axis. Finally, there is a second de-focussing region "d".

At first sight it might be expected that there would be equal and opposite focussing impulses given to the ions in each of the regions and thus that the overall focussing would be zero. However,this is not the case, because the ions have a higher energy in the regions "a,d" and thus are exposed for a shorter time to the defocussing field. The net effect is that the positive focussing forces always dominate and an einzel lens is always a converging element.

MAGNETIC ELEMENTS

Magnetic fields are an important element for the deflection and focussing of charged particles. The dependance of the bending power on the momentum of the ion is exploited in implanters for mass separation.

Experimentally, it is found that the force on ions within a magnetic field is at right angles to the velocity so that no acceleration of the ion takes place as it passes through a magnetic field. In addition, when a charged particle is travelling parallel to a magnetic field there is no deflecting force . A force is only found when there is a velocity component perpendicular to the magnetic field. The force on the ion is proportional to the electric charge on the ion, (q), the ion velocity, (v), and the component of the magnetic field at right angles to the velocity $(B_\perp)$.

In most implanter mass analyzers the velocity is perpendicular to the magnetic field and so the force, F, is given by

$$F = q.v.B$$ with the direction of the force at
                           right angles to the velocity.

Thus, the ions execute a circular arc in the magnetic field and suffer no change of speed; only their direction changes. In a uniform field, the ions will travel in circular orbits with a centrifugal force,

$$F = m.v^2/r$$

exactly balanced by the magnetic force on the ion:

$$F = q.v.B.$$

This balancing of forces can be used to derive relationships between the radius of the circular orbit, the magnetic field strength and the ion momentum or energy. Since the kinetic energy, $\frac{1}{2}mv^2$, equals the charge, q, multiplied by the potential drop through which the charge has been accelerated, V, we have that

$$Br = \sqrt{\left(2mV/q^2\right)}$$

This important quantity, Br, is called the <u>magnetic rigidity</u> of the particle.

Mass Analyzers and their Focal Properties

In implanter systems magnetic fields are often used to provide mass analysis. The need for this analysis stems from the fact the ions extracted from a source include a variety of unwanted particles such as BF^+ when molecular gasses are used in the source, contaminants from residual air, hydrocarbons from the vacuum pumps and contaminants from the solid components of the source. Both uniform and non-uniform fields are used in these analyzers.

When ions traverse a uniform magnetic field, each particle describes a circular orbit whose radius is proportional to the square root of the product of the mass and energy. Normally, the particles with the same charge will have the same energy so that the radius of curvature of the trajectories is proportional to the square root of the mass.

The origin of the focussing effects in the median plane can be seen from figure 9. Here it can be seen that the focussing effects are purely geometrical; upon entering the uniform field region each of the initially parallel ion trajectories becomes a circular arc and it can be see that these arcs intersect after passing through 90°.

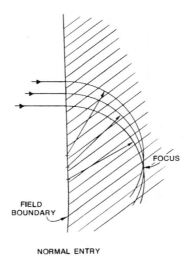

NORMAL ENTRY

Figure 9.   *Origin of the deflecting forces in the median plane.*

If the magnetic field is arranged so that the exit and entry boundaries are at right angles to the central ray, as in figure 10, it can be seen that the trajectory which is displaced a distance, d, from the central ray travels a greater distance within the magnetic field and so is bent through a larger angle than the central ray. By simple geometry it can be seen that the additional distance travelled by the outer ray in the magnetic field is

$$\Delta S = d.\sin\vartheta$$

Thus, the additional angle of deflection of this ray is just

$$\Delta\vartheta = d.\sin\vartheta/R_0$$

This leads to a focal length of

$$f = R_0/\sin\vartheta$$

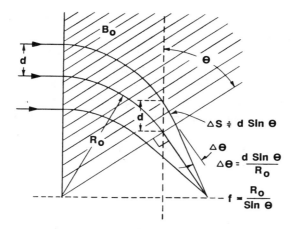

$$\Delta S \div d \sin \Theta$$

$$\Delta \Theta = \frac{d \sin \Theta}{R_0}$$

$$f = \frac{R_0}{\sin \Theta}$$

Figure 10. *Magnetic Field with exit and entry boundaries at right angles to the central ray.*

In the geometry shown in figure 10, the angle of deflection has been chosen to be 45° and it can be seen that the focal point, F, is just a distance R_0 away from the pole boundary. Thus, if we take two identical magnets and set them together back-to-back, as can be seen in figure 11, one has a 90° deflector. Particles leaving the axis at O become parallel in the "pseudo-gap" and are focussed back to the focal point at F. For a 90o magnet with poles that are normal to the central trajectory particles which start from the axis at a distance of R_0 from the pole boundary are refocussed to the axis at a distance of R_0 beyond the exit boundary.

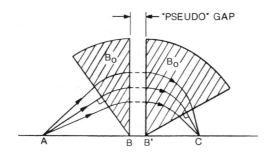

Figure 11.

It should be emphasized that the only focussing present in the geometry of the previous paragraph occurs in the bending plane which it is convenient to assume is the horizontal plane. Ions that enter the magnetic field at an angle to the median plane are not focussed because everywhere the magnetic field is normal to the median plane; in order to produce deflections toward the median plane there must be a horizontal component of the magnetic field. Frequently, this lack of vertical focussing is unacceptable because it is often troublesome to the designer to have strongly astigmatic focal systems. Also, particles that enter the magnetic field at an angle to the central ray will continue to diverge and may be lost by striking the vacuum chamber walls.

The Introduction of Vertical Focussing

In the previous section, the description of the magnet field assumed that the field was zero everywhere outside of the indicated boundaries and jumped to a constant value inside the boundaries. This assumption violates fundamental laws of physics and in reality the magnetic field bulges out well beyond the physical pole boundaries and only becomes effectively zero several pole gaps away.

Figure 12 shows a cross sectional view of the fringe field which exists at the boundary of a real magnet. It can be seen that at any point not in the median plane the fringing field has both a normal component, $B_\perp$, and a horizontal component, B_H. If the ions are travelling in the plane of the paper (so that the magnet pole boundary is at right angles to the ion velocity), the horizontal component of the magnetic field is always directed along the ion velocity and hence introduces no vertical force on the ions.

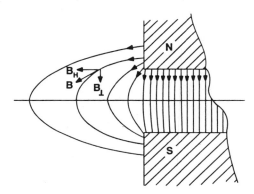

Figure 12.   A cross sectional view of the fringe field (labelled).

On the other hand if the pole edge of figure 12 is rotated so that it is no longer at right angles to the plane of the paper, the horizontal component, B_H, can be itself resolved into two components, one at right angles to the beam. B_T and one parallel to the beam, B_P. This situation is shown as a plan view in figure 13. Clearly, the vertical focussing component B_T increases as the twist angle is increased and it is easy to see that the vertical focussing can be smoothly changed from strongly positive to strongly negative as one changes the angle of the twist.

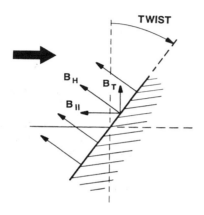

POSITIVE Y FOCUSSING

Figure 13.   Plan View of the Magnet Boundary.

The Effects of Pole Edge Rotation on Median Plane Focussing

The effects of these boundary rotation on the focussing conditions in the median plane can be seen by examining figure 14. A rotation of the pole boundary to increase the focussing strength in the vertical direction simultaneously has the effect of reducing the lens strength in the median plane. Conversely, a boundary twist which introduces a negative lens in the vertical direction will increase the lens strength in the median plane. This coupled effect is an example of a very general rule of ion optics. When one employs transverse fields for focussing, increasing the strength of focussing in one dimension automatically results in a corresponding reduction in the other.

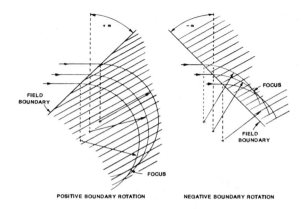

POSITIVE BOUNDARY ROTATION   NEGATIVE BOUNDARY ROTATION

Figure 14.   The effects of pole edge rotation on the median plane focussing.

It is often extremely useful to use a 90° deflection magnet with stigmatic double focussing; namely, the object and the image points are identical in both planes. It is found that this double focussing condition is satisfied when the pole-edge rotation angle, a, satisfies the relation

$$\tan \alpha = 1/2$$

$$\text{or} \quad \alpha = 26.6°$$

Under these conditions, the focal distance to the image points from the pole boundaries is just twice the radius of curvature of the central trajectory, $(2R_0)$. Figure 15 is a ray diagram of the trajectories in the vertical direction, out of the median plane. As expected the vertical focussing takes place only at the entrance and the exit boundaries.

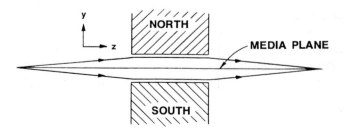

Figure 15.   Ray diagram of the trajectories in a vertical direction.

Non-Uniform Field Magnetic Analyzers and Quadrupole Lenses

Although easy to construct and align, one of the significant limitations of a double-focussing uniform field magnetic deflection system is that the focal strengths in each plane are low. As a consequence, the focal lengths in each plane are long and lead to the requirement of large magnet gaps. This in turn dictates the use of large power supplies because the power needed to drive a magnet scales with the square of the gap.

One way to avoid this problem and strengthen the focal properties of a dipole magnet is to add a non-uniform quadrupole to the normal dipole field. The quadrupole field is a powerful and versatile field configuration which is used as a beam transport component in many ion and electron beam systems. Figure 16 shows the arrangement of a simple magnetic quadrupole. The surfaces of the poles are rectangular hyperbolas which ideally approach the x and y axes asymptotically. By simple mathematics it can be shown that that the magnitude of both field components is accurately proportional to the distance from the axis. The reason quadrupole fields are so highly efficient is that, in contrast to an einzel lens, the fields are transverse to the central trajectory and thus have a maximum focussing effect on the ions.

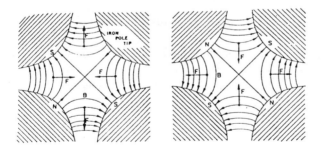

Figure 16 a b.

A positive ion that enters the lens to the right of the center in figure 16a is subject to a focussing force which converges it back towards the axis. In contrast, if the ion enters above the center of the lens the ion is subjected to a defocussing force which deflects it away from the axis. In this case the quadrupole focuses positively in the horizontal plane and negatively in the vertical plane. Clearly, if the magnetic polarities are reversed the focal properties would be reversed; positive focussing in the vertical plane and negative focussing in the horizontal.

At first sight, such lenses may not appear to be very useful because of the strong defocussing which always occurs in one plane. However, when two such lenses are arranged in series, one with positive horizontal focussing and the second with negative horizontal focussing, ions entering the first lens on the horizontal axis will be deflected closer to the axis. When these ions arrive at the second lens they are closer to the axis, the defocussing forces are less and hence the angular deflection is less in the second lens than in the first. Overall, the net deflection is towards the axis and there will be focussing. Figure 17 is two ray plots in the x and y planes for a quadrupole doublet. It can be seen that for this case the magnification is different in the two planes and that the image produced is no longer symmetrical.

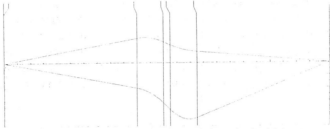

Figure 17.   Ray plots for a quadrupole doublet.

If this asymmetry is unacceptable it is possible to add an additional quadrupole element, as shown in figure 18, and convert the doublet geometry into a triplet. With this extra parameter it is possible to make the image symmetrical and to reduce the maximum size of the envelope within the quadrupole.

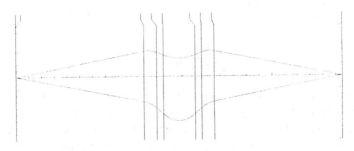

Figure 18.   Ray plots for quadrupole triplets.

Strong-Focussing Dipoles

There are several advantages in combining the properties of a uniform field dipole magnet with that of a quadrupole field distribution. Figure 19 shows the conceptual manner in which one can describe the wanted magnetic field as an addition of dipole, quadrupole and sextupole components[7].

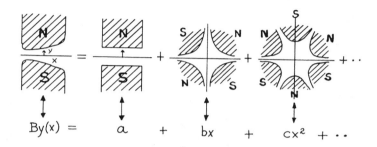

Figure 19. *Conceptual Taylor Expansion of Magnetic Fields.*

Mathematically, the magnetic profile of any magnet can be expanded in the form:

$$B_y(x) = a + b.x + c.x^2 + \cdots\cdots$$

Each of the terms in this Taylor expansion is the mathematical description of the strength of each physical entity shown in figure 19. The coefficients have the following meanings:

a   refers to the dipole strength
b   refers to the quadrupole component
c   refers to the sextupole correction component (see figure 20).

It is possible to shape the magnet poles so that the coefficients, a,b,c have the values necessary to generate any field needed by the system designer to meet his particular goals. In this way the very general focussing and bending elements can be combined into a single structure with considerable savings of space, vacuum efficiency and cost.

Using these techniques makes possible the design of magnets where the aberrations are reduced to levels that are insignificant compared to the required resolution and natural beam size. As an example, the General Ionex 90° injection magnet, whose pole profile is illustrated in figure 20, has sextupole corrections that are added by curving the entrance pole face and by adjusting the depth of the cross-cut at the center. Since the x-extent of the beam is greatest at the center of the magnet and the y-extent is uniform throughout the magnet it is possible to adjust the curvature of the poles and the cross-cut at the center to eliminate the aberrations. This particular magnet achieves a resolving power, m/Dm, of 400 with acceptance angles of ± 20 mrad and a 35.6cm radius of curvature.

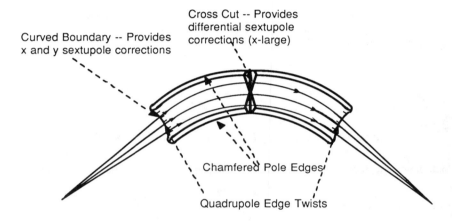

Figure 20.   The IONEX 90 degree inflection magnet.

Another example is the Eaton-Nova NV10 analysis magnet. This magnet combines all of the focussing features we have discussed and is a good example of modern-day magnet design. The magnet has a 70° bend and so incorporates a significant amount of the uniform-field focussing discussed earlier. In addition, strong quadrupole components are introduced by large boundary twists and a fairly large gradient across the magnet pole. This strong focussing makes it possible to use a rather large bending radius (r = 53.8cm) so that at 8 kgauss the magnet is capable of deflecting 80 keV Antimony ions. The entry and exit shims are curved to minimize aberrations giving the magnet a mass resolution of about 65. Rose shims parallel to the beam path and on the edges of the 5cm gap maximize the usable volume of magnetic field and minimizes the amount of steel used in the poles and in the return yoke. The strong focussing is realized with a symmetric triplet of quadrupole elements to ensure equal magnifications in both planes. To emphasize the usefulness of the strong-focussing technique, Glavish points out that the weight of the NV-10 magnet is only 2000 lbs. In contrast a uniform field bending magnet capable of achieving the same imaging would have to operate above 16 kgauss and would weigh over 4000 lbs. The power requirements would be substantially higher.

THE FOCUSSING USED IN THE IONEX IX1500 MeV IMPLANTER

Figure 21 shows the major optical elements of the Ionex IX1500 MeV implanter. The graphical output for the Ionex program OPTICIAN for the complete system is shown in figure 22. It consists of a plot of the beam envelope with the central axis of the plot reduced in scale to fit the plot. The horizontal beam profile is plotted above the axis; below the axis is plotted the vertical profile.

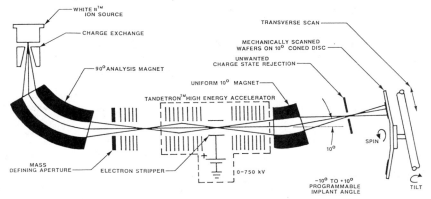

Figure 21.   Major Optical elements of the IX1500.

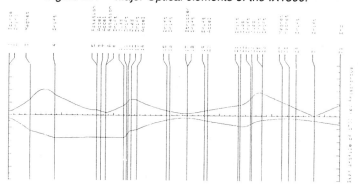

Figure 22.   OPTICIAN plot for IX1500.

In the system negative ions are generated by charge exchanging an intense positive ion beam in magnesium vapor to produce negative atomic ions. The ions are mass analyzed by a sextupole corrected magnetic analyzer which has a mass resolution of approximately 100. Although focussing is introduced in this magnet in both the median plane and in the vertical direction of this magnet it can be seen from figure 22 that the beam is only brought to a focus at the rejection aperture in the horizontal plane.

Before injection into the tandem the ions are accelerated to an energy which can be as high as 200keV by an acceleration tube whose gradient is purposely kept low at the entrance to reduce the focal effects of this element.

At 200keV the focal strength of the lens at the entrance to the accelerator is not sufficient to focus the ions through the stripper canal in the terminal and a quadrupole doublet is introduced to return the beam to a stigmatic shape and to focus the ions through the small diameter stripper. This use anamorphic focussing (different in the two planes) results in an exceptionally compact beamline despite the high resolving power. After this second charge exchange, when the ions are converted from negative polarity to positive, the ions are accelerated another time to a full energy which may be as high as 3.2MeV.

The final focussing and analysis element is a combination of a second electrostatic quadrupole doublet and a 10° magnetic deflection. This combination serves to select the wanted charge state of the ions as they leave the accelerator and also focus the ions to a suitable size for implantation.

REFERENCE LIST

1   White, N.R. NIM B21 , (1987) 339

2   Focussing of Charge Particles vol I and II. Editor Septier, A.
        (Academic Press, New York 1967)

3   The Optics of Dipole Magnets, Livingood, J.J. (Academic Press, New York 1969)

4   Brown, K.L., SLAC-75 Stanford Linear Accelerator Center (1967)

5   Glavish, H.F. Nucl. Instr. and Methods 189, (1981), 45

6   The Transport of Charged Particle Beams, Banford, M. (Spon Books Ltd, London. 1966.

7   White, N.R., Purser, K.H., and Kowalski, S.B. NIMS , A258, (1987), 437

Wafer Cooling, Faraday Design and Wafer Charging

M. E. Mack

Eaton Semiconductor Equipment Division
108 Cherry Hill Drive
Beverly, Massachusetts 01915

1. INTRODUCTION

Ion implantation was first used for device fabrication in the mid 1960's [1]. Since that time, the use of ion implantation for silicon implantation and device fabrication has grown enormously [2,3]. Key factors in the success of ion implantation are its accuracy in its spatial placement of the doping species and its precision and reproducibility in dose. The accuracy in spatial placement of the implanted material is largely due to the development of photoresist masking technology. The photoresist is thick enough that ions penetrate through the resist to the silicon only where windows are developed out of the photoresist. These windows accurately define where the implanted species is to be placed. However, photoresist is subject to damage by ion implantation [4]. Photoresist, which has been subjected to excessive temperature during implantation, will blister and flake off the wafer. At somewhat lower temperatures, flow and a change of critical dimensions in the photoresist will occur. To compound the problem, the photoresist may then be very difficult to remove. Usually, if wafer temperature is held below 100° C during implantion, no damage will occur and the resist will be easily removed.

In addition to the requirements imposed by the use of photoresist, the electrical properties of the wafer may be affected by temperature during implant. At high temperature a partial annealing of the wafer may be produced, altering the sheet resistivity of the product when complete. Beanland et. al. [5] have also reported another temperature-related factor producing a variability in wafer sheet resistivity. This latter effect arises because the dose at which the wafer surface becomes sufficiently amorphized to prevent channeling is temperature dependent. The effect is small for the more usual 7° implants into <100> silicon. Nonetheless, it is clearly prudent to control and minimize wafer temperature during implantation. Wafer cooling is discussed in Section 2.

In order to dope semiconductors with a given material the material is ionized and then accelerated to the energy required to produce the desired penetration into the semiconductor surface. Ionization of the material is necessary in order to achieve the electrostatic acceleration. The fact that the implant species is ionized allows a very simple and accurate measurement of implanted dose, namely by measurement and integration of the beam current. Electrical current easily can be measured to an accuracy of a tenth of a percent or less and, equally important, can be measured over a very large dynamic range, certainly from nanoamperes to amperes. This is fortunate, since the dose range required for successful semiconductor implantion, $10^{10}/\text{cm}^2$ to $10^{18}/\text{cm}^2$, necessitates an equally large range in implanter beam currents. The measurement of beam current is accomplished using Faraday cups. Faraday systems and dose measurement are discussed in Section 3.

While the ionic charge of the implanted species solves one problem, dose measurement, it creates another, wafer charging. Semiconductor device fabrication often entails implantation of insulating structures, such as, for example, MOS capacitors. In such structures the charge, transported by the implanted ions, will accumulate, developing a potential between the substrate and the surface structure. If a sufficient voltage is reached, the charge will arc through the insulator destroying the device. These catastrophic breakdowns produce microscopic craters, similar in appearance to volcanic craters. Of course, the result is reduced product yield. Even if a sufficient potential to generate catastrophic breakdown is not reached other subtle effects can occur. When a voltage is applied to a dielectric, even well below breakdown, a small current will pass through the dielectric due to Fowler-Nordheim tunneling [6,7]. The lifetime of the dielectric can be estimated by integrating this tunneled current [8]. When the integrated current reaches Q_{bd}, the charge to breakdown, a catastrophic failure will occur. That is to say, breakdown in the dielectric is related to passed charge rather than to a specific critical voltage. Of course, the tunneling current will increase rather rapidly with dielectric stress, giving the appearance of a critical voltage. In ion implantation the dielectric is stressed by the accumulated charge. The resulting tunneling current will reduce dielectric lifetime. Thus, if the dielectric is appreciably stressed during implantation and is subsequently tested, evidence of deterioration in the breakdown characteristics may be found even though no punch through craters are observed.

In cases where the properties of the dielectric are unaffected by implantation, wafer charging can still affect the implant. The charge buildup on the wafer surface can perturb the propagation of the ion beam being used to perform the implant [9,10]. The result is a dynamic change in beam profile as the beam sweeps across the wafer to perform the implant. This change in beam profile produces a nonuniform implantation, generally characterized by a low dose at the center of the wafer and higher dose near the edge. The effect is particularly insidious because bare monitor wafers implanted at the same time as the product wafers do not charge and, therefore, show no nonuniformity. The usual technique to prevent excess wafer charging and nonuniform implants due to charging is to flood the wafer surface with low energy electrons. In principle, this compensates for the positive charging due the beam ions and eliminates deleterious effects of charging. Wafer charging and charge neutralization are discussed in Section 4.

2. WAFER COOLING

2.1 Introduction

As noted above, effective cooling is required in order to ensure integrity of photoresist masking and consistency of wafer electrical properties. The type of cooling used and the effectiveness required depends upon the design of the machine being used. Serial machines were the first to be used commercially. These machines implant a single wafer at a time. In the effort to process higher dose wafers in a short period of time, machine beam currents were increased. Eventually, a point was reached when adequate cooling could no longer be provided. Batch machines, which spread beam power over many wafers, were developed as a strategy for coping with the higher beam currents required for high dose implants.

The physics of wafer heating and cooling are discussed in Section 2.2. Section 2.3 discusses serial machine cooling while Section 2.4 outlines batch machine strategies. In these sections serial and batch machines design parameters are related to cooling effectiveness. Section 2.5 summarizes the status of implanter wafer cooling.

2.2. Physics of Wafer Heating and Cooling

2.2.1 Heat Thermalization and Transient Surface Temperature Rise

Fig. 1 shows heat transfer during ion implantation. The ion beam striking the surface stops within a few tenths of a micron of the surface and its energy is thermalized, essentially instantaneously. For a given point on the wafer surface the exposure time will be the time required for the beam to scan a distance equal to its own dimensions. The heat so deposited then will diffuse into the wafer. If some type of conductive cooling is provided, such as elastomeric contact or gas cooling, heat will be conducted from the back of the wafer. Heat will also be radiated away from both the front and the back of the wafer. These various techniques are described in Sections 2.2.2 through 2.2.5 below. The primary interest here is the equilibration of heat through the wafer following the initial heating pulse.

For simplicity, assume that the heat is deposited only at the surface rather than through a finite depth. Of course the surface will be hotter than the bulk of the wafer. The surface temperature rise, $\Delta T_s(t)$, will be [11,12]

$$\Delta T_s(t) = 2(P_B/A_B)(t/\pi k \rho C_p)^{1/2} \qquad -(1)$$

where P_B is the beam power, k is the thermal conductivity of the wafer, A_B is the area of the beam, ρ is the wafer density, and C_p is the specific heat of the wafer. For silicon, $k \approx 1.3$ w/cm/°C, $\rho = 2.33$ gm/cm³ and $C_p = 0.75$ J/gm/°C. Here, it is assumed that the time, t, is short compared to the thermal diffusion time, τ_{TD},

$$\tau_{TD} = \rho C_p L^2 / k$$

where L is the wafer thickness. For a 500 μm thick wafer, $\tau_{TD} \approx 3$ msec. The surface temperature rise after a single scan of the beam across a given point on the wafer will be given

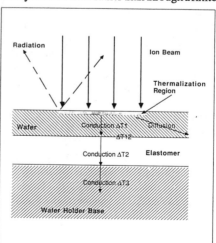

FIGURE 1. Heat Transfer During Implantation

by (1) above with $t=t_{exp}$, the exposure time. In general the exposure time is small compared to the diffusion time, so that (1) is valid. If the beam is rectangular with a height, H, and a width, W, then $A_B=HW$. Assuming the beam is to be scanned in the direction of its width, then the exposure time is proportional to the beam width. In fact, if the period of scan is T and the distance scanned is X, then $t_{exp}=TW/X$ and

$$\Delta T_{max}=2(P_B/H\sqrt{W})(T/X\pi k\rho C_p)^{1/2}. \hspace{2cm} -(3)$$

Beam powers and areas are quite different for serial and batch machines. A typical serial machine produces a beam current of roughly 1.5 mA at 200 kV for a total beam power of about 300 Watts. A typical scan time (half scan) would be about 0.5 msec and assuming a 100 mm wafer, X≈10 cm. Thus, $\Delta T_{max}\approx2/H\sqrt{W}$. Taking H=W=1 cm, $\Delta T_{max}\approx2$ °C. In a medium current machine there may be more than one scan across the same point on the wafer before the beam scans off, but the temperature rise relative to the bulk is still small, perhaps 2° to 5° C.

In the case of the batch machines, the Eaton NV-10 and NV-20 are typical. For the NV-10 at 10 mA and 160 kV (i. e. 1600 Watts of beam power) the transient wafer surface temperature rise is [12] $\Delta T_{max}\approx36/H\sqrt{W}$. Taking H=2.5 cm and W=1 cm, $\Delta T_{max}\approx15°$ C. In the NV-20 the beam current is increased to 20 mA and the beam energy to 200 kV, but, the beam height is also increased from 2.5 cm to approximately 4 cm in height. In addition, disk rotation speed is increased from 950 rpm, to 1500 rpm so that exposure time is proportionately less. Thus, for the NV-20, the transient surface temperature rise will be about 18°C. Note that, in the case of the batch machines, because the exposure time is not negligible compared to the diffusion time, [(2) above], the results here somewhat overestimate actual surface temperature rises [12].

In the case of a photoresist surface coating, a somewhat higher transient surface rise will be experienced because of the lower thermal conductivity of the photoresist. However, the photoresist coating is extremely thin, 1 μm vs 500 μm for the wafer. The thermal diffusion time for this layer is very short because of the reduced thickness of the layer. Steady state will be reached during the scan for the photoresist so that the temperature gradient through the resist will be approximately linear. Worst case photoresist surface temperature rise can be estimated [11] as

$$\Delta T_{max} \approx (P_B/A_B)(L_{PR}/k_{PR}) \hspace{2cm} -(4)$$

where L_{PR} and k_{PR} are the thickness and thermal conductivity of the photoresist respectively. Taking $L_{PR}=1$ μm, $k_{PR}\approx2\times10^{-3}$ w/cm/°C, $P_B=1600$ W and $A_B=3$ cm^2 gives $\Delta T_{max}\approx25°$C. Thus, the surface temperature rise is slightly larger with the photoresist but not greatly larger.

Once the scan pulse passes the point of interest on the surface, heat will diffuse into the wafer and the surface temperature will drop. After many pulses, steady state will be reached. The following sections are primarily concerned with determination of the steady state temperature rise of the wafer. To determine the temperature rise of the surface of the wafer the transient rise must be added to the steady state rise. In the sections below, for simplicity, the bulk of the wafer is assumed to be isothermal. Ignoring the superimposed transient temperature rise, this is an excellent approximation. With backside cooling, the more effective the cooling, the larger the gradient through the wafer. For conductive cooling, for example, the efficiency of cooling is denoted by the cooling coefficient, h, which typically is of the order of 20 mW/cm^2/°C. The corresponding coefficient for conduction through the wafer is k/L = 26,000 mW/cm^2/°C. Thus, the steady state temperature gradient through the wafer is less than 0.1% of the drop across the cooling interface.

2.2.2 Radiative Cooling

Radiative cooling was the first technique used for wafer cooling. It is also the least effective. The case of radiative cooling has been analyzed in detail by Parry [13] assuming, as discussed above, that the wafer is an isotherm. The heat flow equation derived by Parry is:

$$A_B L \rho C_p (dT_w/dt) = P_B(t) - H(T_w - T_{wH}) - \alpha(T_w^4 - T_s^4) - \beta(T_w^4 - T_{wH}^4) \qquad -(5)$$

where T_w is the bulk wafer temperature, T_{wH} the temperature of the wafer holder and T_s is the temperature of the surrounding chamber in which the implant is performed. The coefficient, H, represents conducted heat loss to the wafer holder. In terms of the wafer emissivity, ε_w, the wafer holder emissivity, ε_{wH}, and the surrounding chamber emissivity, ε_s, the factors a and b are:

$$\alpha = \sigma A_B(\varepsilon_w \varepsilon_s)/(\varepsilon_w + \varepsilon_s - \varepsilon_w \varepsilon_s) \qquad -(6a)$$

$$\beta = \sigma A_B(\varepsilon_w \varepsilon_{wH})/(\varepsilon_w + \varepsilon_{wH} - \varepsilon_w \varepsilon_{wH}) \qquad -(6b)$$

Here σ is the Stefan-Boltzman constant, $\sigma = 5.67 \times 10^{-12}$ w/cm^2/°C^4.

In the case that radiative cooling dominates over conduction, a simple solution of the heating equation (5) is not possible. Instead, computer solution of (5) is necessary. The salient features of implanter operation in this regime have been described in the literature [13,14,15]. Fig.2 shows the temperature rise calculated by Glawischnig [15] for a Ferris wheel type batch implanter at 300 Watts of beam power. In this implanter, wafers are mounted to a rotating cylindrical holder and the beam is slowly scanned lengthwise down the cylinder. Fig. 2 shows that each time the wafer is rotated through the beam wafer temperature increases perhaps 15°C. After 7 rotations the beam has scanned off the wafer and the wafer cools radiatively until the slow scan returns again. The peak temperature continues to grow until about the 4th slow scan cycle, when the heating due to the fast scan cycles is approximately balanced by the cooling during the slow scan cycle. By current standards the peak temperature rise reached, slightly over 200° C with 300 W of beam power, is high. This is due primarily to the relatively poor heat transfer which can be achieved by radiative cooling.

Smith [4] has determined the limiting temperature for wafers cooled by radiation alone. This "limiting temperature" is the maximum temperature that would be reached in a very long implant and is shown in Fig. 3 as a function of effective scanned beam power per unit area. The effective

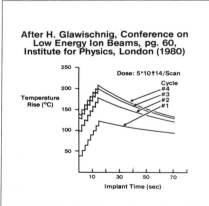

FIGURE 2. Calculated Temperature in Ferris Wheel Type Implanter

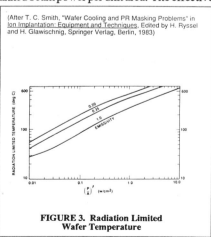

FIGURE 3. Radiation Limited Wafer Temperature

beam power per unit area is the beam power divided by the total area scanned by the beam. The functional dependence of wafer temperature rise on power per scanned area is also shown to be the case for conductive cooling as discussed in Section 2.2.3 below. The fact that the temperature rise is inversely proportional to the total scanned area quantifies what is already intuitive, namely, that if the beam power is distributed over a number of wafers and a proportionately larger area, the wafer temperature rise is proportionately reduced. Of course, this is the principle of operation of batch implanters.

From Fig. 3 it is seen that, because of the T^4 dependence of cooling, wafer temperature rise is not linear with beam power. Thus, there is no simple constant cooling coefficient, h, as there is in conductive cooling. Nonetheless, it is instructive to divide the temperature rise by the scanned power density to determine an effective cooling coefficient. Taking an emissivity of 0.35 and $(P/A)'$=0.5 w/cm², a value typical of many batch machines, h'≈2 mW/cm²/°C. (Note that Smith takes the emissivities of the target holder and the chamber to be equal, so that $\alpha=\beta=\varepsilon\sigma A_B$, where ε is the net emissivity). At $(P/A)'$=3 w/cm², typical of many serial machines, h'≈5 mW/cm²/°C. As noted earlier, h≈20 mW/cm²/°C. is more typical of the conductively- cooled schemes used in batch and serial implanters. It is for this reason that conductive cooling generally has replaced radiative cooling in most modern implanters.

2.2.3 Conductive Cooling

Conductive cooling in a vacuum is not easily achieved. If a wafer is pressed against a metal surface in vacuum, the cooling across the interface is not significantly increased over radiational cooling. Two techniques have evolved which are widely used to effect contact cooling in the vacuum. These are elastomeric contact and gas cooling. Elastomeric cooling is widely used for batch implantation because of its simplicity. Elastomeric cooling is easily implemented on a moving mechanical stage and yet is effective enough to maintain wafer temperature below 100° C. in a batch end station. Gas cooling can be somewhat more effective but is difficult to implement in a moving wafer holder. Because of its greater effectiveness, gas cooling is widely used in serial implanters.

Fig. 4 shows a typical implant geometry. The ion beam is scanned in two dimensions across a wafer. In a high current machine the ratio of the fast to slow scan speeds would be a factor of 100 or more to ensure good beam overlap on successive disk rotations so that the doping is uniform [16]. In medium current machines the ratio is much lower and may even be equal in the two directions but scan patterns are more complex and interleaved in order to ensure good scan overlap and, thus, good uniformity [16,17,18]. The power loading at any given point on the wafer surface is pulsating as shown schematically in Fig. 5. Here, t_1 is the time duration of a single fast-scan exposure, t_2 is the time between fast scans assuming more than one pass over a given point on the wafer and t_3 is the time between slow scans. The temperature of the bulk of the wafer can be

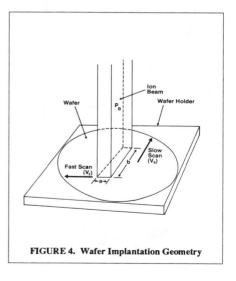

FIGURE 4. Wafer Implantation Geometry

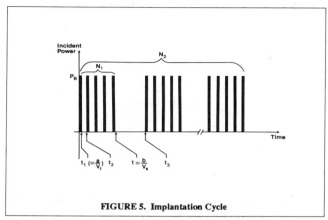

FIGURE 5. Implantation Cycle

determined by solving equation (5) subject to the pulsating power load of Fig. 5. On each pass of the beam over the wafer, the wafer will rise in temperature. Between scans temperature will decay exponentially. For conduction dominated cooling the decay time is

$$\tau = \rho C_p L/h \qquad -(7)$$

where $h = H/A_B$ is the cooling coefficient in watts per cm² per °C. If the characteristic times, t_1, t_2, and t_3, are all very short compared to the cooling time, τ, and if the total implant time is very long compared to τ, then the final temperature rise takes on a very simple form [11],

$$\Delta T_\infty = (P_B/As^*)/h \qquad -(8)$$

where As* is the total scanned area exclusive of the overscan area. The conditions assumed in equation (8) are only approximately valid for many implants and implanters. Nevertheless, (8) is useful for estimating wafer temperature and in many cases correctly gives wafer temperature. Equation (8), as noted earlier, demonstrates why batch processing lowers wafer temperatures. Typical scan areas in serial machines are in the range of a 100 to, perhaps, 400 cm² while the scan areas for batch machines range from 2000 to 4000 cm² or more. Thus, with a beam power of 1500 Watts, and a scanned area of 2000 cm² a batch machine requires $h \geq 8$ mW/cm²/°C to ensure a bulk temperature rise less than 100° C. On the other hand, a serial machine with a beam power of 300 Watts and a scanned area of 150 cm² requires $h \geq 20$ mW/cm²/°C for the same condition.

2.2.4 Elastomeric Cooling

Fig. 6 shows elastomeric cooling. The major difficulty with achieving conductive cooling in a vacuum is that the backside of the wafer is microscopically rough. Consequently, the actual contact area is small if the wafer is simply pressed against a metal surface. Instead an elastomeric covering is used and the wafer is pressed against the elastomer. The elastomer yields and conforms to the microscopic backside structure allowing a much larger contact area than would be the case for a bare metal surface. These conditions have been quantified by Benveniste [12], who has shown that the cooling coefficient, h, is approximately proportional to the ratio kP/M where k is thermal conductivity

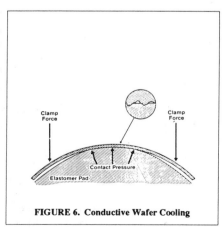

FIGURE 6. Conductive Wafer Cooling

of the elastomer, P is the contact pressure between the wafer and the elastomer and M is the microhardness of the elastomer. The actual design of the elastomeric covering, usually an RTV Silicon compound, is not simple. Additives such as aluminum increase the thermal conductivity of the RTV but also increase the microhardness.

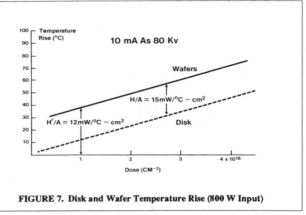

FIGURE 7. Disk and Wafer Temperature Rise (800 W Input)

Optimum thickness must be determined empirically.

The proportionality to contact pressure is important to note. Pressing a wafer onto a flat surface develops a contact pressure only at the contact ring. The portion of the wafer exposed to the beam would then have no contact pressure and, therefore, little or no conductive cooling. If a wafer is subjected to a uniform backside pressure, P, the wafer will form a convex shape. To achieve a uniform contact pressure, P, with a surface for conductive cooling, the wafer must be pressed against a surface having exactly this same convex shape. A circular disk clamped by a ring at the edges and subjected to a uniform backside pressure, P, undergoes a deflection [19]

$$y(r) = \{3PR^2(m^2-1)/8Em^2t^3\}\{ R^2[(5m+1)/(2m+2)]+r^4[1/(2R^2)]$$
$$-r^2[(3m+1)/(m+1)]\} \qquad \text{-(9)}$$

where r is the distance from the center of the disk, R is the radius at which the wafer is "clamped" (edges free), m is the inverse of the Poisson ratio and for silicon is 2.4, E is the elastic modulus which for silicon is roughly 1.5×10^7 PSI, and t is the thickness of the disk (wafer). If a uniform contact pressure is necessary then the pad on which the wafer is clamped must have exactly the shape given in equation (9). Equation (9) is not rigorously correct for a crystalline material such as silicon and, usually, center deflection is measured empirically by hydrostatically loading a wafer. Then the shape of the pad is

$$y(r) = y(0)\{1+[(2m+2)/(5m+1)R^2]\{r^4[1/(2R^2)]-r^2[(3m+1)/(m+1)]\}\} \qquad \text{-(10)}$$

where y(0) is the empirically measured center deflection. For elastomeric cooling, uniform contact pressure is not essential so that it is common to use a spherical approximation to (10).

Elastomeric cooling of the type described above typically uses a contact pressure of 0.3 to 1.5 PSI. For a 100 mm wafer, the center deflection would be of the order of 0.4 to 0.7 mm. Note that for such pads as Fig. 6 illustrates, the wafer is curved so that implant angle is not constant. The implant angle variation due to the pad curvature is of the order of 1° to 3°. Depending on the implant angle and other details of machine design, this can produce a variation in the amount of channelling, which occurs in the implant, and, thus, could give the appearance of a dose non-uniformity [20]. Fig. 7 shows the performance of typical elastomeric cooling pads in a batch machine, the Eaton NV-10. In this machine the disk is not internally water cooled but instead is cooled between implants. Thus, the disk also rises in temperature for larger doses. The cooling coefficient achieved by the RTV alone is h = 15 mW/cm²/°C whereas the effective cooling coefficient at a dose of 1x10¹⁶ including the disk rise is h'= 12 mW/cm²/°C⁻¹.

In batch machines the wafers are typically implanted on a spinning disk. It is then possible to use the normal component of centrifugal force to provide the contact pressure [21,22]. In such instances the pad can be flat so that implant angle need not show a variation. In addition there are no die shadowed by the clamp, reduced sputtering from the clamp and no particulates generated by the clamping. However, the contact pressure achieveable by centrifugal force is much reduced over that achieveable with an actual clamp. Thus, the cooling coefficient is reduced so that the scan area must be increased to maintain wafer temperature. An exception is the Eaton clampless cooling [21] which augments centrifugal force with a surface tension force [23] to achieve cooling coefficients as high as 40 mW/cm²/°C.

2.2.5 Gas Cooling

Gas cooling is another form of conduction cooling. In order to use gas cooling, the wafer is sealed by an O-ring against a platen with provisions for introducing a gas behind the wafer. The gas provides a thermally-conducting path between the wafer and the cooled platen. Note that it is not necessary to flow the gas. Moreover, flowing the gas will not measureably increase the cooling, at least at the usual operating pressures. To minimize leakage of gas past the wafer and to minimize stress on the wafer, a relatively low cooling gas pressure must be used.

The use of gas cooling for ion implantation has been discussed by King and Rose [24]. Two regimes must be distinguished, the viscous flow regime, where the mean free path in the gas is less than the gap between the wafer and the platen and the molecular flow regime, where the gap is small compared to the mean free path [25]. In the viscous regime the heat conductivity, like the viscosity, is independent of pressure. The molecular flow regime is the more important for gas cooling in ion implantation. Heat conduction in this regime has been analyzed by Knudsen [26]. As long as the gap is small compared to the mean free path, heat conduction is independent of the gap and

$$h = \alpha \Lambda P \qquad\qquad\qquad -(11)$$

where α is the accomodation coefficient and Λ is the free molecular heat conductivity. The accomodation coefficient essentially measures to what extent molecules striking a wall and reflecting from the wall adjust their kinetic energy to that corresponding to the temperature of the wall. Completely roughened surfaces have $\alpha \approx 1$ while highly polished surfaces have $\alpha << 1$. Table 1 lists mean free path, the molecular conductivity, Λ, and the high pressure thermal conductivity, k. At high pressure the conduction is independent of pressure but inversely proportional to gap. Lighter gases are seen to be superior for gas cooling, reflecting the fact that molecular conductivity is inversely proportional to the square root of atomic mass [11,26].

Maintaining a gap less than the mean free path requires that the surface be convex in much the same way as for elastomeric cooling. Thus,

Gas	Mean Free Path†	Λ ‡	k˙
H_2	93	60.7	1.62
He	147	29.4	1.35
N_2	46	16.6	0.22
O_2	53	15.6	0.26
H_2O	30	26.5	0.14
Ar	52	9.3	0.16
CO_2	30	16.7	0.14

† μm-Torr
‡ mW/cm²/°C Torr-1
˙ mW/cm²/°C

TABLE 1

for example, if operation at 5 Torr is required then the convex pad must have the shape given by (9) with P=5 Torr. At pressures higher than 5 Torr the wafer would then move away from the convex pad opening the gap. When the gap becomes larger than the

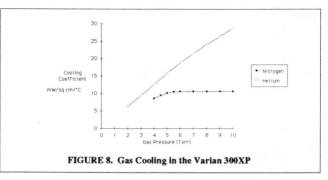

FIGURE 8. Gas Cooling in the Varian 300XP

mean free path the cooling becomes viscous, and the cooling coefficient initially becomes constant and then decreases. In the case of gas cooling, using a spherical approximation to the correct pad shape, equations (9) and (10), is not acceptable because the difference between the correct shape and the spherical approximation is comparable to the mean free path. At 5 Torr, for example, mean free path is only about 20 μm for most of the gases of interest.

Fig. 8 shows gas cooling in the Varian 300XP [27]. With nitrogen the cooling coefficient increases with pressure and saturates. Because of its relatively small mean free path, the nitrogen cooling becomes viscous dominated shortly after reaching the contact pressure. Helium has a much longer mean free path and consequently, does not saturate for the range of pressures shown. Note that the effective accomodation coefficient is relatively low, on the order of α=0.1.

As in the case of elastomeric cooling the curvature of the platen results in a variation in implant angle across the wafer. Consequently, undesireable channelling may result. The use of helium or hydrogen minimizes the required contact pressure and therefore, minimizes the angular variation across the wafer. Well designed gas cooling systems can achieve cooling coefficients of order of 25 mW/cm²/°C with a platen angle variation only about 1°. Cooling coefficients more than twice this can be achieved with higher angle variations [28].

2.3 Serial Machine Cooling

Most serial machines are gas cooled, the reason being that the average power per unit scanned area is high, between 2 and 5 Watts/cm² and thus, the highest possible cooling coefficient is required. The majority of the commercial serial implanters use electrostatic scanning in both scan directions. The scan time, t_1, (section 2.2.3) and the time between scans, t_2, are both short compared to the wafer cooling time, τ, of equation (7). In this case the beam can be treated as a uniform beam covering the entire scan area and

$$\Delta T(t)=(P_B/As\ h)(1-\exp(-t/\tau))\qquad -(12)$$

For long implant times this reduces to the equation (8). Dose and time are related according to

$$t=1.6\times10^{-19}DAs/I_B\qquad -(13)$$

where D is the dose, and I_B is the beam current. Combined with (12) this gives

$$\Delta T(t)=(P_B/As\ h)(1-\exp(-1.6\times10^{-19}DAs/I_B\tau))\qquad -(14)$$

Taking the case of 625 μm thick wafers and h=25 mW/cm²/°C, the cooling time constant is t=4.4 seconds. Assuming a scan area of 160 cm², which would be typical for 125 mm wafers, and a beam power of 300 W, the limiting temperature rise (equation 8) would be about 75° C.

However, for many implants, because of the heat capacity of the wafer, itself, the actual temperature rise is less than the limiting temperature. If beam current is 1.5 mA, temperature rise would be less than half the limiting value for doses below $2 \times 10^{14} cm^{-2}$. In practice, to ensure optimum uniformity, one would probably reduce beam current and extend implant time rather than attempt implants shorter than 5 seconds, so actual temperature rise would vary between 68% (5 second implants) and 100% of the limiting temperature.

In at least one case, the recently developed ASM medium current machine, the scanning is not electrostatic in both dimensions but is mechanical in one dimension and electrostatic/ magnetic in the other. In this case the analysis above is not valid because the time between scans is no longer negligible compared to the cooling time. The procedure outlined below for batch machines is more appropriate in this case. Temperature will be higher than that given in (12) and, as explained below, will be a minimum in the center of the wafer and higher near the mechanical scan turn around.

2.4 Batch Machine Cooling

Fig. 9 illustrates long term heating in the case of a batch implanter. Scanning in the fast scan direction results in a heated stripe which then is scanned across the wafer. Just past the edges of the wafer (top and bottom in Fig. 9), the beam will turn around and restart the scan. Near the edges of the wafer the silicon experiences back to back heating pulses with little time to cool. On the other hand, in the center of the wafer cooling time between heating pulses is a maximum. As a result it would be expected that the wafer edges should exhibit a higher temperature than the center and in fact, this is the case (see also reference [12]). An analytical solution of (5) is possible at the wafer edge if cooling during the scan turn around time is neglected and if lateral heat diffusion in the wafer is ignored. The results are rather complicated and, are not reproduced here. As noted by Wauk [22] wafer temperature rise is inversely proportional to disk radius and, of course, directly proportional to beam power. Dependencies on beam height, and slow scan speed are more complicated. In fact, slow scan speed is quite critical. Fig. 10 shows results of the analysis for the Eaton NV10. Here the scanned beam area is about 2400 cm² and the beam power is 1600 Watts. The effect of scan speed is very important. In the case of a cooling coefficient of 15 mW/cm²/°C⁻¹, typical for clamped elastomeric cooling, decreasing the slow speed scan time from 30 seconds to 6 seconds decreases the wafer temperature rise by about 50° C. The impact is less with better cooling, but still important. With clampless cooling pads having a cooling coefficient of 40 mW/cm²/°C, the corresponding temperature decrease is 35° C. Comparing Fig. 10 with the result of equation (8) shows, as would be expected, that the temperature approaches that predicted by (8) as the scan speed is increased. Thus, for 40 mW/

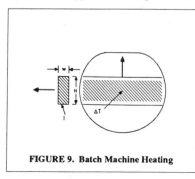

FIGURE 9. Batch Machine Heating

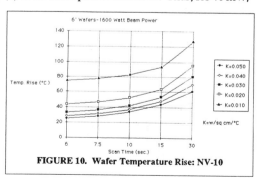

FIGURE 10. Wafer Temperature Rise: NV-10

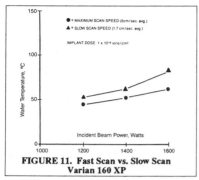

**FIGURE 11. Fast Scan vs. Slow Scan
Varian 160 XP**

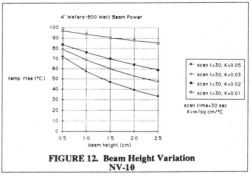

**FIGURE 12. Beam Height Variation
NV-10**

cm²/°C, equation (8) predicts a temperature rise of 18° C vs about 28° C at the fastest scan time of Fig 10. The results of Fig. 10 are analytic but have been confirmed experimentally. Fig. 11 shows similar results for the Varian 160XP. For comparison the scan times of Fig. 10 correspond to a scan velocity range of from about 1.3 cm/sec to 6.5 cm/sec.

Fig. 12 shows the effect of beam height variation in the Eaton NV10. (From Fig. 9 it is clear that beam width plays no role except for the transient surface temperature rise-Section 2.2.1.) It must be noted that two effects play roles in determining the variation of wafer temperature with beam height. Increasing beam height decreases the heat flux in the moving stripe, but also increases exposure time for any point on the wafer. In the case of the NV-10 ensuring that beam height exceeds 1.5 to 2 cm does reduce temperature.

One factor which is different among various batch machines is whether the implant disk is water cooled or not. Water cooling eliminates any wafer holder temperature rise during the implant but complicates disk interchange. In the Eaton NV-10, for example, the implant disk is not water cooled. Instead, when the disk is exchanged and unloaded, it is cooled down in preparation for the next implant. This reduces the cooling capability slightly but allows ready interchange of disks so that, if desired, disks can be dedicated to species, and implant angle or size can be changed quickly. The heat capacity of the disk and holders limits disk temperature rise as noted in Fig. 7 to about 10° C for a 1x10¹⁶cm⁻² implant at 800W. Other systems such as the Eaton NV-20, the Varian 160XP and the AMT Precision 9000, have internally water cooled disks to eliminate holder temperature rise and, in fact, do require additional time for disk interchange. Both approaches are workable and under the conditions for which they are designed, both effectively cool wafers.

2.5 Conclusions

The results given in sections 2.3 and 2.4 above give the bulk temperature rise of the wafer. To this must be added the transient surface temperature rise of section 2.2.1. In addition, any temperature rise in the wafer holder, itself, must be added. However, if it is intended that photoresist temperature be held below 100° C, the initial holder temperature must be added to the temperature rise to give final wafer surface temperature. It is not uncommon for initial holder starting temperatures to be as high as 35° C or more. High starting temperatures place greater demands on the wafer cooling capability. End station chillers or low temperature machine cooling can be used to help in this regard. Properly maintained and used, modern implanters do allow implantation into positive photoresist over a wide range of implant conditions without undesireable damage to the resist.

3. FARADAY SYSTEMS AND DOSE CONTROL

3.1 Introduction

As noted in Section 1, the fact that dosing is accomplished using ions permits dose measurement by measuring beam current. Beam current in an ion implanter is sensed using a Faraday cup [29, 30]. Simplisticly, it might be assumed that to measure beam current one need only insert a target in the beam as shown in Fig. 13. However, energetic ions impinging on a target generate secondary electrons and ions [30, 31]. The most copious secondaries leaving such a target would be electrons. The metering thus measures two currents, the impinging ion beam and the escaping secondary electrons. The two currents add giving the effect of a significantly larger ionic current than is actually present. In addition external sources of charged particles other than the implanted ions exist in the implanter. Electron flooding (or showering) to minimize wafer charge buildup (Section 4) is common. It is important that these particles be excluded from striking the measurement target. Two techniques have been developed to prevent the escape of secondaries and to prevent unwanted particles from being measured. Both involve separation of the sensing target from the implanter beamline by an electric or magnetic field, which prevents the transmission of particles other than the ion beam. Electrostatic Faraday suppression is discussed in Section 3.2, while magnetic suppression is discussed in 3.3. There are a number of factors beside suppression which can affect the dose measurement [29,31,32]. Pressure is probably the greatest concern and is discussed in Section 3.4.

Having designed a Faraday, it must then be set up for dose control. Two different philosophies have evolved for use of Faraday systems. For want of better terms these are referred to here as wafer internal and wafer external Faradays. In wafer internal Faraday systems the wafer and its holder are mounted within the Faraday cup, itself. As the name implies external Faraday systems incorporate dose measurement outside and away from the wafer holder. There are advantages and disadvantages to each, but with proper implementation, both are effective and reliable. Faraday implementations are discussed in Section 3.5. The use of Faraday systems for dose control is discussed in Section 3.6.

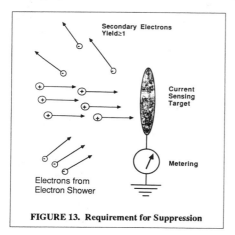

FIGURE 13. Requirement for Suppression

3.2 Electrostatically Suppressed Faradays.

Fig. 14 shows a complete electrostatically suppressed Faraday system. The flat target of Fig.13 is replaced with a closed cylindrical chamber. In front of the sensing cylinder is placed another cylinder biased at a negative voltage, typically of the order of 1000 V. This bias cylinder is intended to return secondary electrons ejected from the inside of the sensing cylinder back into the sensing cylinder. In addition, the negative bias prevents externally generated electrons from entering the Faraday. A grounded defining aperture completes the system. The defining aperture terminates the electric field from the bias ring so that the bias field will not affect upstream beam propagation. In addition, the aperture assures that stray beam components will not strike the walls of the Faraday system.

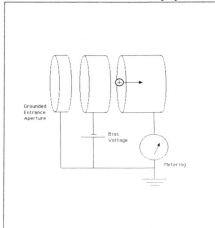

FIGURE 14. A Complete Electrostatic Faraday

Fig. 14 demonstrates another important property of the Faraday cup. An entering positive charge is actually sensed before the charge strikes the rear wall of the sensing cylinder. The entering charge induces a charge on the cylinder as it enters. That is to say, the positive charge of the entering ion attracts electrons to the sensing cylinder from ground. The electron current measured by the metering system is exactly the same as the ionic current. When the ion strikes the rear surface of the sense cylinder it simply neutralizes the electrons already present. It is for this reason that Faraday systems still function even if the target struck is insulating. This is often the case where insulating wafers are implanted within a wafer internal Faraday system.

The suppressing field provided by the bias electrode must penetrate to the center of the bias electrode. A general rule of thumb design has been provided by Jamba [30]. However, at high beam currents the potential of the beam itself may shield the suppression field. The latter effect has been quantified by McKenna [31,33] through computer modeling, which includes not only the Faraday electrodes but also the field of the beam, itself. Fig. 15 shows the Faraday system, which McKenna modeled, while Fig. 16 shows the equipotential distribution for the case of negligible beam current. The dimensioning on the two axes is in arbitrary (computer) units whereas the equipotential curves are labeled by voltage with respect to ground. The potential distribution shows that the bias scheme is adequate to contain secondary electrons up to electron energies of about 100 eV. This is sufficient, since secondary electron energies are quite low, generally 1 to 10 eV [33]. Fig. 17 shows the modification of the potential field by the presence of a 1 cm diameter, 1 mA arsenic beam. On axis bias potential has dropped from 100 V to nearly 0 volts. To emphasize the problem Fig. 18 shows the trajectories of 100 eV secondary electrons ejected from the back of the Faraday at angles of 0 and 45°. While this is an extreme case, it does emphasize that, under such circumstances, the beam can serve as a conduit to allow secondary electrons to escape from the Faraday system. Modern electrostatically suppressed Faraday systems [34] are designed by techniques similar to that developed by Mc Kenna.

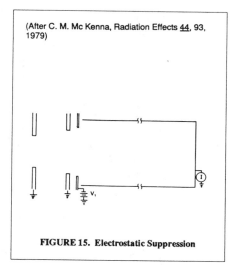

(After C. M. Mc Kenna, Radiation Effects 44, 93, 1979)

FIGURE 15. Electrostatic Suppression

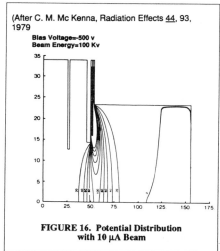

(After C. M. Mc Kenna, Radiation Effects 44, 93, 1979)

Bias Voltage=-500 v
Beam Energy=100 Kv

**FIGURE 16. Potential Distribution
with 10 μA Beam**

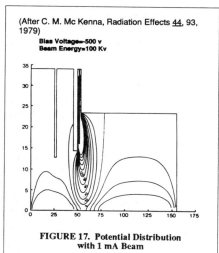

(After C. M. Mc Kenna, Radiation Effects 44, 93, 1979)

Bias Voltage=-500 v
Beam Energy=100 Kv

**FIGURE 17. Potential Distribution
with 1 mA Beam**

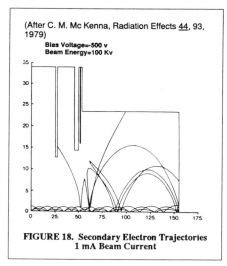

(After C. M. Mc Kenna, Radiation Effects 44, 93, 1979)

Bias Voltage=-500 v
Beam Energy=100 Kv

**FIGURE 18. Secondary Electron Trajectories
1 mA Beam Current**

3.3 Magnetically Suppressed Faraday Systems

Electrostatic suppression systems require power supplies operating at relatively high voltage and are somewhat complex. Insulator leakage currents can adversely affect dose measurement accuracy [35], although in carefully designed systems this difficulty can be avoided (see 3.5 below). A somewhat simpler alternative is magnetic Faraday suppression [30,36]. Fig. 19 shows schematically a magnetically suppressed Faraday. Here the magnetic field acts as a barrier to prevent either the escape of secondaries or the entrance of externally generated

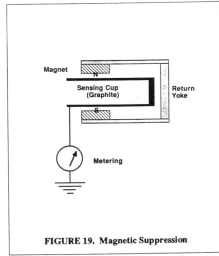

FIGURE 19. Magnetic Suppression

charged particles. Fig. 20, for example, shows an externally generated 100 eV electron reflected by the magnetic field near the entrance of the NV-10 magnetic Faraday. Note that magnetic suppression is effective for both positive and negative secondary particles.

Magnetically suppressed Faradays can suffer a problem similar to the beam potential shielding encountered in electrostatically suppressed Faradays. In Fig. 19 the magnetic field is vertical. A high current beam entering the Faraday will produce an electric field which has a horizontal component out of the plane of the figure. The orthogonal electric and magnetic fields produce forces, which oppose each other. In the case of a uniform magnetic field, at one velocity, v=E/B, the forces cancel and the particle will move in a straight line. This is the principle of operation of an E x B velocity filter. Thus, if secondaries are ejected from the inside of such a Faraday, electrons having a velocity near E/B could escape and, similarly, charged particles could enter from the outside. Of course, the electric field from the beam is not

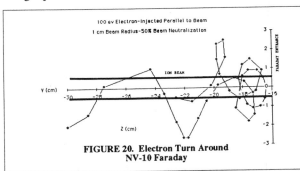

FIGURE 20. Electron Turn Around NV-10 Faraday

constant but depends on position relative to the beam. Computer modeling shows that the E x B condition must be taken into account, just as beam potential shielding must in electrostatic Faradays. In magnetically suppressed Faradays the solution to the E x B problem is to tailor the magnetic field so that

ejected secondaries and entering particles cannot maintain a constant or near constant E/B condition. As in the case of electrostatically suppressed Faradays, computer design is the key to successful operation.

Properly designed, magnetically suppressed Faradays, in fact, are as effective or more effective than electrostatically suppressed Faradays and are simpler.

3.4 Pressure Effects in Faraday Systems

One of the major pitfalls in Faraday design is the effect of ambient pressure [36]. Collisions between entering beam ions, A^+, and residual gas atoms, X, generate additional charged particles which must be managed properly to accurately measure ion beam current. Three types of collision reactions have high probability,

$$A^+ + X \longrightarrow A^+ + X^+ + e^- \quad \text{(Ionization)} \qquad -(15a)$$

$$A^+ + X \longrightarrow A + X^+ \quad \text{(Charge Exchange)} \qquad -(15b)$$

$$A^+ + X \longrightarrow A^{++} + X + e^- \quad \text{(Stripping)} \qquad -(15c)$$

The probabilities of these are roughly in the order given. It is important to note that Faraday systems are likely to be close to the wafers under implant so that pressure bursting to pressures of 10^{-4} Torr or more may be encountered. (This is certainly true for wafer internal Faraday systems.)

Fig. 21 shows measured beam current as a function of pressure for the Faraday shown in Fig. 14. Initial beam current at the start of the experiment had been 10 mA [36]. A rather pronounced drop in measured beam current with increasing pressure is observed. This Faraday would overdose the wafers at high implant pressure. To the current integrator, the current appears to have decreased and therefore, the integrator will increase the implant time. The explanation for the drop off is shown in Fig. 22. The beam in the Faraday system ionizes residual gas {(15a) above}. Electrons generated during the ionization are driven into the sensing cup, but the ions are pulled to the bias electrode and are not sensed. The path length within the Faraday used for the test was 15 cm. For an ionization cross section of $\approx 2 \times 10^{-15}$ cm^2 a signal reduction of about 10% would be expected with a pressure increase of 10^{-4} Torr. This is in reasonably good agreement with the observed result. Ryding et. al. [36] examined a number of Faraday configurations and demonstrated that two Faraday designs give pressure independent results, the magnetically suppressed Faraday of Fig. 19 and the doubly-biased electrostatic Faraday of Fig. 23. The remaining drop in measured current with pressure represents charge exchange (15b) in the implanter beamline rather than an actual Faraday effect.

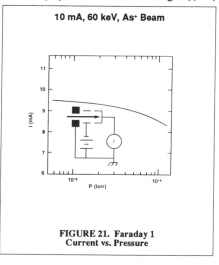

FIGURE 21. Faraday 1
Current vs. Pressure

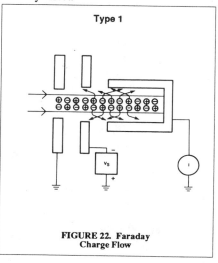

FIGURE 22. Faraday
Charge Flow

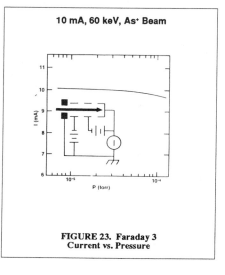

FIGURE 23. Faraday 3
Current vs. Pressure

Not all commercial implanters use pressure neutral Faradays. However, pressure effects in the dose may not be apparent. For example, if the bias supply in Fig. 14 is connected so the bias current flows through the metering system in an effort to improve the pressure response, it is found that the Faraday gives higher current readings at higher pressures [36]. In contrast to the Fig. 14 Faraday, this latter Faraday would then underdose at high pressures (i.e. the measured current is greater so that the dose controller will decrease the implant time). However, if boron, which has a high charge exchange cross section, is implanted, charge exchange in the beamline can offset the inherent pressure dependence of the Faraday and give a result seemingly independent of pressure. Because of their smaller charge exchange cross sections, arsenic and phosphorous would show the high pressure underdosing inherent to the Faraday, itself. This example underscores the interdependency of the Faraday design and the remainder of the implanter in determining successful dose control.

3.5 Faraday Implementations

3.5.1 Introduction

As noted in the introduction on Faraday systems two schemes have evolved regarding the implementation of Faradays. These are wafer internal Faradays, where the wafer and its holder are contained within the Faraday cup, itself, and the wafer external Faradays, where the Faraday is at some distance from the wafer and its holder. Wafer external Faradays are discussed in Section 3.5.2 while wafer internal systems are discussed in 3.5.3. Each type of Faraday system purports advantages and each has disadvantages. The two system types are compared in Section 3.5.4.

3.5.2. Wafer External Faraday Systems

The Eaton high current implanters such as the NV-10 and NV-20 are good examples of external Faraday systems. With magnetic suppression, the Faradays are simple, reliable and pressure neutral. However, it is instructive to examine an electrostatically supressed wafer external Faraday system. The new Varian 300XP serial implanter is a good example [37]. The beam line arrangement is shown in Fig. 24. The quad aperture Faraday has 4 Faraday cups located in the overscan region of the beam. The wafers and the electron flood for charge neutralization are both located well behind the Faraday system. The separation of the Faraday from the electron shower and the suppression provided in the Faraday both ensure that electrons from the flood gun do not enter the Faraday and affect dose. The fact that the wafers are not part of the Faraday system allows a relatively simple, grounded wafer platen.

Fig. 25 shows a cross section of one of the corner cups of the Faraday, itself. The Faraday is a pressure neutral design similar to that shown in Fig. 23. The 300XP Faraday has been designed with attention to minimize insulator leakage affects [35]. The insulator supporting the 300 V bias plate is connected to the ground plate. Leakage across this insulator cannot affect the Faraday reading. The insulator supporting the 180 V bias ring is connected to the beam collector ring, which in this case is the actual current sensing element. Because the 180V bias voltage is reference to the sensing element, leakage across this insulator also will not affect dose. Only leakage across the insulator supporting the beam collector ring can influence the reading. However, the maximum voltage on this collector ring is only 1 V, thus, leakage is truly minimized. Almost any other arrangement of insulators would increase the potential for leakage and spurious beam current measurement. Leakage is a particularly important issue for serial implanters, because the currents are frequently very low, often nanoamperes, in order to perform low dose implants such as threshold shifts.

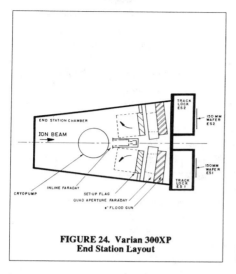

**FIGURE 24. Varian 300XP
End Station Layout**

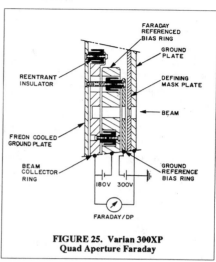

**FIGURE 25. Varian 300XP
Quad Aperture Faraday**

3.5.3 Wafer Internal Faraday Systems

Wafer internal Faraday systems are used in the Eaton serial products such as the NV-6200 and the NV-8200 as well as the Varian batch products. The design of the Varian 160XP is instructive and is shown in Fig. 26. Note that not only the wafer but also the electron flood (shower) is located within the Faraday. Otherwise it would not be possible for electrons from the shower to reach the wafer. Since the flood is part of the Faraday, it is important that electrons from the flood gun not be able to leave the Faraday. If electrons leave the Faraday, dose errors will occur. A tight seal between the rotating implant disk and the rear Faraday is not possible, so the -40 V bias ring helps to ensure that shower

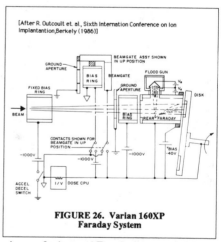

**FIGURE 26. Varian 160XP
Faraday System**

electrons do not escape by such a route. Note that in a wafer internal Faraday system the wafer holder, in this case the implant disk, must be electrically isolated and connected to the dose control system. All other Faraday components must likewise be connected to the dose control system.

3.5.4 Comparison of Wafer Internal and Wafer External Faradays

Wafer internal and wafer external systems each have their proponents. Electrical isolation of the wafer holder and a large number of Faraday components is not easy to achieve and maintain, particularly where rotating components are involved. On the other side of the ledger, wafer external systems, of necessity, sample beam current while wafer internal systems are able to measure beam current continously. This would seem to be an advantage for wafer internal

Faradays. In wafer external systems it is necessary to keep electron shower electrons out while in wafer internal systems it is nessary to keep them in, a more difficult task where 300 V primary electrons are involved. Which is the better system? Large numbers of both types of systems are in usage for ion implantation and both give satisfactory results. The problems involved are mostly engineering related and they are solvable. Thus, it is possible to engineer electrically floating rotating disks and to keep the electron flood electrons within the Faraday. It is also possible to engineer beam sampling systems that sample at a fast enough rate so that beam fluctuations do not measureably affect dose.

One area of difference between internal and external Faraday systems involves implanter response to wafer charging. As discussed in the following section, there are two effects of wafer charging, implant non-uniformity and dielectric punchthrough. Wafer external Faraday systems typically have a relatively long, field free region ahead of the wafer. Ion collision with residual gas will generate space charge neutralizing electrons in sufficient numbers that a charged wafer can draw on these electrons and prevent dielectric punchthrough. However, the long field free drift path does make beam blowup and nonuniformity more likely. In contrast, the bias field in a wafer internal system strips all of the space charge neutralizing electrons out of the beam, making dielectric punchthrough more likely. The same electric field prevents wafer charging from drawing space charge neutralizing electrons from the beam, thus, eliminating beam blowup and nonuniformity. In both systems the problem is corrected using electron flooding and in both cases results in general will not be completely satisfactory without some form of charge neutralization.

3.6 Dose Control by Faradays

3.6.1. Introduction

Successful Faraday design is not the end of dose control but only the beginning. It is beyond the scope of the present work to discuss dose electronics in any detail. However, some system effects are worth discussion. Charge neutralization in the beamline is an issue which affects every implanter to a greater or lesser degree. Charge neutralization effects are discussed in Section 3.6.2. Two modes of dose control have been developed, these are closed loop and open loop control (Section 3.6.3). Finally, section 3.6.4 summarizes the status of dose control by Faraday systems.

3.6.2. Beamline Charge Exchange

The influence of charge exchange on dose has been discussed by Wittkower [38]. Two effects can be distinguished, a dose nonuniformity and a dose shift. A dose nonuniformity can occur in machines utilizing electrostatic or magnetic scan and not having a beamline offset. Neutrals are not scanned and strike the wafer at a fixed position leading to a nonuniformity. The solution is an electrostatic offset of the beam in conjunction with the scanning. In such a case, no part of the wafer is in a direct path with the undeflected neutral beam. However, problems can still occur with multiply charged implants [39,40] and the distance from the deflector to the wafer is not negligible, so that some charge exchange still occurs [38]. Nonetheless, dose uniformities of from 0.5% to 1% are routinely achieved in commercial implantation equipment.

In machines which are fully mechanically scanned, beam neutrals cannot produce a nonuniformity, but like the ions, the neutrals resulting from charge exchange are scanned over the entire wafer. These neutrals are not detected by the Faraday system and, thus, give a dose offset. A dose offset will also occur in electrostatically scanned machines due to charge exchange after deflection. It is important to note that the magnitude of neutral contamination in the beam is proportional to the product of pressure times path length so that even though pressure is low, if path length is long beam neutralization can be significant (a few percent). This dose offset is one area which may contribute to the differences in absolute dose between different implanters (see 3.6.4).

3.6.3 Closed Loop vs. Open Loop Dose Control

Two rather different philosophies of dose control have been developed by the manufacturers of commercial implant equipment. Most commercial ion implanters use open loop dose control. In such a system the beam is scanned across the wafer and the total wafer dose is integrated until the preset dose is reached. In contrast, closed loop systems such as the Eaton NV-10 and NV-20 [41] and the ASM medium current implanter use closed loop feedback to the scanning system to compensate for spatial variations in implant across the wafer. Thus, in the NV-10 compensation for beam fluctuations and for the i/r effect of the spinning disk [41] is automatic. Similar compensations can be achieved in open loop systems [17,18,42,43] but the compensations must be anticipated and built in. While it would certainly seem that the closed loop system is superior, the true implications of this difference are not clear. In practice both types of systems are capable of achieving uniformities of 1% or better and both are capable of achieving batch to batch reproducibilities of 1% or better. Moreover, factors which adversely affect uniformity and reproducibility in practical applications, such as pressure bursting and beam blowup are not sensed by the dosing system alone and, thus, require additional sensing to provide protection in either type of system. However, with such sensing, dose correction is readily provided in close loop systems, since the feedback path is already present.

3.6.4. Dose Control Status

The status of dose control in ion implantation is summarized by the dosing round robins conducted by Current and Keenan [44] and more recently by Larson [45]. Commercial implanters control dose uniformity and batch to batch consistency very well [44-47]. However, absolute dose can differ by as much as 15% among the different commercial implanters. This result is at first surprising considering that current measurement is NBS traceable. However, pressure response is different among the various Faraday systems. Beamlines and beamline charge exchange are different and dose control philosophies are also different. In practice, where different machines must work side by side, the differences in absolute dose are usually controlled by modifying the set dose or altering dose processor scaling constants[48].

4. WAFER CHARGING AND CHARGE NEUTRALIZATION

4.1 Introduction

Semiconductor implantation frequently involves implantation into insulating structures. As noted in the introduction, by its nature ion implantation transports charge to the implanted target. The presence of space charge neutralizing electrons in the positive beam potential will reduce the total charging. Nevertheless, significant surface charging still can occur with insulating structures. Charge may accumulate until catastrophic breakdown occurs or damage may be subtle, resulting in a permanent alteration of the dielectric breakdown properties. Wafer charging can also affect beam propagation, resulting in beam blowup when the beam is scanned over an insulating wafer. The resulting implant nonuniformity can be quite large, as high as 20% in some cases. This problem is particularly troubling, since bare monitor wafers implanted at the same time as insulating product wafers will show no such effect. The need for some means of maintaining target neutrality during implantation has long been recognized [49-52].

Section 4.2 below discusses wafer charging while Section 4.3 discusses beam propagation. Particular emphasis is given to rotating disk batch implanters. Although both show detrimental charging effects [53], the high beam currents and relative long single pass wafer exposure times make batch implanters more subject to these effects than serial implanters. Charge neutralization is discussed in Section 4.4. Section 4.5 reviews the status of charge neutralization. The results indicate that with proper beam optics and adequate electron flooding, wafer uniformity to ±0.5% can be achieved. Wafer charging can also be held to ±10 V.

4.2 Wafer Charging

Fig. 27 illustrates some of the physics of wafer charging during ion implantion. As discussed in Section 4.3, the beam contains positive ions and electrons, which, in the absence of electron flooding, will be generated by collision with residual gas ions. These electrons will neutralize the beam to a degree dependent on the beam optics of the specific implanter under consideration.

Ions impinge on the wafer at high energy, ejecting from the surface primarily electrons but also ions. The wafer surface is not homogeneous but includes various conducting and insulating structures. Secondary emission from adjacent structures or the lack thereof, can stongly affect the survival of a gate at a particular site on the wafer. Obviously, this problem is extremely complicated and not amenable to direct analysis.

To evaluate the importance of wafer charging during implantation, it is instructive to consider a much simpler problem, namely, the voltage reached on an insulating device of area, A and thickness, d, in a single pass of the beam. For simplicity, secondary emission from the wafer surface is ignored. A rectangular beam of width, W, and height, H,

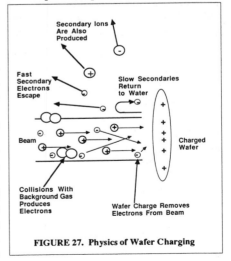

FIGURE 27. Physics of Wafer Charging

will transfer a charge, Q, to the device. The charge Q is

$$Q= \int i \, dt \quad -(16)$$

Where i is the net current into the device. If the total beam current is I_B and the beam neutralization is ß, then

$$i= (1-ß)I_B A/HW \qquad -(17)$$

The charging of the device produces a voltage, V=Q/C, where C is the capacitance of the device,

$$C=k\varepsilon_0 A/d. \quad -(18)$$

Here k is the dielectric constant of the insulator and $\varepsilon_0 = 8.85 \times 10^{-12}$ Farads/m. For a spinning disk machine the exposure time to beam will be,T,

$$T=W/2\pi Rf \, -(19)$$

where R is the radius of the implant site on the disk and f is the frequency of rotation. Combining (16) through (19) gives for the dielectric stress,

$$S = V/d = (1-ß)I_B/2\pi Hk\varepsilon_0 Rf \qquad -(20)$$

This dielectric stress is plotted in Fig. 28 for the Eaton NV-20 as a function of beam height, and as a function of beam current, assuming ß=0 (which is not the case). The results underscore the importance of charging. High quality thermal oxide will break down at stresses of the order of 10 MV/cm and Fig. 28 shows that on a single pass through the beam the device charges at least 1/3 of the way to destruction. Of course, the results here should be recast in terms of charge to breakdown, Q_{bd} [8,54]; however, the tunneled current would be dependent on the stress in equation (20).

Equation (20) is of limited value because it neglects important surface physics and because it represents only a single pass in a multipass process. The result is useful in highlighting machine parameters important to charging. Equation (20) suggests that increasing beam height in the slow scan direction, increasing disk radius and increasing disk rotation frequency might all be helpful in reducing device charging. However, a real implant will entail many passes through the beam and, for example, increasing disk radius reduces the single pass exposure time but lengthens the implant total implant time. Since no charge decay processes have been included in the derivation of equation (20), is not clear to what extent the larger disk will truly help. Even so, taking (20) literally suggests that the machine designer has perhaps a factor of 2 to 6 at his disposal by increasing disk radius, increasing disk rotation rate and increasing beam height. However, the most critical factor by far in equation (20) is the beam neutralization. For wafer internal Faraday machines with no electron shower, ß≈0 and 1-ß≈1. With electron shower or in a machine with a long field free beam path, ß≈0.98 and 1-ß≈0.02. Thus, beam neutrality has a factor of 50

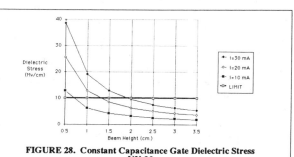

FIGURE 28. Constant Capacitance Gate Dielectric Stress NV-20

range. It is clear where the efforts of the machine designer must go.

It is also instructive to examine the potential of the ion beam, itself. In the absence of space charge neutralization (i. e. ß≈0), a substantial potential exists on the beam axis. For a circular beam, the potential well, V_w, from the edge of the beam to its axis is,

$$V_w = (I_B/4\pi\varepsilon_0)(m/2qE)^{1/2} \tag{21}$$

where m is the ion mass, q is its charge and E the potential through which the ion was accelerated. For a ribbon beam the factor 4π is replaced by the approximate factor 8W/H. In addition to the potential drop across the beam, there is a potential drop, V_b, from the edge of the beam to the surrounding walls of the beamline. Again assuming cylindrical symmetry

$$V_b = (I_B/2\pi\varepsilon_0)(m/2qE)^{1/2}\ln(r/R) \tag{22}$$

The total potential drop is the sum of (21) and (22) and will depend on the beam radius, r, and the beamline radius, R. Note that for a 10 mA arsenic beam at 80 kV, the total potential can exceed 1 kV. When it is remember that this potential also appears at the surface of the device being implanted, it again puts into perspective the difficulty of the problem. Of course, with space charge neutralization the beam potential will be much reduced. If the electrons, which provide the space charge neutralization, are distributed uniformly throughout the beam, then I_B above is replaced by $(1-ß)I_B$, so that the total potential drop will be decreased by a factor of 50 or more. This is a rather idealized result, since if some type of external source of electrons is used to provide the space charge neutralizing electrons, the electrons will not be uniformly distributed throughout the beam [55].

As noted above, the true situation is very complex. When the beam strikes the wafer, secondaries are ejected. Secondary current and unneutralized ion beam current both result in wafer charging. The charging will draw space charge neutralizing electrons from the beam and will pull back some of the ejected secondaries. Space charge neutralizing electrons will be replaced by ion-gas collisions and by external injection, such as by electron flooding. The final voltage reached will depend on a complicated balance of factors. It is important to note that both positive and negative device charging can occur and both can cause device damage. Positive charging is associated with beam ions and secondary electron ejection. Negative charging is due to space charge neutralizing electrons, excess flooding electrons and secondary ion emission.

Wafer topography also plays an important role [56]. Fig. 29 shows substrate dependence. Positive charging in p-type substrates and negative charging in n-type substrates result in the formation of a depletion layer below the insulating oxide [57]. This layer reduces the dielectric stress experienced in the device oxide. In contrast, negative charging in p-type material and positive charge in n-type, places the full stress across the dielectric. Fig. 30 shows implantation on oxide adjacent to bare silicon. The bare silicon ejects secondary electrons which can assist in neutralizing the oxide. Photoresist, on the other hand, ejects few secondaries and it has been observed that devices surrounded by photoresist show a lower yield [58]. The photoresist may also affect charge leakage paths. Fig. 31 shows a frequently used charging test structure. In Fig. 31a, the polysilicon (crosshatched region) covers the entire gate structure region. This is the most commonly used test site structure for charging. The gate oxide is usually thermally grown, while the thicker field oxide may use water vapor to speed growth rate. The structure in Fig. 31b is not a satisfactory test site because implantation into and through the oxide occurs in region C, damaging the oxide, and because the geometry results in extra electrical stress at the edges of region B, when voltage is applied.

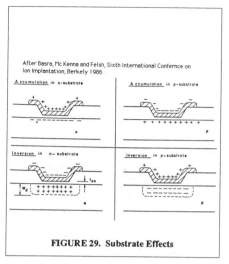

FIGURE 29. Substrate Effects

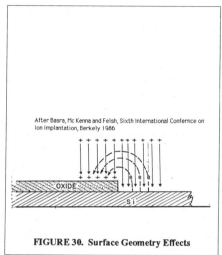

FIGURE 30. Surface Geometry Effects

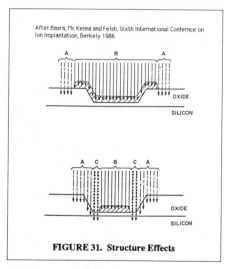

FIGURE 31. Structure Effects

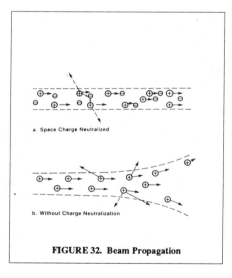

FIGURE 32. Beam Propagation

4.3 Beam Effects

One means of wafer neutralization is by drawing space charge neutralizing electrons out of the beam. As shown in Fig. 32 this impacts the beam propagation in the implanter. In Fig. 32a, the beam is fully space neutralized. The electrons responsible for the space charge neutralization usually are generated by ion-residual gas atom collision, equation (15a) above. The electrons are then trapped by the beam potential and the gas ions are repelled. The electrons drift in the potential, which the fast moving ions produce. Note that the probability for direct capture of these electrons by the energetic ions is extremely small. The space charge neutralizing electrons offset the mutual repulsion of the ions and allow the ion beam to propagate long distances without undue expansion. If these electrons are drawn off, mutual repulsion of the ions will cause a rapid increase ("blow up") of beam diameter (Fig. 32b), resulting in poor implant uniformity.

The time constant for beam neutralization by gas collisions is

$$T = (P_0/\sigma PL)(m/2qE)^{1/2} \qquad\qquad\qquad\qquad -(23)$$

where P is the pressure in the beam, P_0 is atmospheric pressure, σ is the cross section for ionization and L is the Loschmidt number. With an ionization cross section of 2×10^{-15} cm^2 and an implant pressure of 2×10^{-5} Torr, this time constant is only about 15 μseconds. This is shorter than the transit time constant across the wafer in commercial ion implanters. Thus, individual wafers on an implant disk are isolated from beam interaction effects produced by neighboring wafers. The dynamic beam enlargement suffered as the beam scans across a charged wafer will rapidly disappear as the beam scans off this wafer onto the disk. The combination of secondary electrons generated by ion collision with the disk and by gas ionization will quickly restore the space charge neutralizing electrons in the beam.

The degree of beam blowup can be estimated for the idealized case of a fully stripped (ß=0) uniform beam with zero emittance. In this case the increase in beam diameter, ΔD, due to the effects of space charge in a short distance, z, is

$$\Delta D = z^2 I_B (m/q)^{1/2}/(8\pi\sqrt{2}\ \varepsilon_0 D E^{3/2}) \qquad\qquad\qquad -(24)$$

Note that the spreading increases as the square of the path length, z. For an 80 kV, 10 mA As$^+$ beam having an initial diameter of 6 mm, the spreading calculated amounts to a fourfold increase in beam diameter in a propagation length of only 25 cm. Obviously, the expansion is significant.

Wafer charging will not withdraw all beam electrons from the beam so that actual beam enlargement will be less than predicted by (24). However, electrons will be withdrawn down the beam length until either an electric or magnetic field blocks the passage of electrons. This field may be encountered well past the resolving aperture. In such a case, two factors will contribute to the nonuniformity, the dynamic beam enlargement, itself [9], and dynamically reduced beam transmission through the resolving aperture [10]. These effects are readily controlled by electron flooding. The Eaton NV-10, for example, uses both an electron shower and an electron gate incorporated into the shower to ensure uniform implantation.

4.4 Charge Neutralization

Techniques for ion beam charge neutralization have been reviewed by Holmes [55,59] and include secondary electron generation, the plasma bridge, and elevated beamline pressure. The technique most commonly used in commercial ion implantation is secondary electron generation. The secondary electrons are generated either by higher energy primary electrons or by beam impact [60]. In the systems using primary electrons, an electron beam of perhaps 200 mA current is accelerated to 300 volts and strikes either a target or a section of the beamline or Faraday wall (see, for example, Fig. 26). The wall then emits low-energy secondary electrons with a yield near one and in principle, at least, these electrons shower over the wafer, neutralizing it. For the purposes of illustration, suppose all of the secondary electrons have a kinetic energy of 10 eV. Then, if a copious supply of secondary electrons were to flood the wafer with electrons, the wafer would charge to -10 volts. At any other higher voltage the 10 eV electrons would be repelled from the surface. If sufficient electrons could be provided, any positive charging by the beam would be offset.

The reason for using secondary emission to generate the low energy electrons is the limitation imposed by space charge at the emitting surface. For a planar diode the space charge limited emission current density, J, is given by the Langmir-Child equation as

$$J = 4\varepsilon_0 (2q/m)^{1/2} V^{3/2}/9d^2$$ -(25)

where V is the voltage across the diode and d is the gap. Thus, if one attempts to create 10 volt electrons directly, for example, by using a heated filament biased at -10 volts, and assuming a filament to wafer gap of d=10 cm, then the space charge limited current density is 74 μA/cm^2. If 30 mA is required, then the filament area must be at least 400 cm^2, a rather large filament. If it is required that electron energies be 5 eV, then the required area grows to nearly 1200 cm^2. By using secondary electrons, a relatively small filament can produce the 300 V primaries. A large area then can be used to generate the secondaries. In actual practice the area used to generate the secondaries is seldom large enough to be below the space charge limit so that the secondary emission is also space charge limited. Indeed, in showers where the secondary target is close to the beam, if the beam becomes unneutralized, the beam potential will assist in removing electrons from the secondary target. While this is extremely useful, users may be concerned that measured secondary current is dependent on species, current and beam size {equations (21) and (22)} as a result.

In actual practice not all secondary electrons are emitted at a fixed energy. Instead there is a distribution of energies with emission decreasing rapidly above 20 eV but rising again above 60 eV (Fig. 33). The increase above 60 eV is due to "back diffused" primary electrons. Unlike the refected electrons these are scattered out of the target in exactly the same way as the secondary electrons and are indistinguishable from true secondaries. The consequences of Fig. 33 are far reaching. It suggests that direct flooding of the wafer may be undesireable, if sensitive wafers are involved. In fact, the work of Hall et. al. appears to confirm this inference [54]. Clearly some type of energy discrimination is required to prevent excessive negative charging by the fast electrons.

Relatively little has been reported regarding the use of other techniques for wafer neutralization during ion implantation. Holmes [55,59] found increased beamline pressure to be most effective. Hall et. al.[54], in fact, have combined increased implant pressure with electron flooding to minimize wafer charging. This is not a panacea, however, because charge exchange collisions (Sections 3.6.2) at the higher pressure can give significant dose errors and implant nonuniformities.

An important adjunct to charge neutralization is the measurement of wafer charging. For nonuniformity studies, a special wafer has been developed [9] which charges well and shows any resultant nonuniformity. This wafer consists of 1200Å of undoped polysilicon over 1000Å of oxide. The wafer charges to relatively high voltages and can be annealed and mapped in just the same way as bare monitor wafers. Because of the beam blowup, charging is indicated by a low dose

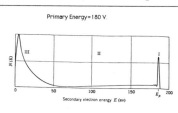

After H. Massey, E. Burhop, and H. Gilbody, *Electronic and Ionic Impact Phenomena*, Oxford University Press, London (1969).

FIGURE 33. Typical Energy Distribution for Secondary Electrons

(high sheet resistivity) in the center of the wafer and higher dosing near the wafer edges. As a result of aperture transmission effects, average dose is usually reduced over that for wafers, which do not charge.

For direct charging, MOS capacitors such as shown in Fig. 31a can be used, but the fabrication of high quality test sites is difficult and the measurement slow and laborious. Measurement usually entails measuring breakdown for a number of the test sites and plotting histograms of the breakdown data. Implant damage usually is manifested as a shift in the histogram to lower breakdown voltages and a marked increase in the failures at or near zero voltage. It is also possible to monitor charging indirectly in the implanter. Fig. 34 shows an early capacitive charge sensor. A positively-charged wafer approaching the sensor induces a negative charge in the sensor. The sensor charge is a maximum when the wafer is directly beneath the sensor. The charge bleeds away when the wafer moves away. Integrating the current from the sensor gives a measure of the total charge on the wafer. Although simple in concept, sensor design is complicated by the presence of a weak plasma in the implant chamber and by the moving geometry. Since the development of this simple sensor, the charge sensor has been further developed to allow examination of small sample charging as that sample passes under the ion beam [61].

Fig. 35 shows charge sensor measurements in the Eaton NV-10. The channel number indicated in the figure is proportional to time. Voltage measured by the sensor initially becomes negative, then positive and finally negative again. The negative charging is due to the electron cloud surrounding the beam [55]. When the wafer sweeps into the beam it first encounters this cloud, then the beam, itself, and finally the cloud again. If the beam is centered on the measurement spot, charging is positive. Figs. 35b and 35c show the results of using an electron flood specially modified to minimize negative charging. The positive charging of the sensor is decreased but negative charging increases somewhat at higher flooding. The electron flood does increase the number of electrons within the area of beam but produces an even greater increase outside the beam [55].

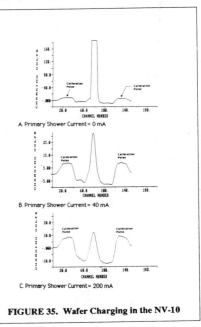

A. Primary Shower Current = 0 mA

B. Primary Shower Current = 40 mA

C. Primary Shower Current = 200 mA

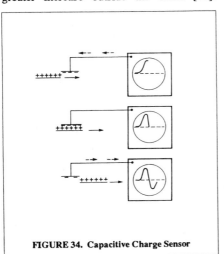

FIGURE 34. Capacitive Charge Sensor

FIGURE 35. Wafer Charging in the NV-10

Because of the presence of high energy electrons (fig.33), in many flood gun systems negative charging to several tens of volts can occur at higher flood currents. These flood currents may be needed to control the positive charging, which the beam produces. Flooding can be combined with higher beamline pressures. However, testing in the NV-10 shows that this combination decreases peak positive charging but does not affect negative charging.

4.5 Effectiveness of Charge Neutralization

Electron flooding can be very effective in reducing wafer charging. The spatial nonuniformities are most readily corrected. Fig. 36 shows wafer uniformity in the Eaton NV-10 using the polysilicon-oxide wafers discussed above. The implant was performed with arsenic at 50 kV to a dose of 2×10^{16}. The use of a heavy species at low energy accentuates the beam blowup and produces a larger nonuniformity {equation (24)}. Fig. 36a shows the result without electron flood. Contour lines represent a 5% change in sheet resistivity. Sheet resistivity is high at the center and low near the edges. Standard deviation in the sheet resistivity for the polysilicon-oxide wafer is about 24%. Fig. 36b shows the result with electron shower at a 95 mA primary shower current and with the electron gate discussed above. Contour lines in Fig. 36b represent a 1% sheet resistivity change and the standard deviation of the resistivity is about 0.6% in this case. Use of the electron gate alone will allow a reduction of the standard deviation from 24% to perhaps 1%. However, use of the gate alone would be similar to the use of a wafer internal Faraday machine without flooding, i.e. the implant would be uniform but actual charging of the wafer would increase.

Reduction in device charging is a more difficult problem. One reason for this is that device requirements are changing and, in general, are becoming more stringent. The trend to thinner and thinner gate oxides demands proportionately better control of device charging. Increasing device density necessitates that yield per gate be increased. Equipment manufacturers have responded to these needs with considerable success but it is clear that this will be an ongoing area of development in an effort to keep up with industry requirements. In terms of present status, Fig. 37 shows peak positive and peak negative charging in the NV-10. With proper electron flood design peak charging is limited in this implanter to about ±10 volts.

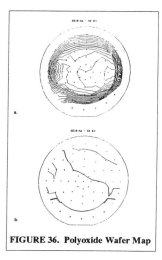

FIGURE 36. Polyoxide Wafer Map

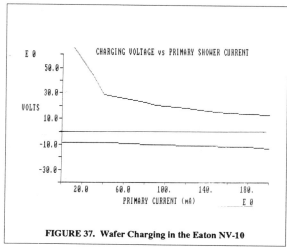

FIGURE 37. Wafer Charging in the Eaton NV-10

REFERENCES

[1]. James F. Gibbons, in *Ion Implantation Technology*, 83, edited by M. I. Current, N. W. Cheung, W. Weisenberg and B. Kirby, North Holland Physics, Amsterdam, 1987

[2]. P. H. Rose, in *Ion Implantation Equipment and Techniques*, 1, edited by J. F. Ziegler and R. l. Brown, North Holland, Amsterdam, 1985

[3]. M. I. Current, in *Ion Implantation Equipment and Techniques*, 9, edited by J. F. Ziegler and R. l. Brown, North Holland, Amsterdam, 1985

[4]. T. C. Smith, in *Ion Implantation Equipment and Techniques*, 196, edited by H. Ryssel and H. Glawischnig, Springer-Verlag (1983). See also this volume.

[5]. D. G. Beanland, W. Temple and D. J. Chivers, *Solid State Electronics 21*, 357 (1978)

[6]. M. Lenzlinger and E. H. Snow, *J. Appl. Phys. 40*, 278 (1969)

[7]. T. Poorter and D. R. Wolters, in *Insulating Films on Semiconductors*, 266, edited by J. G. Verwey and D. R. Wolters, North Holland, Amsterdam (1983)

[8]. D. R. Walters and J. J. van der Schoot, *Philips J. of Research 40*, 164 (1985), *Philips J. of Research 40*, 137 (1985), *Philips J. of Research 40*, 115 (1985)

[9]. M. E. Mack, G. Ryding, D. H. Douglas-Hamilton, K. Steeples, M. Farley, V. Gillis, N. White, A. Wittkower and R. Lambracht, in *Ion Implantation Equipment and Techniques*, 9, edited by J. F. Ziegler and R. l. Brown, North Holland, Amsterdam, 1985

[10]. N. White, M. E. Mack, G. Ryding, D. H. Douglas-Hamilton, K. Steeples, M. Farley, V. Gillis, A. Wittkower and R. Lambracht, *Solid State Technology 28*, 151 (1985)

[11]. M. E. Mack, in *Ion Implantation Equipment and Techniques*, 221, edited by H. Ryssel and H. Glawischnig, Springer-Verlag (1983).

[12]. V. Benveniste, in *Ion Implantation Technology*, 366, edited by M. I. Current, N. W. Cheung, W. Weisenberg and B. Kirby, North Holland Physics, Amsterdam, 1987

[13]. P. D. Parry, *J. Vac. Sci. Tech. 15*, 111 (1978)

[14]. R. J. Stocker, *Nucl. Inst. Meth. 189*, 281 (1981)

[15]. H. Glawischnig, in *Low Energy Ion Beams*, edited by I. H. Wilson and K. G. Stevens, 60, Institute for Physics, London (1980)

[16]. D. M. Jamba, *Nucl. Inst. Meth. 189*, 253 (1981)

[17]. N. Turner, "Improved Uniformity of Implanted Dose by a Compensated Scan Pattern Generator", 5th International Conference on Ion Implantation, Ontario, July 1980

[18]. D. D. Myron and D. R. Zrudsky, in *Ion Implantation Technology*, 410, edited by M. I. Current, N. W. Cheung, W. Weisenberg and B. Kirby, North Holland Physics, Amsterdam, 1987

[19]. R.J. Rourk, *Formulas for Stress and Strain*, 217, McGraw-Hill (1954)

[20]. M. I. Current, N. L. Turner, T. C. Smith, and D. Crane, in *Ion Implantation Equipment and Techniques*, 336, edited by J. F. Ziegler and R. l. Brown, North Holland, Amsterdam, 1985

[21]. C. Taylor, P. Splinter, A. Weed, J. Grant, and S. Holden, in *Ion Implantation Technology*, 224, edited by M. I. Current, N. W. Cheung, W. Weisenberg and B. Kirby, North Holland Physics, Amsterdam, 1987

[22]. M. T. Wauk, in *Ion Implantation Technology*, 280, edited by M. I. Current, N. W. Cheung, W. Weisenberg and B. Kirby, North Holland Physics, Amsterdam, 1987

[23]. E. Rabinowicz, *Friction and Wear of Materials*, Chapts. 2 and 3, Wiley and Sons, New York (1965)

[24]. M. King and P. H. Rose, Nucl. Inst. Meth. 189, 169 (1981)

[25]. D. C. Evans, in *Ion Implantation Technology*, 385, edited by M. I. Current, N. W. Cheung, W. Weisenberg and B. Kirby, North Holland Physics, Amsterdam, 1987

[26]. S. Dushman, *Scientific Foundation of Vacuum Technique*, Wiley and Sons, New York (1949)

[27]. J. Jost and S. Dinaro, unpublished

[28]. M. Sieradzki, in *Ion Implantation Equipment and Techniques*, 237, edited by J. F. Ziegler and R. l. Brown, North Holland, Amsterdam, 1985

[29]. P. L. F. Hemment, *Institute of Physics Conference Serial 54*, 77, Institute for Physics, London (1980)

[30]. D. M. Jamba, *Rev. Sci. Instrum. 49*, 634 (1978)

[31]. C. M. McKenna, in *Ion Implantation Techniques*, 73, edited by H. Ryssel and H. Glawischnig, Springer-Verlag, Berlin (1983)

[32]. G. Ryding, *Microelectronics Manufacturing and Testing*,1, Feb 1983

[33]. C. M. McKenna, *Radiation Effects 44*, 93 (1979)"

[34]. R. Outcault, C. McKenna, T. Robertson and L. Biondo, in *Ion Implantation Technology*, 354, edited by M. I. Current, N. W. Cheung, W. Weisenberg and B. Kirby, North Holland Physics, Amsterdam, 1987

[35]. C. B. Yarling, *Semiconductor International*, 110, Aug 1983

[36]. G. Ryding, M. Farley, M. Mack, K. Steeples, and V. Gillis, 10th International Conference of Electron and Ion Beam Science and Technology, Montreal (1982)

[37]. P. Lundquist, C. McKenna, R. Brick, and P. Corey, in *Ion Implantation Technology*, 414, edited by M. I. Current, N. W. Cheung, W. Weisenberg and B. Kirby, North Holland Physics, Amsterdam, 1987

[38]. A. B. Wittkower, *Solid State Technology*, 77, Sept. 1982

[39]. P. Spinelli, J. Escaron, A. Soubie, and M. Bruel, in *Ion Implantation Equipment and Techniques*, 283, edited by J. F. Ziegler and R. l. Brown, North Holland, Amsterdam, 1985

[40]. C. R. Kalbfus and R. Milgate, in *Ion Implantation Technology*, 400, edited by M. I. Current, N. W. Cheung, W. Weisenberg and B. Kirby, North Holland Physics, Amsterdam, 1987

[41]. G. Ryding and M. Farley, *Nucl. Inst. Meth. 189*, 319 (1981)

[42]. C. Ehrlich, A. Delforge and R. Liebert, in *Ion Implantation Equipment and Techniques*, 228, edited by J. F. Ziegler and R. l. Brown, North Holland, Amsterdam, 1985

[43]. R. Eddy, A. Long, S. Smith and J. Tkach, in *Ion Implantation Technology*, 424, edited by M. I. Current, N. W. Cheung, W. Weisenberg and B. Kirby, North Holland Physics, Amsterdam, 1987

[44]. M. I. Current and W. A. Keenan, in *Ion Implantation Equipment and Techniques*, 418, edited by J. F. Ziegler and R. l. Brown, North Holland, Amsterdam, 1985

[45]. L. Larson, Greater Silicon Valley Ion Implant Users Group Meeting, June 1986

[46]. K. Steeples, in *Ion Implantation Equipment and Techniques*, 412, edited by J. F. Ziegler and R. l. Brown, North Holland, Amsterdam, 1985

[47]. G. Ryding, *Mat. Res. Soc, Symp. Proc. 45*, 367 (1985)

[48]. L. A. Larson and G. L. Kennedy, in *Ion Implantation Technology*, 421, edited by M. I. Current, N. W. Cheung, W. Weisenberg and B. Kirby, North Holland Physics, Amsterdam, 1987

[49]. G. V. Spivak, A. I. Krohnika, T. V. Yovorshaya and Y. A. Durasova, *Dok. Akad. Nauk. SSR 114*,1001 (1957)

[50]. R. L. Hines and R. Wallor, *J. Appl. Phys. 32*, 202 (1961)

[51]. R. L. Hines and R. Arndt, *Phys. Rev. 119*, 623 (1960)

[52]. A. I. Akishin, S. S. Vasilev and L. N. Isaev, *Izves. Akad. Nauk. SSR 26*, 1379 (1962)

[53]. P. E. Bakeman and A. F. Puttlitz, in *Ion Implantation Equipment and Techniques*, 399, edited by J. F. Ziegler and R. l. Brown, North Holland, Amsterdam, 1985

[54]. J. M. Hall, H. Glawishnig and W. Holtschmidt, in *Ion Implantation Technology,* 350, edited by M. I. Current, N. W. Cheung, W. Weisenberg and B. Kirby, North Holland Physics, Amsterdam, 1987

[55]. A. J. Holmes, *Rad. Effects 44*, 47 (1979)

[56]. V. K. Basra, C. M. McKenna and S. B. Felch, in *Ion Implantation Technology*, 360, edited by M. I. Current, N. W. Cheung, W. Weisenberg and B. Kirby, North Holland Physics, Amsterdam, 1987

[57]. B. G. Streetman, *Solid State Electronic Devices*, Section 8.3, Prentiss -Hall, New Jersey (1980)

[58]. R. Tong and P. McNally, in *Ion Implantation Equipment and Techniques,* 376, edited by J. F. Ziegler and R. l. Brown, North Holland, Amsterdam, 1985

[59]. A. J. T. Holmes, *Phys. Rev. A 19*, 389 (1979)

[60]. M. L. King and S. E. Sampayan, in *Ion Implantation Technology*, 396, edited by M. I. Current, N. W. Cheung, W. Weisenberg and B. Kirby, North Holland Physics, Amsterdam, 1987

[61]. V. Benveniste, unpublished

PHOTORESIST PROBLEMS AND PARTICLE CONTAMINATION

T. C. Smith
Motorola Semiconductor Group
Mesa, AZ, 85202, USA

CHAPTER OUTLINE:

1.0 INTRODUCTION

This chapter treats two important subjects related to the successful application of ion implantation techniques to IC wafer fabrication. First, the ability to use photoresist (PR) as a "room temperature" mask for the selective doping of substrates is a very important technological advantage of ion implantation over diffusion methods. However, this benefit has some problems associated with it, especially in high current type implanters. Second, continuing trends toward smaller circuit geometries impose stringent requirements for low particulate levels in virtually all IC manufacturing processes. The topic of monitoring and controlling particle contamination in implanters is all the more important because implantation is a key process and maintaining low particle performance in implanters is critical for achieving high yields for VLSI devices.

The practical aspects of using PR as part of the overall process flow will be discussed in detail. The minimum thickness of any given PR required for effective masking depends not only upon the implant and PR parameters but also the particular application of the implant step in the process flow. Consideration of how much of the implanted dopant might be tolerated in the regions being masked determines the meaning of "effective" masking. The additional factors which should be considered in order to preserve the integrity of the resist image, ensure proper operation of the implanter, and provide for complete stripping of the implanted PR will be covered. The effect of pre-treatments as they influence what happens in the implanter and the interdependent parameters which determine the wafer temperature under different operating conditions are reviewed. The impact of outgassing and its possible effect upon the accuracy of the dose measurement is also discussed. An explanation of the damage mechanisms in the PR will show that outgassing of the PR layer in the vacuum system of the implanter and the "difficulty of PR stripping" are phenomena which are intimately related to each other. An understanding of these factors in different applications and systems is important to ensure the correct incorporation of PR masking into the manufacturing flow and the efficient utilization of equipment in both the photolithography and implant processing areas.

Contamination of wafers by particulate matter in any process step leads to lower yields of fully functional devices at final test. In the case of implantation steps, particles which are on the wafers at the time of the actual implantation may cause "blocked implants" because the depth of penetration for most implants is in the sub-micron range. Some of these blocked implants may be "killer defects" which lead to device or circuit failure. In addition to this effect, particles which are deposited on wafers in the implanter can be transferred to other wafers or wafer carriers and carried on to subsequent processes, such as diffusion or photolithography operations, where they may become the cause of further yield loss. The generation and redistribution of particles in implantation systems will be discussed, along with the techniques used to measure their size and quantity. Operating procedures for making statistically significant measurements, qualifying equipment, and establishing control limits will be reviewed. A description of machine modifications which have led to reductions in particle counts will be given. With the trend toward increasing the number of implant steps used in leading edge technologies, these considerations are increasingly important for realizing high yields.

2.0 PHOTORESIST PROBLEMS
2.1 PHOTORESIST MASKING CONSIDERATIONS

The use of PR masking allows great processing flexibility and is a major reason for the successful incorporation of implantation into IC process flows.   The alternate masking technique, which was traditional in the diffusion doping technology previously used, employed patterned layers of silicon dioxide or silicon nitride.   Although these materials can block diffusing dopants and withstand high furnace temperatures, these films must be grown or deposited, and then the wafers must still be processed through a photolithographic step and also an etching step.   With implantation, only the photo process is necessary, because the dopant is introduced by means of the ion's kinetic energy while the wafer is at a temperature which is low enough to preserve the original PR image.   The doping and the thermal redistribution processes can thus be separated using the implant technique.   In any case, for emerging technologies, implantation is the preferred doping technique for shallow junctions and low thermal budgets.

First, the details of the processing must be arranged so that for each specific implant, the thickness of the masking PR is sufficient to ensure that the ions do not penetrate through to the region being protected from the implant.   Estimates of t, the required PR thickness, can be made from a knowledge of R_p, the mean projected range, and ΔR_p, the straggle for the particular ion/PR combination.   In one of the earliest papers on the subject of PR masking, the energy dependence of R_p and ΔR_p for boron implanted into KTFR negative resist was experimentally determined (72). Tabulations of calculated values of these parameters as a function of energy for various common ions are available for AZ111 positive resist (75) and for KTFR negative resist (77).   The range parameters for boron in polymers have also been obtained experimentally using secondary ion mass spectrometry (SIMS) techniques and compared to published values (85e). As a practical matter, for the one to two micron PR thicknesses commonly used, the principal instance where masking effectiveness should be checked is for boron at energies above about 100 keV.   When using MeV implants, masking effectiveness is a matter of concern for a much broader range of ions and energies.

If the implant profile is assumed to be gaussian in shape, the resulting fraction of the implant dose which is beyond the lower boundary

348   T. C. SMITH

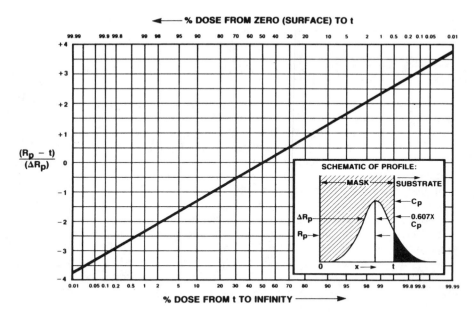

Figure 1.   Masking effectiveness for a Gaussian implant profile.

of the mask is described mathematically by the "complementary error
function".   This dependence results in a straight line when plotted on
probability graph paper, as shown in Fig. 1.   The inset of that figure shows
the implant profile schematically, with the portion of the implant which
penetrates beyond the masking layer shown as the solid black region.   The
vertical axis shows the figure of merit, $[(R_p-t)/\Delta R_p]$, which determines the
masking effectiveness.   The lower horizontal scale indicates what
fraction (as a %) of the implanted dose is in the substrate while the upper
horizontal scale shows what fraction is stopped in the masking layer.
This graph is useful for estimating the penetration of any masking layer,
as long as R_p and ΔR_p of the masking material are known.   A conservative
"rule of thumb" would state that to be certain that less than 0.001% of the
dose gets through the mask, then the thickness of the mask must be
greater than $[R_p+4.3x(\Delta R_p)]$.   In some cases, one part in 10^5 of the implant
dose might not be anything to be concerned about, as when implanting a
dose of $10^{12}/cm^2$.   In the case of a much higher dose, such as an MOS
source/drain implant at a dose of $10^{16}/cm^2$, if 0.001% of the dose gets
through to certain regions of the substrate, it could change a device's
threshold voltage significantly.   After checking these considerations, it

may be necessary to decrease the implant energy or increase the masking thickness.

2.2 PROBLEMS AND PERSPECTIVES IN IMPLANTERS

In addition to the problem of whether or not the PR film can effectively mask the implant, a number of other difficulties with the process have been encountered. This was especially true when production applications moved to higher doses and equipment with higher beam current capabilities became commercially available. These problems include destruction or deformation of the PR film due to excessive wafer heating, severe outgassing, difficulty in stripping PR after high dose implants, and increased frequency of cleaning residues from end station chambers and Faraday cup assemblies. Regardless of their exact formulation, most PRs consist of polymers and solvent compounds containing carbon, hydrogen, and oxygen along with a photoactive compound and possibly other organic additives. In most cases, the PR film will not decompose during the implantation process if the wafer temperature does not exceed the manufacturer's recommended maximum safe temperature. For negative PRs, more commonly used in the past, this temperature typically might be in the range of 150°C to 200°C. For positive PR's, used in advanced technologies with fine line geometries, a maximum temperature of only 100°C to 150°C can be tolerated. At temperatures slightly above these maximum safe limits, the developed line image flows and the pattern gets distorted. At still higher temperatures, the PR may crack and form blisters. Hence, much effort in implanter equipment development has been directed toward providing some means of cooling wafers during implant.

A second problem is that excessively high pressures in the end station and beamline regions of the implanter can cause neutralization and other effects which lead to overdosing and dose non-uniformity. In some machines, when outgassing produces high pressure, vacuum system interlocks cause the controller to put the implant run on "hold" and operator assistance is required to resume the process. In most cases, an understanding of the relationship of implant parameters to the difficulty of resist strip was lacking, especially since this latter effect seemed difficult to quantify. Perspectives on the problem of PR stripping and methods of dealing with it are varied because users have worked with a wide variety of operating conditions in both implantation systems and PR stripping equipment. The various implanters used may have differing wafer cooling capabilities, depending upon when they were manufactured. Some results first presented at the 4th International Conference on Ion Implantation seemed contrary to common sense (83b). For instance, it was shown that outgassing is *not* caused by wafer heating and that the difficulty of PR stripping is *not* a function of the temperature of the wafer during the implant itself.   Explanations of these statements are given below when discussing outgassing and PR damage mechanisms.

2.3 WAFER HEATING AND COOLING CONSIDERATIONS

Because these topics are covered in depth in another chapter of this book, the present discussion will simply deal with some general features of the thermal response of the wafer under different conditions.   In the general case, the temperature of the wafer is governed by the power balance, where the power put into the wafer is equal to the rate at which energy is stored plus the power which is lost from the wafer.   In the vacuum system of the implanter, one means for heat to be lost is by heat transfer via black body radiation, which is more effective at elevated temperatures.   In addition, for implanters equipped with some conductive cooling scheme, heat is rejected from the back side of the wafer to a cooled platen when processing single wafers or to the disk or wheel in the case of batch machines.

When considering the effective power per unit area into the wafer, the simplest form of the basic power balance equation is:

$$(P/A)' = (\rho\ C\ thw)\ dT/dt + 2\ \sigma\ \varepsilon\ (T^4 - T_{or}^4) + C_j\ (T - T_{oc}) \qquad (1)$$

Estimates of the response of wafer temperature versus time can be made from considering the limiting forms of this equation under various implanting conditions.   On the left hand side of this equation, the effective power per unit area is equal to the power in the beam divided by the area over which the beam is scanned.   The power in the beam is simply the arrival rate of ions times the energy per ion.   This is equivalent to the beam current times the total voltage through which the ions have been accelerated.   It is convenient here to express the effective power density, $(P/A)'$, in milli-Watts/cm^2.   The first term on the right hand side is the rate at which energy is stored in the wafer per unit area.   The factors included are: ρ, the density of the wafer, C, the heat capacity per unit mass of the wafer, thw, the thickness of the wafer, and dT/dt, the rate of rise of wafer temperature.   For 0.05 cm (0.020") thick silicon wafers, the term $(\rho\ C\ thw)$ has a value of 84 milli-joule/(cm^2-deg).   The second term on the right is the rate at which energy is radiated from the wafer, at temperature T, to the surroundings at a temperature T_{or}, with these temperatures expressed on the absolute temperature scale in deg. K.   The factor σ is the Stefan-Boltzmann constant, [5.67E-9 mW/(cm^2-deg^4)], ϵ is the effective emissivity of the wafer and the surroundings (typically 0.35), and the factor of two in front applies if one assumes that *both* sides of the wafer radiate heat to the surroundings. The third term on the right is the power flux conducted from the wafer to the platen or disk which is at a temperature T_{oc}. The factor C_j, called the surface conductivity of the interface and expressed in mW/(cm^2-deg), is the figure of merit for heat transfer from the wafer to the platen or disk through a thermally conductive elastomer or conductive gas. Depending upon the specific cooling design, values of C_j typically range from 4 to 80 mW/(cm^2-deg).

A lot of information can be gained by considering Eqn.(1) in its limiting forms for the various kinds of wafer cooling which may be employed. First, knowing the implanting time, the temperature rise of the wafer with *no* cooling mechanism can be estimated from neglecting the last two terms on the right hand side of the power balance equation. Under these conditions, the heat is stored in the wafer and its temperature rise above ambient temperature is less than 76°C for doses up to 2.0E14/cm^2 at energies up to 200 keV. In the case where only radiative cooling is available in a system, the radiation limited temperature, when dT/dt is zero, can be calculated by setting the first and third term on the right side of Eqn.(1) equal to zero. Similarly, for the case of conductive heat transfer, the conduction limited wafer temperature rise above ambient temperature can be derived by neglecting all but the third term. The "time constant" for the temperature response can be estimated by dividing the equilibrium temperature rise above ambient temperature by the original rate of rise of wafer temperature.

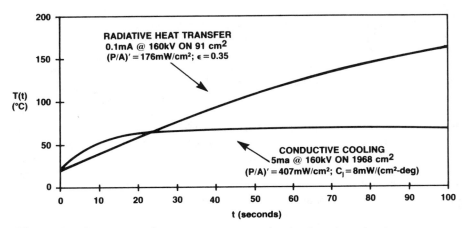

Figure 2. Average wafer temperature vs. implanting time for two cases.

Fig. 2 shows the average wafer temperature versus implanting time for two cases of interest: radiative heat transfer in a single wafer implanter (i.e. no conductive cooling) and conductive cooling in a high current batch machine. The parameters for each implant are given in the figure. For the top curve, at the end of 100 seconds the wafer temperature is 164°C and the dose is $6.9E14/cm^2$. The time constant is 88 seconds, and if this wafer remained in the implanter for an infinitely long time, its final equilibrium temperature would be 204°C. For the bottom curve, the time constant is 10.5 seconds, so the final equilibrium temperature is reached much earlier in the implant. In this latter case, the dose after 100 seconds is $3.2E15/cm^2$. So, the actual wafer temperature depends upon what the effective power density is and what kind of provisions are made for cooling the wafer.

In these examples, the high current machine has 50 times more power in the beam, but it is distributed over a much larger scan area, thus the average power density is only 2.3 times higher than in the single wafer implanter. If more efficient conductive gas cooling were used in the single wafer implanter, with a C_j of 50 mW/(cm^2-deg) the temperature rise of the wafer would be only 3.5°C. If the maximum beam current of this machine were 1.5 mA at 200 KV, the wafer temperature rise with this conductive cooling would be 66°C. So, high values of C_j are required when the effective power density is high. In any case, for PR coated wafers, it is generally true that as long as the macroscopic wafer temperature does *not* exceed the manufacturer's recommended maximum temperature, the PR will survive the implant without blistering, flowing, or cracking.

2.4 PHOTORESIST OUTGASSING

From the very beginning of using PR masking in implanters, it was known that the pressure in the vacuum system would rise due to outgassing of the thin film of PR. This was an area of concern, because if the ion beam were neutralized to a significant extent in high pressure regions of the beamline and end station, then dose errors would occur. Initially, the relationship of outgassing to the implant parameters was not well understood, since users' perspectives were very different. In addition, it was thought that wafer temperature, which was poorly controlled, was probably an important parameter in outgassing. Studies of end station pressure and measurements of the total quantity of gas evolved under various implant conditions led to the development of an empirical model of PR damage which explains the observed outgassing and several related phenomena (83b).

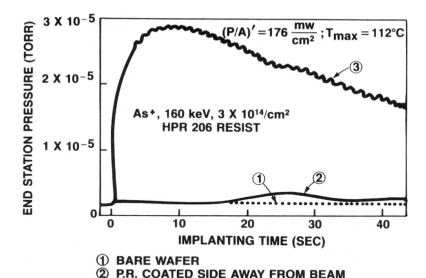

Figure 3.   PR outgassing: end station pressure versus implanting time for single wafer implanter (after Smith, Ref.83b).

End station pressure versus time, in a single wafer implanter system, is given in Fig. 3.   Three cases of interest are shown, with the implant parameters as stated.   In this system, which was equipped with cryogenic pumps, the beam current was 100 μA, and the scan area was 91.0 cm², somewhat larger than the area of the 100 mm diameter wafers used.   The bottom curve, which corresponds to a bare silicon wafer, shows no evidence of outgassing.   The top curve shows P vs. t for a wafer covered with unbaked HPR 206 positive photoresist, and is a typical response curve for this kind of implant, where the pressure reaches a maximum value and then falls off for the remainder of the implant.   (The oscillations in the pressure will be explained below).   In all three cases, with no conductive wafer cooling, the wafer temperature response is the same, increasing continuously as shown by the top curve in Fig. 2.   This is because the power density was the same, but in Fig. 2 the implanting time was longer. For the middle curve in Fig. 3, the wafer was reversed, so that the PR coated side was facing away from the incoming beam.   Since there was no clamp ring, any gas generated could freely escape into the vacuum system to be pumped away.   In this case, the end station pressure showed a slight rise when the wafer temperature was about 75°C, which may be attributed to the evolution of volatile solvents.

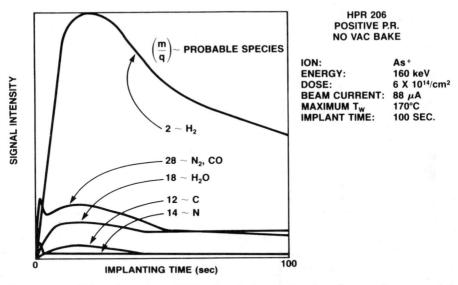

Figure 4.   PR outgassing: RGA signal (in arbitrary units on linear scale)
versus implanting time for single wafer implanter.

When a residual gas analyzer (RGA) was connected to the end station
vacuum system, the intensity on a linear scale of the principal peaks
observed was recorded as a function of time. The results for an arsenic
implant at 160 keV and a dose of $6.0E14/cm^2$ are shown in Fig. 4. It is
seen that for this positive PR, about 90% of the gas evolved was hydrogen,
with water vapor and mass 28 each accounting for about 5%. (With the
same conditions, for HNR 120 negative PR, 78% of the gas evolved was
hydrogen, while 8% was due to mass 28 and 14% was due to all other
masses.) The ripple evident in the total pressure for the upper curve of
Fig. 3 was not seen in the RGA study in the curves for the component gases.
That was because there was a power failure in between the times when
the data for these two graphs was recorded, and the cryogenic pumps in
the system were regenerated! This indicated that the end station pump
was saturated with hydrogen. The oscillations of total pressure seen on
the end station ion gauge were synchronized with the motion of the
displacers of the cryo pump. Since cryo pumps can not condense hydrogen
on their cold heads, hydrogen is only pumped by being adsorbed on an array
of activated charcoal surfaces. Depending upon the implant parameters,
this condition of being saturated with hydrogen can be reached after
implanting several thousand PR coated wafers. Consequently, it has
become standard practice in implantation systems to use cryo pumps
having "enhanced hydrogen capacity" and to regenerate the cryo pumps

more frequently to avoid saturating their charcoal arrays with hydrogen.

Additional observations on outgassing were helpful in understanding the mechanisms at work in the system. For instance, if the beam were switched off, the pressure in the system would fall to the level of base vacuum in a few seconds. When the beam was switched on again, the pressure rapidly recovered to its previous level and then proceeded with the same response as before. This seemed to be clear evidence that outgassing was not due to wafer temperature effects, because the wafer temperature could not respond so quickly. Besides, the wafer outgassing was *decreasing* while the temperature was *increasing* in this case.

At low doses, less than $1.0E13/cm^2$, pressure measurements made at various energies and beam currents showed a different behavior. Whenever the beam current was adjusted to a different value during the implant, the pressure in system would quickly change and after a few seconds remain constant without falling off over time. In this dose region, the pressure changed linearly with beam current. When the energy was changed, with the beam current held constant, the end station pressure changed with a less than linear response. In the low dose region, one could say that the rate at which gas molecules evolve is equal to a "gas multiplier" (M_g) times the arrival rate of ions. That is:

$$(dN/dt)_{gas} = M_g \ (dN/dt)_{ions} \qquad (2)$$

For singly charged ions, the arrival rate of ions is related to the beam current by:

$$(dN/dt)_{ions} = (I_{beam})/e \qquad (3)$$

Here, e is the charge of the electron. The rate at which gas evolves and is pumped away by the end station vacuum system is:

$$(dN/dt)_{gas} = (S \ P)/(kT) \qquad (4)$$

Where S is the pumping speed for hydrogen, P is the end station pressure, k is Boltzmann's constant, and T is the ambient gas temperature in the end station vacuum system. Knowing that at room temperature (20°C), one Torr-liter of gas corresponds to $3.3E19$ molecules, then M_g can be calculated if S is known. From these considerations, for arsenic ions into HPR 204 positive PR, it was found that with a measured pumping speed for hydrogen of 2100 l/sec, the values of M_g were about 900, 1400, and 2600 at energies of 40, 80, and 160 keV respectively.

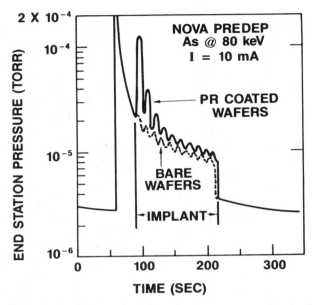

Figure 5.   PR outgassing: end station pressure versus implanting time for
batch type implanter (after Ryding, Ref.82b).

        In the case of a batch type implanter, the P vs. t characteristic is
that of a single wafer modulated by a function describing what fraction of
the area being scanned at any instant consists of PR coated wafers.   As
shown in Fig. 5, the pressure rise is greatest on the first pass of the beam
over the wafers (82b).   For the implanting conditions in the figure, the
dose per pass is 3.8E14/cm^2.   The implanting time is slightly longer for PR
coated wafers than for the bare silicon wafers because of charge
neutralization which causes the intensity of the incident beam to decrease
when the pressure is high.   This can produce overdosing in a band through
the center of the wafers in a direction parallel to the rotational motion
ofthe spinning disk.   This occurs because when the electrically indicated
beam current is lower, the feedback circuitry slows down the radial scan
response accordingly, but the flux of doping ions is actually constant
throughout the run.   The sensitivity of indicated beam current to system
pressure has been investigated for a variety of Faraday cups with
different biasing arrangements (83a).   For 3.0E15/cm^2 arsenic at 80 keV,
the resulting effect of dose errors upon sheet resistance when implanting
PR coated wafers has been estimated to be less than 2% (84b).   At
increasingly higher doses, the overall dose error would be lower because
the fraction of the implant during which the pressure was high would

decrease.   Some implanter systems incorporate "pressure compensation" circuitry which makes appropriate corrections to the dosimetry based upon instantaneous readings of end station pressure.   Some other systems only take beam current readings when the wafers are completely out of the beam.

In the previously described experiments (83b), it was noticed that if for the same dose, the beam current were reduced by a factor of two, so that the implanting time was doubled and the maximum wafer temperature was lower, the area under the curve for P vs. t was the same. This area is a measure of the quantity of gas which evolved from the PR film. When the implant was performed with one fourth the beam current, with four times the implanting time and still lower maximum wafer temperature, *again* the area under the curve for P vs. t was the same.   When the integral of pressure with time for arsenic ions into HPR 206 PR is plotted as a function of dose over a broad range of dose, the response is as shown in Fig. 6.   At a dose of about 2E14/cm^2, the response changes from a linear characteristic and then increases slowly at higher doses.   The effect upon outgassing of pre-treatments of the PR coated wafers, including vacuum bake, plasma processing, and U.V. cures was measured and compared to the

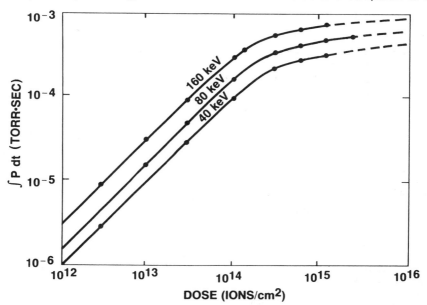

Figure 6.   Time integral of pressure versus dose for arsenic at various energies into HPR 206 positive PR (after Smith, Ref.83b).

case of no PR pre-treatment. The quantity of gas evolved was the same within 5 or 10 percent in most cases, with pre-treated wafers sometimes exhibiting greater amounts of gas evolution!

Similar results have been observed in other RGA studies in which ion energy and beam current were varied (85d). In that work, pressure integral curves were obtained for arsenic and other ion species as well. There, it was also noted that the quantity of gas evolved is a function of dose and energy, but is not a function of beam current and therefore wafer temperature. Studies of outgassing in batch machines lead to identical conclusions about the dependence upon energy, beam current, and dose (86a). These last two studies confirmed the fact that PR pre-treatment schemes do not significantly reduce the amount of outgassing observed. In instrumentation for dosimetry, measurements of optical density versus dose at various energies show a similar kind of linear response, followed by a saturation region at doses above about $2E14/cm^2$ (87c).

2.5 DAMAGE MECHANISMS

Some characteristics of heavily implanted PR were determined by Okuyama et al (78). They investigated the changes in optical density and improvements in thermal, physical, and chemical resistances of various PRs caused by high dose implants. The thickness change due to the implant was also measured, along with the stripping characteristics in an oxygen plasma. They confirmed by infrared absorption spectrum analysis and gas chromatograph analysis that the implant converted the organic film into a material resembling disordered graphite. It was found that the reaction had more to do with the base polymer than with the photoactive sensitizers. These changes in the PR were also found to be independent of the beam current used and therefore thermal effects during the implant were excluded as the cause of this "graphitization". They concluded that the resist was damaged by the ion bombardment, with incident ions breaking chemical bonds, releasing hydrogen and oxygen atoms, thus leaving the film richer in carbon. However, no quantitative data on outgassing was reported.

The above observations on the outgassing characteristics of PR were consistent with a mechanism of radiation damage by ion bombardment and resulted in the proposal of a simple model of damage to PR (83b). The diagram in Fig. 7 schematically shows the implant profile as well as the profile of damaged PR. The ion concentration is highest at R_p, but the ion breaks bonds in PR while losing energy all along its path. The maximum concentration of hydrogen atoms, C_{max}, depends upon the exact formulation of the base polymer of the PR, but is probably about $7E22/cm^3$. A single

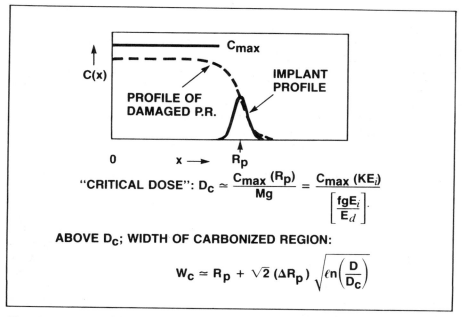

$$\text{"CRITICAL DOSE": } D_C \simeq \frac{C_{max}\,(R_p)}{Mg} = \frac{C_{max}\,(KE_i)}{\left[\dfrac{fgE_i}{E_d}\right]}.$$

ABOVE D_C; WIDTH OF CARBONIZED REGION:

$$W_C \simeq R_p + \sqrt{2}\,(\Delta R_p)\,\sqrt{\ell n\!\left(\frac{D}{D_C}\right)}$$

Figure 7.   Schematic diagram of implanted ion profile and damaged PR profile. Model for D_C and W_C (after Smith, Ref.83b).

incident ion can cause many gas molecules to be released in collisions of the primary ion and recoiled atoms in the PR. The energy deposition is not constant along the ion path as the dotted line near the surface in Fig. 7 might imply. To first order, the gas multiplier, M_g, is taken to be some fraction, f_g, of the incident ion energy, E_i, divided by the average bond dissociation energy, E_d. If the dose were increased to a "critical dose", D_c, one might say that all the hydrogen will be depleted from the carbonized portion of the PR layer to a depth of R_p, as depicted in the figure. From the terms on the right, this simple damage model would predict that since both R_p and M_g are proportional to energy, D_c should be independent of energy as seen in Fig. 6. In addition, rough estimates of the various factors would predict that D_c is approximately $1E14/cm^2$.

This model assumes as a first approximation, that below D_c, fractional carbonization exists to a depth of about R_p. As the dose increases, the PR becomes 100% carbonized to that depth when the dose is equal to the critical dose. At doses above D_c, one could estimate the "width of the carbonized PR layer", W_c, as the coordinate of the point beyond the peak concentration at which the implanted ion concentration is equal to its value at D_c. If the implant profile is assumed to be Gaussian, W_c is related to energy (through R_p and ΔR_p) and to dose by the expression given in the lower part of Fig. 7. This expression is somewhat inexact, because above D_c, the energy loss of the incident ions is *dominated* by the characteristics of amorphous carbon, since the PR has been substantially modified by that point! (It is interesting to note that many experimentally determined values of R_p and ΔR_p for PR have been obtained with doses of about $5E14/cm^2$, and therefore may be somewhat in error). In addition, from considerations of the density of typical PR resins and that of amorphous carbon, the thickness of the layer of amorphous carbon, $W_{\alpha c}$, would be expected to be about one half of W_c, the original width of the PR layer which has been carbonized. This thickness change due to the implant has been observed in some experiments.

In this model, the critical dose is the break point between linear behavior and the slowly varying dependence upon dose described by the "square root of $\ln(D/D_c)$" term. Both the pressure integral curves and the curves of optical density versus dose show these characteristics because they are indications of the same carbonized layer formation. From experiments with AZ1350 positive PR, in the dose range above D_c, the correlation coefficient for a linear regression fit of these two parameters was 0.995. Therefore, the quantity of gas which evolves and the optical density each serve as a "tag" on the value of W_c. In this case, optical density, defined as $OD=\log(I_{out}/I_{in})$, is a measurement of the attenuation of a beam of light transmitted through opaque and transparent layers in the PR and the underlying glass substrate. The change in OD, from before implant to after implant, would be expected to be proportional to $W_{\alpha c}$, the thickness of the layer of amorphous carbon. If a few simple estimates are made for R_p and ΔR_p in PR, and if one assumes a characteristic decay length for the light beam passing through amorphous carbon, the trends in the expected results are identical to those reported in Ref.(78).

In Fig. 3 and Fig. 4, the end station pressure reaches a peak value and begins to decrease at a time when the dose is equal to the critical dose. In the case of the high current machine as seen in Fig. 5, D_c was *exceeded* on the first pass of the beam over the disk. At low doses, well below D_c, the pressure remains constant because each incident ion releases many gas molecules, with a ratio given by M_g. The pressure in the system is proportional to beam power because the arrival rate of ions is related to the beam current and the depth of penetration is related to the voltage through which the ions have been accelerated. That is, the range of the ion in the PR, or the path length along which gas is liberated, is proportional to energy. However, this outgassing is *not* caused by wafer heating.

Regardless of the rate at which ions arrive, for a given energy and dose, the total quantity of gas which evolves, the degree of carbonization (or the values of W_c and $W_{\alpha c}$), and the optical density *are always the same* because the final state of the damaged PR is the same. Also, for these basic reasons, none of these effects depend upon the temperature of the wafer during the implant, because the damage is not caused by thermal effects. Nor would these properties be expected to be significantly affected by various pre-treatment schemes. The empirical formula for W_c, given in Fig. 7, shows that higher energies would produce greater changes in W_c than higher doses would. Fig. 8 schematically shows the progressive degradation of the upper surface of the PR film for increasing doses, expressed as multiples of the critical dose. The degree of PR carbonization and the PR film thickness change rapidly at doses near D_c. However, at increasingly higher doses, the changes in PR film thickness and the width of the layer of amorphous carbon, $W_{\alpha c}$, occur much more slowly. Similar effects are described in work directed at improving the dry etching stability of electron-beam and other resists by an ion exposure technique (82a).

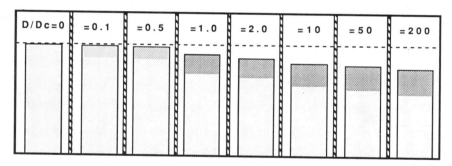

Figure 8.   Schematic diagram of progressive PR carbonization and film thickness change as a function of dose relative to critical dose, D_c.

2.6 DIFFICULTY OF STRIPPING IMPLANTED PHOTORESIST

In an earlier reference, a matrix of experimental implants with arsenic was described for which data was obtained in an attempt to quantify wafer temperature, outgassing, and the "difficulty of resist strip" as a function of PR type, pre-treatment schemes, and implant parameters (83b). These studies were done in two kinds of implanters: a single wafer machine with only radiative cooling and a batch machine with conductive cooling capabilities. The experiments focussed on implants at equal energy densities: 6.0E14/cm² at 160 keV and 1.2E15/cm² at 80 keV. The basic validity of the temperature modeling was confirmed and outgassing characteristics described above were explained satisfactorily. An understanding of the damage mechanisms, as presented in the previous section, also led to insights into the problem of stripping implanted PR.

The usual means of stripping PR from wafers include wet chemical and plasma "ashing" techniques. These may be used either separately or in combination depending upon the layer being stripped. Wet chemical methods involve oxidizing solutions, such as a heated mixture of sulfuric acid and hydrogen peroxide commonly called "piranha", while ashing usually implies processing in an oxygen plasma. Continued emphasis on "all dry" processing has lead to the development of "downstream" or "afterglow" ashers in which the wafer is separated from the region where the plasma is generated in order to minimize damage to oxide layers in MOS circuits. Studies of an all dry ashing process have shown that non-volatile compounds of the implanted species and the resist, as well as sputter deposited metal from the implanter, cause surface residues (87b).

A suitable figure of merit for quantifying the difficulty of resist strip may be the ratio of ashing time for implanted PR divided by the ashing time for non-implanted PR of the same thickness. For the wafers in the experimental matrix mentioned above, it was found that in a single wafer plasma asher, the ashing time for all wafers implanted at 160 keV with $6.0E14/cm^2$ dose was 1.33 times greater than the ashing time for PR without implant. Similarly, this ratio was 1.20 for every wafer implanted at 80 keV with $1.2E15/cm^2$ dose *regardless* of the other implant related parameters, such as beam current, wafer temperature, PR pre-treatment scheme, and even which machine was used. This trend was to be expected because the mechanism of stripping implanted PR obviously consists of first ashing the layer of amorphous carbon on top, of thickness W_{ac}, at a reduced ashing rate relative to virgin PR, and then ashing the remaining undamaged PR underneath at a normal ashing rate. In this instance, the ashing rate for the layer of amorphous carbon seemed to be about one third that of virgin PR. The result that the high energy/low dose implant is actually "harder to strip" than the low energy/high dose case, with ratios close to the experimental values, could be predicted from estimates of W_c and W_{ac} using the pressure integral data from outgassing studies.

Considerations of the implant parameters which determine the degree of PR carbonization and the width of the carbonized PR layer, W_c, would show that for doses below about $5E13/cm^2$ there is no problem. Also, above D_c, the difficulty of resist strip would be influenced more strongly by energy and increase slowly with increasing dose. As depicted in Fig. 8 by the heavily shaded region, W_{ac}, the thickness of the layer of amorphous carbon, is about half of W_c, the width of the portion of the PR layer which

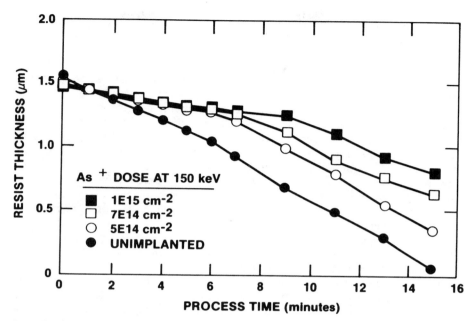

Figure 9.   Remaining PR thickness versus processing time in asher for various total ion doses (after Orvek and Huffman, Ref.85b).

has been carbonized.   In another study made with a variety of implants (85b), cross-section SEM micrographs of carbonized PR lines confirmed many of the same observations about the dependence of the damage upon implanter operating parameters.   This technique provided direct measurements of W_{ac} from the micrographs.   It could also be seen that on sloped sidewalls, the resist was carbonized to a lesser extent as measured from the outer surface.   Data on the PR thickness as a function of ashing time in an oxygen/argon plasma is presented in Fig. 9.   As expected, the ashing proceeds through the layer of amorphous carbon at a reduced rate (about 30% of the rate for virgin PR), and then the undamaged PR is ashed at the normal rate.   As the simple model of PR damage predicts, the film thickness after implant is reduced and W_{ac} is only slightly different for the doses used in this case.

2.7 SUMMARY

Consideration of the implant parameters are important in determining the effectiveness of the PR masking layer and for ensuring that the wafer temperature does not exceed the maximum allowable value recommended for the specific PR being used. However, it is now quite clear that outgassing and the damage to the PR which makes it difficult to strip are the *inevitable* result of ion bombardment. The mechanisms of radiation damage which cause outgassing and the formation of a carbonized layer on the top surface of the PR do not depend upon the wafer temperature and are not significantly affected by PR pre-treatments. A simple model for PR damage satisfactorily explains the phenomena of outgassing on either side of D_c, the critical dose. It also explains the observed trends in optical density, the thickness change of the PR, the directly measured thickness of the layer of amorphous carbon, $W_{\alpha c}$, and the subsequent difficulty of resist strip. These effects are all the result of the conversion of the base organic compounds into amorphous carbon. The relationship of these effects to implanter parameters have been sorted out in a variety of complementary experiments which have come to some of the same conclusions.

A review of some details of existing process flows may be worthwhile to enhance productivity. Otherwise, many of the ideas of the past which seemed quite reasonable, such as the notion that outgassing of PR is caused by wafer heating (and could therefore be avoided by running with low beam current), will continue to impact the wafer throughput or utilization of equipment. It may still be thought that wafers processed in a low beam power mode are easier to strip. Similarly, productivity in some ashing operations might be improved. If a single process employing a very long ashing time were used after *all* implants regardless of the implant parameters, this may actually be "overkill" for a large portion of the material being processed. In addition, some pre-treatments might be eliminated after reconsidering their utility.

3.0 PARTICLE CONTAMINATION
3.1 IMPACT ON YIELD IN IC WAFER FABRICATION

In virtually all phases of processing wafers, the control of particle contamination from the equipment itself is every bit as important as the proper management of the clean room facilities and the people working in them. Particles on the surface of wafers can cause randomly distributed defects in photolithographic processes and in other critical steps. As mentioned above, in the case of implantation, this could lead to blocked implants, some of which could be killer defects in terms of degrading device and therefore circuit performance. Yield is defined as the number of good die tested divided by the total number of available die on a wafer or in a lot, and is usually expressed as a percentage. In any given processing technology, the number of critical masking steps is fixed and the circuit design rules determine the die size. Continuing trends toward lower critical dimensions imply that smaller particles will become killer defects. At the same time, trends toward higher levels of circuit integration imply that die sizes will continue to increase, limited only by packaging technology. Each generation of technology may require a factor of five reduction in defect densities attributable to particles. Thus, the pressure to reduce contamination due to particles in all processing equipment is unrelenting.

A number of empirical yield models have been developed (83c). Overall yield at final test is basically a function of die area and defect density. One model which is said to be useful whenever yields are greater than 33%, for circuits with similar processing, is Murphy's yield model given by:

$$Y = [(1-e^{-AD})/AD]^2 \qquad (5)$$

where A is the die area and D is the effective cumulative defect density or number of defects per unit area. The circuit yield according to this model is shown graphically in Fig. 10, where it can be seen that for a die size of 50 mm^2 (77,500 mil^2), decreasing the effective defect density from 2.0 to 0.5 defects/cm^2 will increase the circuit yield from 40% to 78%. On 150 mm diameter wafers, an effective defect density of 1.0 defect/cm^2

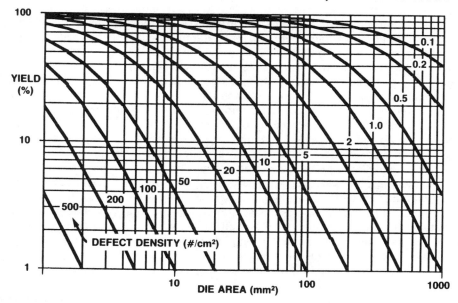

Figure 10. IC yield versus die area for various effective defect densities, according to Murphy's yield model.

represents 177 cumulative particles per wafer. This corresponds to adding about 10 particles per wafer per pass on each of 18 critical layers. For particles of diameter >0.5μm, defect densities in state-of-the-art implanters are typically less than 0.1 defect/cm^2. This is near the defect density required for cost effective manufacture of 1M DRAMs (Dynamic Random Access Memories). Yield models which appropriately describe the defect sensitivity of memory chips have been treated by Stapper (86b).

Although other processing equipment involves similar vacuum system operation and wafer handling mechanisms, some of the best quantitative information on particle contamination in these areas has been obtained by the manufacturers and users of implanters. Improvements in particle performance have been gradual, usually involving redesign or modification of hardware and/or software. Many of the comments below about procedures and practices for monitoring and maintaining the particle performance of implanters apply to both serial processing and batch processing machines, regardless of differences in design features.

3.2 SOURCES OF PARTICLES IN IMPLANTERS

The major sources of particles will be described along with a discussion of the forces which act on particles and lead to their redistribution in these systems. Particles may be deposited on wafers processed in implanters both in the external regions contiguous to the clean room and inside the vacuum environment of the end station. One principal source of particles is the silicon wafers themselves, which can generate microscopic particles from being chipped at their outer edges or from abrasion due to handling. Catastrophic wafer breakage which may be due to mechanical stresses in wafers or may result from malfunction of a handling or clamping mechanism generates a copious supply of particles! It is safe to say that by *eliminating* wafer breakage, any processing operation will go a long way toward keeping particle counts low. In addition, particles can be generated by moving parts in the vicinity of the wafer handlers or end station. Residues from cleaning wipes used on end station parts may be present. Deposits on the walls inside the vacuum system, from condensed films of source material or backsputtered material in the beamline, may flake off and migrate onto wafers. PR films on wafers may get chipped by handling or clamping mechanisms. In addition, fragments of blisters in PR may be generated inside the implanter when PR coated wafers are insufficiently cooled (84b).

The types of forces which act on particles and cause them to be transported about in implanters include aerodynamic, centrifugal, gravitational, electrostatic, and impact forces. Quantitative estimates of the magnitude of these respective forces have been made for a high current machine (85a). For a particle of 1µm radius, aerodynamic drag forces at pressures above 10^{-3} Torr exceed the force of gravity. The gravitational force is also exceeded by the centrifugal force in machines which hold the wafers on a disk which spins at high RPMs. Aerodynamic forces vary strongly with particle size and increase sharply at high system pressures. These aerodynamic drag forces can be eliminated by moving the wafers only in sufficiently good vacuum. In batch machines, these same facts are put to good use to provide a periodic cleanup of the end station by alternately injecting gas through the venting port and pumping the gas away through the roughing port while spinning the disk during both sequences. This particle purging or auto-clean cycle is effective for reducing particle counts to their baseline values. It is generally thought that particles adhere to surfaces by chemical bonding or electrostatic forces and that they are transported by aerodynamic forces in turbulent gas flow during the roughing and venting sequences and by centrifugal forces, in the case of machines which use a spinning disk.

3.3 MEASUREMENT TECHNIQUES

In cleanrooms, it is standard practice to take periodic measurements of airborne particles using aerosol particle detectors, especially in the vicinity of the end stations of implanters, where wafers are exposed. This checks the general level of cleanliness and verifies the effectiveness of the laminar flow equipment. In addition, the principal method of monitoring particle performance related to each individual implanter is to make measurements with a surface inspection system, obtaining particle counts on wafers before and after the implant step. The "before" counts are subtracted from the "after" counts, so that the "delta", or difference in particle counts, is used to characterize the condition of the system being tested. The figure of merit is usually the "particles per wafer per pass" (85f). Because of the high variation in measured values which is often observed, proper statistical techniques must be applied in order to draw inferences from the results. The output measurements are usually presented as maps showing the locations of detected particles, or in the form of tables or histograms corresponding to preset intervals of particle size.

Surface inspection systems capable of making measurements on patterned product wafers have become commercially available, but to date, virtually all measurements of particles added have been made using bare silicon test wafers or oxide coated wafers. These systems illuminate the surface with a high intensity collimated light source or a scanned laser spot and detect the scattered light using a photomultiplier

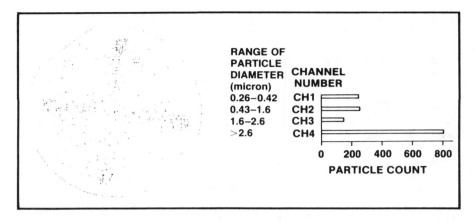

Figure 11.   Output from surface inspection system: map showing locations
            of particles and histogram of counts for various channels.

tube.   The particle counts within about 7 mm of the edge of the wafer are
usually excluded from analysis.   Various response levels correspond to
different particle sizes; systems can be calibrated using latex spheres of
known diameter or standard reference wafers having patterns of etched
defects of known quantity and size.

A map of an implanted wafer and its associated histogram is shown in
Fig. 11, with the ranges of particle size for the different channel numbers
given in the figure.   This particular pattern of high particle counts is
related to the location of four pumping ports connecting the volume on the
front side of the wafer with the back side region.   This result indicates
contamination of the pump-out manifold, probably when a wafer was
broken in the vacuum lock.

An optical system which provides a real-time readout of particle flux
inside vacuum processing equipment has recently become available.   Its
use for monitoring particle counts in the end station of a high current
implanter has been described (87a).   The compact sensor, which is mounted
inside the vacuum chamber, uses a laser beam which is reflected back and
forth between two mirrors and is finally collected on a beam stop.   The
signal from photocells mounted directly above the mirrors detects light
which is scattered from particles.   The operation of the sensor is
independent of system pressure and provides a time resolution of better
than 0.1 second.   Particle flux counts (down to 0.5 micron diameter size)
can thus be obtained as a function of time during events such as roughing,
spinning the disk, venting, etc.   Cleaning cycles can be initiated on the

basis of changes in real time data instead of being scheduled at arbitrary intervals based on time or number of runs made.

3.4 EQUIPMENT QUALIFICATION

The procedures used for qualifying the particle performance of implanters and other processing equipment are not yet standardized throughout the various wafer fabrication facilities. However, many IC manufacturers with multiple processing areas may apply their own internal standards. Because of the different wafer sizes in use, the figure of merit used for comparison is usually the number of particles per unit area greater than a certain size. For instance, a vendor might guarantee fewer than 0.05 particles/cm^2 greater than 0.5 μm diameter. To date, numerous different types of surface inspection systems with different characteristics have been employed by users and vendors alike for testing the particle performance of equipment. Not only do the actual distributions of particle sizes encountered vary greatly, but also the number and width of the "bins" or "channels" of particle sizes reported are often dissimilar from one particle detector set-up to another. The sensitivity to particle shapes and the underlying films may vary, giving different results from one type of particle detector to another for particle counts on a given set of wafers, even when using the same model of surface inspection system! For these reasons, it is often difficult to compare systems (both implanters *and* particle detectors) and to compare processing areas.

A whole new industry has grown out of the need for contamination control in the design, construction, and operation of clean room facilities which are filled with a wide variety of processing equipment. In most existing areas, efforts to reduce particles have been ongoing, with gradual improvements being made in response to more stringent requirements. In the clean room, the responsibility for maintaining specified particle performance levels is ultimately shared by everyone in all operations. Comparable particle count specifications and common procedures should be established and controls put in place so that appropriate actions are taken whenever limits are exceeded.

The proper calibration and maintenance of surface inspection systems is mandatory, and the trend toward lower limits of allowable particles per wafer per pass requires greater measurement accuracy and precision. The results of any program depend strongly upon the processing area's environment, its clean room culture, and cleaning procedures. In the initial phases, the design of experiments and the planning and execution of subsequent measurements are important factors. Care must be taken to obtain statistically significant results in processing which are comparable to those obtained at the vendor's factory (85f, 87d). An evaluation procedure for locating and identifying any sources of particle contamination throughout the wafer handling operations of a serial processing machine has been developed in flow chart form (85e).

Those sources of variability which can be controlled should be identified and the testing procedure standardized accordingly. Wherever possible, the operating conditions for running monitors should be the same as for running product wafers. Later, it is desirable and more efficient to run routine particle count monitors through the entire implant operation along with production lots. This may raise concerns as to whether or not the number of particles added is significantly effected by the actual implant itself or by the presence of a PR coating on the accompanying product wafers. The author has not observed any correlation in results to these factors in either a serial processing machine or a batch processing machine, both of which run material with PR masking being used at least 75% of the time.

3.5 OPERATING PROCEDURES

Once the required particle performance has been verified, a periodic testing schedule using suitable monitors should be established. The results are usually displayed on process control charts which are maintained with realistic "warning" and "shut-down" limits. These limits *must* be rigorously enforced as part of the operating specification of the implanter. Complete details of the testing procedure should be prescribed with step-by-step instructions. Adherence to schedules and procedures for periodic cleaning is mandatory. This includes the wafer loading areas and the interior of the end station vacuum chambers, especially the Faraday cup assemblies. The periodic use of particle purging cycles is strongly recommended, if the machine is equipped with this capability.

If a wafer breaks or end station maintenance is performed, the corrective actions to be taken by either operators or maintenance personnel to restore acceptable performance should be documented. Recommended cleaning procedures for areas at atmosphere and surfaces inside the vacuum chambers include using a vacuum cleaner suitable for cleanroom service, preferably one which also effectively traps toxic material. This is followed by a thorough wiping with lint-free cloths soaked in Freon TF and then an isopropyl alcohol rinse (87d). In many instances, for example following a major system disassembly, cleaning, and reassembly, it may be appropriate to run a cassette of clean dummy wafers through a full implant cycle, since these may serve to "getter" particles which might still remain after cleaning. Following these steps, the machine particle qualification procedure should be performed and the implanter should not be released to run production material unless the added particle counts are below the warning limits.

For implant operations where a high percentage of wafers run use PR masking, the incorporation of "edge bead removal" techniques as part of the photolithography process has been reported to reduce particle counts in an implanter by 50% (84a). The bead of PR which forms at the edge of the wafer in the spin coating step can be removed to a point several millimeters in from the edge in the edge bead removal step. This portion of the wafer might be contacted by wafer handlers or clamping mechanisms in the end station of some implanters. Since baked PR is brittle, contacting the wafer edge tends to generate particles when edge bead removal is not used (85g).

3.6 MACHINE MODIFICATIONS

Particle contamination studies of implanters have usually led to redesigned hardware or modified operating features which in turn have led to lower particle counts. The resulting improvements have been brought about through the efforts of both the users and the manufacturers of various systems. In many cases, the manufacturers have made retrofit kits, which are available for upgrading the particle performance of older systems. Some of the modifications or changes in operation can be adapted to existing systems, even though they are not identical to the system configuration involved in the original study. For example, an early report pointed out that particle counts were reduced by a factor of 2 to 5 when the vent valve in the vacuum locks of the end station was relocated so that it was opposite the location of the roughing valve, instead of adjacent to it (85g). This arrangement produced a uni-directional flow of gas when pumping or venting the lock, and could easily be adapted to other similar systems. Another observation was that, even though final filters were used on the gas inlet lines, when the delivery pressure of the venting gas was too high, the particle counts on wafers would increase by an order of magnitude because of greater turbulence in the vacuum lock.

The careful selection of components such as filters, valves, vacuum fittings, and tubing has been documented (85c). The basis of comparison was the measured counts of particles generated under conditions of repeated cycling. Any possibility of metal scraping on metal should be eliminated and teflon coating of parts which are likely to rub together is recommended. Wafer handling mechanisms of the "pick and place" type, which avoid any sliding or abrasion of wafers by the use of mechanisms which provide for backside contact only, are preferred.   Positive sensing of wafer position throughout the sequence of motions is desirable to avoid wafer breakage.   Automated wafer handling also reduces the contact by operators with the material being processed, further reducing contamination from that source.   Process gas used for venting the vacuum chambers should be filtered at its point of use and the compressed air used for pneumatic actuators should be exhausted into non-critical areas well away from the regions where wafers are exposed.

Since the gas turbulence inside vacuum locks during the roughing and venting cycles is generally conceded to be a major cause of added particle counts, the use of restricting orifices on the roughing and venting lines has been investigated (87e).   It was found that the contamination level was a strong function of the venting orifice size and was relatively insensitive to the roughing orifice size.   Due to the longer pumping and venting times, the wafer throughput is reduced by approximately 10%. This could probably be tolerated if it results in lower particle counts and therefore increased wafer yield.   It was also found that it is important to balance the flow of gas over the front and back sides of the wafer, so that particles on the back of the wafer do not migrate to the front side.   When this balance is achieved, the observation of higher particle counts on the "first use" ceased to occur.

3.7 SUMMARY

The subject of particle contamination in implanters has been reviewed in relation to concerns for reducing defect levels in all wafer fabrication processes.   Continuous improvement has been necessary in order to achieve high yields and to be successful in cost effective IC manufacturing.   The maximum allowable levels of defect densities for state-of-the-art ICs are constantly changing, but examples of present day (1988) requirements were discussed with reference to one of the time-tested yield models in order to put the numbers in perspective.   The generation and transport of particles and the forces which act upon them in implanters were briefly described.   Measurement techniques are not standardized, making comparisons of equipment very difficult.   In spite of this, improved performance has been demonstrated in the past few years, especially in newer equipment.   Whether purchasing new systems or operating existing implanters, the verification of specific particle performance is now considered to be a key part of the overall system qualification.

Later discussions outlined various particle studies, which have led to modifications to equipment or changes in operating procedures.   These studies showed how implanters and their environment can be adapted to meet the goals of low particles per wafer per pass. In the processing area, after the required baseline level of particle performance is initially achieved, continuous monitoring of particle counts on wafers is required to verify that each system is qualified to run product. For any system, compliance to the particle count specification is just as important as providing the right ion at the proper energy and correct dose! Realistic warning and shut-down limits must be established and rigorously enforced. The schedules for periodic cleaning should be adhered to with the same discipline as other preventive maintenance measures. Achieving and maintaining low defect densities is necessary for survival in present day manufacturing and indispensable in the future.

4.0 REFERENCES

72) G. Baccarani and K. A. Pickar, "Range and Straggle of Boron in Photoresist", Solid-State Electronics, Vol. 15, (1972), p. 239.

75) J. F. Gibbons, W. S. Johnson, and S. W. Mylroie, *Projected Range Statistics- Semiconductors and Related Materials,* 2nd Edition, (Dowden, Hutchinson, and Ross Inc., Stroudsburg, PA, 1975).

77) B. Smith, *Ion Implantation Range Data for Silicon and Germanium Device Technologies,* (Research Studies, Forest Grove, OR, 1977).

78) Y. Okuyama, T. Hashimoto, and T. Koguchi, "High Dose Implantation into Photoresist", J. Electrochem. Soc., Vol. 125, No. 8, (1978), p.1293.

82a) K. Mochiji, Y. Wada, and H. Obayashi, "Improved Dry Etching Resistance of Electron-Beam Resist by Ion Exposure Process", J. Electrochem. Soc., Vol. 129, No. 11, (1982), p. 2556.

82b) G. Ryding, "Evolution and Performance of the Nova NV-10 Predep Implanter", in *Ion Implantation Techniques*, (Eds. H. Ryssel and H. Glawischnig, Springer-Verlag, New York, 1982), p. 319.

83a) G. Ryding, "Dosimetry and Beam Quality", in *Ion Implantation Equipment and Techniques*, (Eds. H. Ryssel and H. Glawischnig, Springer-Verlag, New York, 1983), p. 274.

83b) T. C. Smith, "Wafer Cooling and Photoresist Masking Problems in Ion Implantation", in *Ion Implantation Equipment and Techniques*, (Eds. H.

Ryssel and H. Glawischnig, Springer-Verlag, New York, 1983), p. 196.

83c) C. H. Stapper, F. M. Armstrong, and K. Saji, "Integrated Circuit Yield Statistics", Proc. IEEE, Vol. 71, No. 4, (1983), p. 453.

84a) N. Durrant and P. Jenkins, "Defect Density Reduction Utilizing Wafer Edge Resist Removal", Kodak Microelectronics Seminar, Interface '84.

84b) K. Steeples, "Dose Control with High Power Ion Beams on Photoresist Masked Targets", J. Vac. Sci. Technol., Vol. B2(1), (1984), p. 58.

85a) D. H. Douglas-Hamilton and C. Taylor, "Particles and Particle Transport in Ion Implanters", Nuc. Instr. & Methods in Phys. Res., Vol. B6, (1985), p. 196.

85b) K. J. Orvek and C. Huffman, "Carbonized Layer Formation in Ion Implanted Photoresist Masks", Nuc. Instr. & Methods in Phys. Res., Vol. B7/8, (1985), p. 501.

85c) J. Pollock, N. Turner, R. Milgate, R. Resnek, and R. Hertel, "Particulate Performance of a Wafer Handler for Serial Process Ion Implantation", Nuc. Instr. & Methods in Phys. Res., Vol. B6, (1985), p. 202.

85d) D. Roche, J. F. Michaud, and M. Bruel, "Outgassing of Photoresist During Ion Implantation", Mat. Res. Soc. Symp. Proc. Vol. 45, (1985), p. 203.

85e) D. M. Tennant, A. H. Dayem, R. E. Howard, and E. H. Westerwick, "Range of Boron Ions in Polymers", J. Vac. Sci. Technol., Vol. B3 (1), (1985), p. 458.

85f) B. J. Tullis, "A Method of Measuring and Specifying Particle Contamination by Process Equipment", Microcontamination, Vol. 3, No.11, (1985), p. 67.

85g) W. H. Weisenberger and M. McCullough, "Particle Generation in Medium Current Implanters", Nuc. Instr. & Methods in Phys. Res., Vol. B6, (1985), p. 190.

86a) B. N. Mehrotra and L. A. Larson, "An FTIR and RGA Study of the Outgassing of Photoresist During Ion Implant", Proc. SPIE Vol. 623, (1986), p. 78.

86b) C. H. Stapper, "The Defect-Sensitivity Effect of Memory Chips" IEEE J. Solid-State Circuits, Vol. SC-21, No. 1, (1986), p. 193.

87a) P. G. Borden, Y. Baron, and B. McGinley, "Monitoring Particles in Vacuum-Process Equipment", Microcontamination, Vol. 5, No. 10, (1987), p. 30.

87b) S. Fujimura, H. Yano, J. Konno, T. Takada, and K. Inayoshi, "Study on Ashing Process for Removal of Implanted Resist Layer", Extended Abstracts of The Electrochem. Soc., Vol. 87-2, (1987) p. 1051.

87c) J. R. Golin, N. W. Schell, J. A. Glaze, and R. G. Ozarski, "Latest Advances in Ion Implant Optical Dosimetry", Nuc. Instr. and Meth. in Phys. Res., Vol. B21, (1987) p. 542.

87d) J. Jost, G. Angelo, and J. Turner, "0.5 µm Particulate Performance of a Batch Process Implanter", Nuc. Instr. & Methods in Phys. Res., Vol. B21, (1987), p. 372.

87e) R. Milgate and R. Simonton, "Recent Improvements in the Particle Performance of a Varian 350D Ion Implanter", Nuc. Instr. & Methods in Phys. Res., Vol. B21, (1987), p. 381.

Ion Implantation Diagnostics and Process Control

M. I. Current, C.B. Yarling,
Applied Materials, Santa Clara, CA
and
W.A. Keenan,
Prometrix Corp., Santa Clara, CA

A. Introduction

The goal of this chapter is to introduce the basic tools and procedures which are routinely used for diagnosis and process control of ion implantation. Although the need for consistent characterization of implanter function is clearly seen for the IC production environment, one also needs to emphasize the importance of similar characterization and control procedures for research and process development functions. Many a research project in ion-solid interactions has been seriously compromised, or at least confused, by undetected, "anomalous" performance of an ion implantation system operated under "routine", but unverified, conditions [1].

The emphasis of this chapter will be on methods which verify the performance of an ion implantation system by analysis of implanted materials, either test surfaces or selected regions of in-process IC device wafers. Complementary analytical methods, which emphasize direct characterization of the implantation system itself, are discussed in Ref. 1. A key feature of many of the methods described in this chapter is the use of various techniques to display the spatial variation of properties of implanted surfaces in order to verify proper performance and trouble-shoot system malfunctions. This is commonly refered to as "wafer mapping".

The range of problems that can cause serious failure of an ion implantation process is broad and constantly changing. One group of implant technologists, the Silicon Valley (CA) Implant Users Group, has been meeting monthly since 1983 to discuss practical issues in ion implantation. The distribution of topics which have been chosen for discussion over the last four years is shown in Fig. 1. The emphasis on wafer charging reflects the critical importance of these effects to the yield and reliability of advanced CMOS devices. Figure 1 also demonstrates the need for continued attention on such issues as dose uniformity, resist integrity, contamination by particles and sputtered films, maintenance and safety procedures in order to be assured of "routine" ion implantation processing.

ION IMPLANTATION:
SCIENCE AND TECHNOLOGY

377

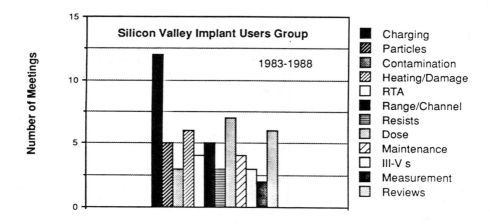

Figure 1.   The distribution of meeting topics for the Silicon Valley Implant
             Users Group for the years mid-1983 to mid-1988.

     For general references on diagnosis and process control of ion implantation
systems, the reader is directed to previous implant school texts [2,3], the classic book
by Wilson and Brewer [4], the recent text by Ryssel and Ruge [5]. The proceedings
from the conference series on Ion Implantation Technology is also a valuable
resource [6].

B. Process Control Concepts

As process requirements for advanced IC manufacture continue to become more demanding, the need for monitoring and controlling the implantation process as a defense against operational, equipment, and process errors becomes increasingly important. The use of automation, combined with proper training of implant operators, can reduce "man-made" errors. System self-diagnostics and adaptive control mechanisms which are now being introduced in advanced ion implantation tools will augment, and ultimately replace, the process control techniques described in this chapter.

Comprehensive and consistent process control programs not only ensure proper operation of the machine itself, but also monitor interactions between the ion implantation equipment and the wafer. The goals of a process control program are higher device yield, better utilization of manufacturing equipment and personnel and a balanced product flow within the production line.

The ability to solve implant process problems can be a complicated and tedious task if the entire measurement procedure is not fully characterized.

A useful definition of process control is [7]:

The measuring of particular critical variables within the manufacturing process for the purpose of (1) gathering information about the process and processing equipment; (2) analyzing that information; and (3) making a decision about the status of the process. If the status is unacceptable or if disturbances or malfunctions exist, corrective action will be initiated .

The steps in a basic process control procedure for ion implantation are implanting and annealing a test monitor wafer, measuring the results on a four-point probe, and plotting the results on a process control graph. This approach, outlined in Fig. 2a, verifies the operation of the implanter and it's interaction with the wafer. A more complete control procedure should include independent verification of the implant monitoring equipment, as shown in Fig. 2b. This verification may be implemented by periodic: (1) probe calibration; (2) tests by software-driven self-diagnostics; (3) processing of independent test monitors and (4) daily processing and subsequent evaluation of samples from implanted batches of wafers.

The tangible results from such an approach should include a process control (PC) chart for the daily implant test monitor, PC charts for the daily probe qualification and test monitor of the four-point probe, and a PC chart for a daily/periodic implant test monitor which is processed in the annealing system, (either furnace or rapid thermal processor). These types of charts and daily checks ensure the integrity of the implanter as well as the implant monitoring process. Then, when anomalies are obsereved in the implanted test wafer, attention may be focused directly upon the implanter, since a high degree of confidence has already been placed in the functionality of the process used to produce that test wafer.

An Integrated System Of Process Control

The implant process control program considered up to now has focused upon the indirect method of measuring sheet resistance upon a test wafer. A much more thorough program would supplement the sheet resistance technique with regular capacitance-voltage (C-V) analysis, optical analysis that includes both particle measurement and dose evaluation systems, and integration of periodic feedback from parametric test data of implanted device structures. Such an optimum program is outlined in Fig. 3. In this example, the central core of process control is still the sheet resistance technique, where sheet resistance monitors are implanted, annealed and measured with the four-point probe. In addition, there are monitor wafers which are implanted and measured on the particle measuring system and dose monitors which are implanted and processed by an optical system using a focused white-light, laser scanning or ellipsometry. Also included in the integrated process control program illustrated in Fig. 3 is capacitance-voltage analysis (C-V), a useful technique for monitoring low-dose implants [8,9].

Parametric Test Data

Parametric test data is collected from test structures which are located on a product wafer. Electrical data from a well-designed test array can yield useful information about the numerous processes (including implant) which occurred during wafer manufacturing. This data is particularly useful when it is combined with the implant-area control data on the performance of individual implantation tools. This is especially valuable when standard process control techniques lack sufficient sensitivity to monitor significant IC device yield effects [10].

The ultimate goal of process control procedures is to be able to measure the performance of each system in the process control loop *in-situ*, or within the system itself, in real time. One can expect to see continued efforts towards the development of the necessary measurement and adaptive control technology.

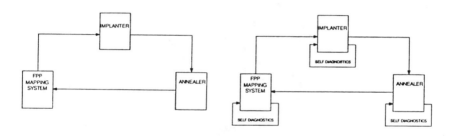

Figure 2: Methods of Process Control (a) Historical method verifies only the operation of the implanter; (b) Modern approach includes independent verification of the implant monitoring equipment.

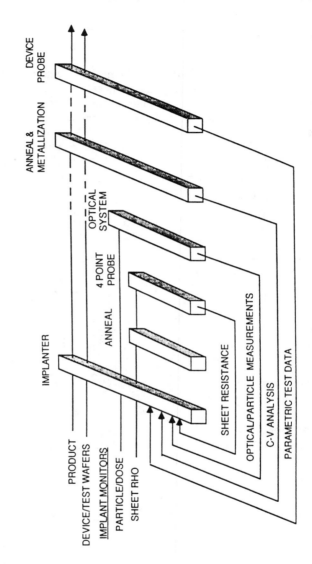

Figure 3: Process Control Flow Chart for Ion Implantation.

C. Measurement Tools

Process Requirements

The choice of measurement tools for implant diagnostics and process control is strongly driven by the requirements for advanced integrated circuit fabrication. The general requirement is that the sensitivity of the measurement tools be an order of magnitude better than the levels of variation which give significant process impact. For instance, to control dose uniformity to better than 1%, the probe noise levels should be less than 0.1%. Modern ion implantation tools are capable of operation with dose uniformities of less than 0.75% and day-to-day repeatability in the 1% range. The relative dose accuracy, which will be discussed in detail later in this chapter, is less impressive; ranging between 5 and 20 %.

Dose accuracy requirements can vary considerably between processes and individual process steps. The most sensitive implants are those that control the transistor gain or switching voltage, such as the base and emitter implants in bipolar devices and the threshold implants in MOS devices. Any variation in dose or implant depth in these implants is directly reflected in transistor characteristics.

High-speed bipolar transistors require base doping control in the range of $\approx$ 0.3% and base widths (the distance between the emitter and base junctions) of a few hundred Angstroms. Increases in device area, such as large gate-array devices and efforts towards wafer-scale integration, drive additional requirements for process uniformity at spatial dimensions of the order of scale of the IC device. Increases in wafer size, with 200 mm wafers now appearing in pilot-line production, and decreases in device feature sizes to below 1 µm, extend the spatial range for process control measurements to over 5 orders of magnitude.

Increased attention is being given to the control (and elimination) of "foreign materials" which are deposited in the device surface during processing. For the case of ion implantation, the principal items for concern are solid-particle "dust" and sputtered films of other dopants and metals [11-13]. Present day requirements for advanced devices call for particle contamination of less than 0.1 particles per cm^2 for particle sizes greater than 0.3 µm. Estimates of particulate contamination for 16 Mb DRAM devices call for improvements in the levels of cleanliness to nearly 10^{-3} particles per cm^2 [14]. Enhanced diffusion effects associated with the sputter deposition of "cross"-dopants, such as the remnants of beams previously implanted in the same machine, have been observed at levels as low as one part in 10^4 of the primary ion dose [12,13].

Measurement Tools

The choice of measurement tools for routine diagnostics and control of implantation operations is related to the tools used for research and process development, but is subject to the additional considerations of convenience, cost, simplicity of analysis, and the relationship between measured test parameters and device perfomance. Dose measurements can be based on direct measures of the number of incident atoms, such as electrical activity (sheet resistance, threshold voltage, capacitance-voltage) or indirect effects associated with implant damage such as the optical density changes in organic films (mylar film burns, Ionscan), or damage in semiconductor layers (ellipsometry, Therma-Wave, double-implant/sheet resistance).

Electrical Measurements

Sheet resistance measurement is widely accepted as a general purpose tool for routine process control measurements. The direct connection between the sheet resistance of an implanted layer and many key aspects of IC device performance and the availablity of resistivity standards for probe calibration are strong points in its favor. Direct measurements with four-point probes can cover an implant dose range from 5×10^{12} to well above 10^{16} ions/cm^2. Surface passivation with sulfuric peroxide can extend the range of direct measurements to $\approx 2 \times 10^{11}$ ions/cm^2.

The double-implant technique, which uses the effect of implant damage on an electrically active layer to measure the implant dose, can extend the use of sheet resistance probing down to 10^{10} ions/cm^2 [15]. An initial implant is done to form a conducting layer on the wafer. This layer is annealed and then implanted with the low dose implant to be tested. The sheet resistance of the initial layer is raised by the damage caused by the second implant. A special advantage of the double-implant technique is that the sensitivity (the percent change in the measurement divided by the percent change in dose) can be tailored by the choice of the dose and energy of the initial implant. The sensitivity of the double implant method decreases rapidly for implant doses above 10^{14} ions/cm^2, which is well within the range of normal operation of the single, direct measurement technique.

Additional electrical techniques which have been used to monitor dose levels and uniformity include measurements of spreading-resistance (two-point probe), Van-der-Pauw resistance, capacitance-voltage curves and threshold voltage shifts in MOS devices [16]. These methods have not been as widely used for process control purposes because of the time delays and expense connected with the additional processing steps required to implement these techniques.

Optical Measurements

Optical measurements offer the significant advantage of a "non-contact" probing method which can, in many cases, be focused to the dimensions of a IC device structure (nominally 1 μm). Direct measurements of damage effects on semiconductor materials show practical changes in optical constants (ellipsometry) and thermo-acoustic properties (Therma-wave) [17,18]. The Therma-wave technique has demonstrated spatial resolution of 1 μm on IC device structures, sensitivity to implant dose over a range of 10^{10} to 10^{15} ions/cm^2 as well as sensitivity for implants into both Si and GaAs crystals. Another optical technique in general use for implant diagnostics measures the change in optical density of an implanted layer of photoresist on a transparent glass substrate (Ionscan) [19].

Time Dependence of Measured Values

One important area of implant diagnostics which is often overlooked is the time-dependent properties of the measurement itself. Time variations in repeated measurements of implanted surfaces have been reported for sheet resistance, double implant/ damage, Therma-wave and optical density (Ionscan) techniques [20-22]. Many of these time-dependent effects can be controlled by treatment of the implanted surfaces, but the implant engineer is wise to treat a "reference standard" wafer which is held in a lab drawer as the "keystone" to a process control system with a measured degree of caution.

Display of Data

In many instances, the key to the diagnosis of an implant malfunction lies in the comprehension of the spatial variation of a measured parameter over the surface of an individual wafer or batch of wafers. The direct method is to print the numerical value of a measured parameter on a spatial grid which approximates the locations of the measurement sites. An example is shown in Fig. 4 for a low-dose, C-V measurement [23]. While the tabular listing of numerical data has the advantage of directly recording a large amount of information, patterns are often difficult to recognize.

A way of presenting spatial measurements in a clearer visual fashion is to divide the range of measured values into a small number of intervals and to diplay the values with a spectrum of colors or shades. A basic example of this technique is shown in Fig. 5, where sheet resistance values are displayed using a range of shadings from a standard printer array. This technique provides an efficient display of the data, although the resolution is limited by the number of available shading levels.

The most commonly used technique for display of implantation data is a contour map, where measured data from an array of test sites are interpolated to show the location of the average value (usually denoted by a darker line) and the locations of measured values which differ from the average by a selected scale (usually shown in terms of a percent variation). Figure 6 shows variations in a low-dose $(6 \times 10^{11}$ ions/cm$^2)$ implant as measured by a Therma-wave optical probe. These dose variations were associated with a "locked-up" (non-uniform) pattern of beam scanning on a electrostatic-scan implanter. The probe sites which measured doses higher than the average value are marked with a (+) and those less than the average with a (-). The scale of variations is shown with the interpolated contours. The contour interval for this map was 1%. The data shown in Fig. 6 indicates that the peak-to-peak variation in dose over this wafer was as high as 6% in certain locations. Contour-interval maps are an efficient and flexible mode for display of a wide range of data. The spatial resolution of the display is limited by the probe site density and the contour interval selected to display the data.

A more dramatic method of displaying large-scale variations of dose over a wafer surface is to convert the contour-interval data into a perspective view in which the parameter values are combined with the spatial location of the probe sites on a flat surface to produce a side-on view of a three-dimensional surface. An example is shown in Fig. 7 where sheet resistance variations over a wafer implanted with a disk-type, mechanical-scan system are shown as a three-dimensional surface. The variations in dose are related to irregular motion of the axis of the spinning disk across the ion beam. Three-dimensional surfaces are useful for focusing attention on a large-scale variation, but are less useful for process control and data storage since recovery of the measured values from a 3-D map is a tricky and unreliable process.

For particle contamination measurements, the *location* of a particle of a given size can be sufficient information for a useful attack on a problem. An example is shown in Fig. 8, where light scattered from particles illuminated with a scanned laser beam is analyzed to show the location of particles on the surface of a flat Si wafer [24]. The evidence of a dirty vent port can be seen near the upper right side of the wafer.

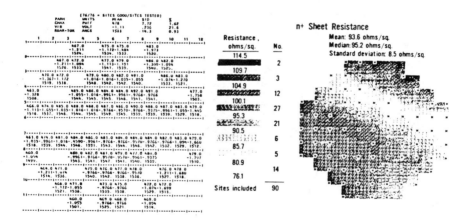

Figure 4. Numerical tabulation of low-dose, C-V data in a form which approximates the location of the probe sites over a 100 mm wafer [23].

Figure 5. A shaded-interval, or gray-tone, map of sheet resistance variations caused by implantation through a non-uniform thickness oxide.

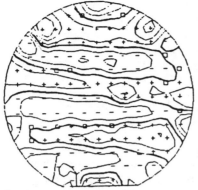

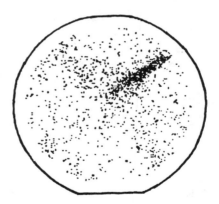

6.48%

−6.04%

Figure 6. Contour-interval map of dose variations measured with a Therma-Wave optical probe. The location of the interpolated average dose value is shown as the darker lines. The lighter lines show the locations dose values which vary from the average in intervals of 1%.

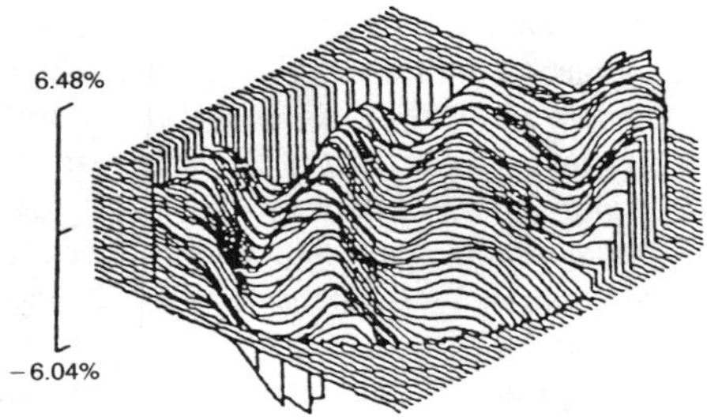

Figure 7. A three-dimensional map where interpolated sheet resistance variations are shown as the surface height along the vertical axis and location on the implanted wafer shown in the tilted plane.

Figure 8. Particle location map from a scanned laser tool showing the contamination resulting from a dirty vent port located near the upper right side of the wafer [24].

D. Spatial Resolution

As device dimensions become smaller and circuit densities increase, the question of small-scale uniformity becomes more important. The functionality of a chip will be determined not only by the average dose for that chip but also by the variation of dose across the chip. For measurement of small scale uniformity, the spatial resolution becomes important. The spatial resolution depends on the physics and geometry of the technique.

The spatial resolution of optical techniques depends on the diameter of the light spot and the angle of incidence of the spot on the sample. For normal incidence, the resolution is equal to the diameter of the spot. For Ionscan, the spot diameter is ≈3mm and for Thermawave, it can be ≈1μm. For optical techniques where the light is incident at an angle ø and the diameter of the spot size is d, the resolution is approximately d / cosø. For the case of the GRQ ellipsometry technique [25] this would be approximately:

$$0.5mm/\cos 70^{o}, \quad \text{or } 1.5\ mm$$

The determination of spatial resolution of the four-point probe is much less straight forward. It is clear from the sheet resistance radial correction factors [26] that the voltage is effected by the presence of the wafer edge at a distance of 4-5 probe spacings. This is the most drastic situation where symmetric current flow is terminated at the wafer edge and more current flows under the voltage probes. The use of the dual current-voltage configuration [27] to calculate the geometrical correction factor for each site eliminates the error introduced by this current crowding, but it is clear that the voltage is influenced by current flow through an area of diameter equal to 8-10 probe spacings. For most of the four point probe mapping systems, the probe spacing is ≈1mm. The distance between the voltage sensing probes is 1mm and the current flows through an area of 8-10 mm in diameter. The voltage generated between the voltage sensing probes is determined by the current flow in the material between the probes. This current flow, in turn, is determined primarily by the resistivity between the probes and secondarily by the resistivity of the rest of the material through which the current is flowing. The four point probe thus provides a weighted average of the resistivity of the material through which the current is flowing, with the material between the voltage probes being the dominant factor. This raises the question as to how small a feature could be detected by the four-point probe.

An experimental determination of spatial resolution was performed using a photo mask to generate 10^{13} ions/cm^2 implanted stripes on a 10^{14} ions/cm^2 uniform implant [28]. The stripes were 2.0, 1.0, 0.5, and 0.25 mm wide and were spaced 5.0 mm center to center. The four point probe could detect the presence of 0.25mm stripes using a diameter scan or a map of appropriate test site density. The mask layout shown in Fig. 9 provides each stripe-width with its own quadrant. The map shown in Fig. 10 was made using 625 sites on a stripe-implanted 100mm wafer. The diameter scan shown in Fig. 11 was done using 0.25 mm steps on the quadrant of the wafer with the 0.50 mm stripes. The presence of the 0.50 mm stripe is clearly indicated. Table 1 summarizes the results of the diameter scans for the four quadrants. The average sheet resistance increases and the percent standard deviation becomes smaller as the width of the stripes become smaller. The peak-to-valley difference (delta) also becomes smaller, but the 3 ohms/square delta for the 0.25 mm stripes is easily detected. The detection limit of the four-point probe is smaller than the probe spacing and also depends on the measurement density. For conventional four-point probe mapping systems, the detection limit is ≈0.25 mm. For the Solid State Measurements system, which uses four spreading resistance-type probes, the resolution is improved to ≈0.1 mm through the use of a 0.5mm probe spacing. Direct spreading resistance measurements have a lateral spatial resolution of ≈ 5 µm.

Table1:Sheet Resistance Values from a Spatial Resolution Study.

Stripe Width (mm)	Average R_S (ohm/square)	Sigma/ R_S (%)	ΔR_S (ohms/square)
2.0	505.2	2.02	25
1.0	508.2	1.36	13
0.5	507.7	1.00	7
0.25	514.4	0.88	3

In mapping the uniformity of a wafer, it is important to use the appropriate density of test sites; too few sites may miss important dose variations and too many will consume valuable measurement time. A commonly used figure for the density of test sites is approximately equal to the density of IC devices, or one probe site for each 1-2 cm^2 [29]. This ensures that device-to-device dose variations can easily be detected. It is also important that the distribution of test sites be uniform and not favor any one region of the wafer. One such uniform-density test pattern is a set of equally-spaced concentric circles.

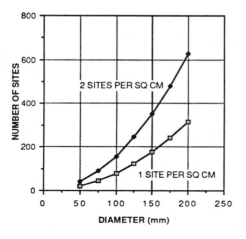

Figure 9.   Plot of number of sites versus wafer diameter to achieve a density of 1 and 2 test sites per square cm.

Figure 10. A 361 site contour map of a 125mm wafer implanted with 1E13 ion/sq cm boron at 600Kev. The mean sheet resistance is 1012 ohm/sq and the uniformity is 0.38%.

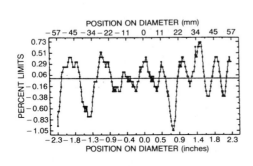

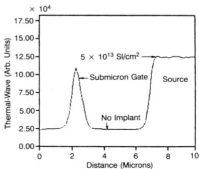

Figure 11. A 361 site diameter scan of the same wafer going from top-right to bottom-left and showing a serious scan lock-up problem.

Figure 12.   A 10μm   linear scan across the gate area of a GaAs FET device using the Therma-Probe 200.

Starting with one test site at the center of the wafer, the first circle has eight sites distributed equally along its circumference. Each additional circle has eight additional test sites. This distribution provides equal area per test site and also preserves the symmetry of the wafer as shown in Fig.12. Fig. 13 shows the number of sites necessary to achieve a density of 1 or 2 test sites per cm^2 as a function of the wafer diameter. For 150mm wafers the required number of probing locations is 180 and 350 sites, respectively.

A diameter scan is often very useful in detecting or resolving problems as shown in Fig. 10 and 11, above. Because the test sites are all in one dimension, the equivalent "density" would be that of the same number of sites squared. The advantage is that the diameter scan can be measured much more quickly. For a 100 site diameter scan of a 100 mm wafer, the resolution would be equivalent to 10,000 site map or 100 sites per cm^2. Fig.14 is a 361 site contour map of a 125 mm wafer implanted with 1E13 ions/cm^2 boron at 600 keV. The sheet resistance is 1010 ohms/square and the apparent uniformity is excellent at 0.38%. The density of sites on this map is sufficient to indicate hints of a scan problem of some kind. The 361 site diameter scan in Fig.15 clearly shows the presence of a scan lock-up problem on this wafer. The chip-to-chip variation in dose is greater than 1% and could cause a yield loss for some device chips. For good process control, it is necessary to carefully study the ion implant maps and routinely perform diameter scans to help identify problems. Merely charting the mean and standard deviation is not enough, as shown in Fig. 14 and 15.

The 1 μm spot size of the Thermaprobe 2000 [30] allows for direct measurements on device wafers. This capability is shown in Fig 16, which is a line scan over a 10 μm distance in the gate area of a GaAs FET. The peak at 2.0 μm corresponds to a submicron gate line that has been implanted with 5 x 10^{13} ions/cm^2 silicon. The signal beginning at 4 μm at the right of this peak corresponds to a broader implanted source structure.

Table 2 compares the various electrical and optical techniques available to the implant engineer for process control. The choice of measurements is difficult. Consequently, as the implant specifications become more stringent, it is advisable to use more than one technique. It should be remembered that in the early days of implantation, the only choice was whether or not to perform a mylar burn. Today a number of viable choices exist.

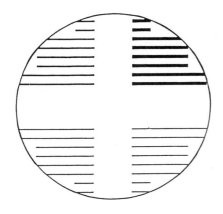

Figure 13.   Layout of stripe pattern on the spatial resolution test mask. Stripes are (a) 2.00, (b) 1.00, (c) 0.50 and (c) 0.25mm wide with a center to center spacing of 5.0mm.

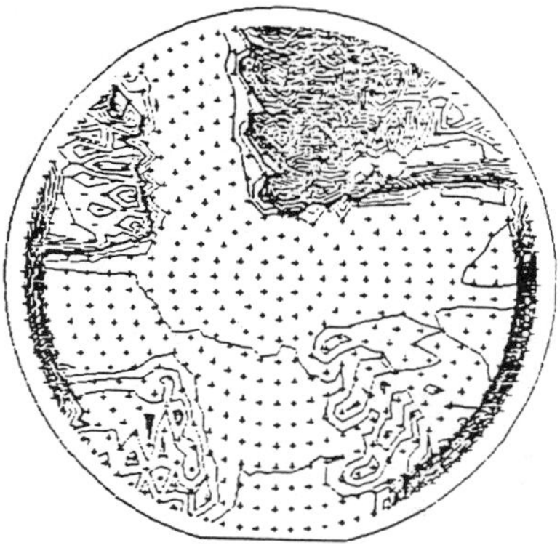

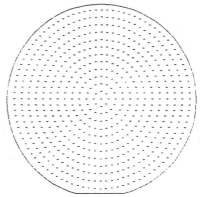

Figure 14.   A 625 site contour map of a 100mm wafer implanted with 1E13 ions/sq cm using stripe mask. Uniform background implant was 1E14 ions /sq cm.

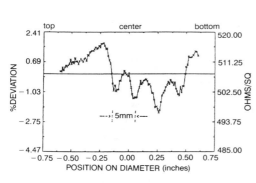

Figure 15.   Linear scan across quadrant (c) of stripe implanted wafer using 0.25mm steps.

Figure 16.   Layout of 625 test site pattern of 12 equally spaced concentric circles, prodviding an equal area per test site.

Table 2. Selected Characteristics of Wafer Mapping Techniques

	Rs Single I/I	Rs Double I/I	Spreading Resistance	Pulsed C-V	Ion Scan	Therma Wave	GRQ Ellipsometry
Type		ELECTRICAL				OPTICAL	
Measures	Sheet Resistance	Damage	Spreading Resistance	Depletion Capacitance	Photoresist Damage	Crystal Damage	Crystal Damage
Resolution	.25mm	.25mm	5μm	.10mm	3mm	1μm	1.5mm
Species	Active	Both Active and Inactive	Active	Active	Both	Both	Both
Sensitivity	.7-.8	.5-1.0	.7-.8	.7-.8	.5	.2-.6	.5
Dose Range	5E11-5E16	1E10-1E14	2E11-1E17	1E10-5E12	1E11-1E14	1E10-1E15	1E11-1E14
Results	Direct	Calibration	Calibration	Direct	Calibration	Calibration	Calibration
NBS Standard	Yes	No	Yes	No	No	No	No
Relaxation	Minor	Serious	Minor	Minor	Serious	Serious	Serious
Substrate	Opposite s Type	Si	Si Either s Type	Si Same s Type	Glass v P/R	Si Either Type	Si
Require	Anneal	Initial I/I	Anneal	Anneal + Oxide + Metal Dots	Measure Before and After	—	Measure Before and After
Advantage(s)	Whole S/C Process Industry Std	Reuse Wafer No Post-Process	Resolution Either Substrate Type	Profile Available	No Post-Process	No Post-Process Contactless	No Post-Process Contactless

E. Examples of Implant Process Diagnostics

Once one is in possession of a process measurement (a sheet resistance contour map, diameter scan, particle map, etc.), the next challenge is the interpretation of this data to detect and diagnose any problems. The aim of this section is to briefly discuss a wide variety of examples of process diagnostics in order to give a feel of the range of problems which can be encountered and the use of wafer mapping to display lateral process variations.

Silicon Wafer Effects

Measurements on Si wafers, either test or product materials, are basic indicators of the quality of an implant process. These can be used to measure dose, sheet resistance, surface contamination and particles. Si test wafers are also sensitive to certain channeling and charging effects.

Surface Contamination
Studies of surface contamination usually focus on mechanisms which transfer contaminants from within the ion implanter to the implanted wafer surface. However, direct contamination induced by wafer handling or other types of physical contact with the wafer is often overlooked. Figure 17 shows a case where anomalous sheet resistance values, indicated by the round contours at the top of the wafer, were linked to metal contamination from handling wafers with tweezers. Another example of wafer contamination is shown in Fig. 18 where unusual circular contours appear on the outer edge of the wafer map. Visual examination of the wafer from which this map was made, showed white, cloudy rings at the approximate location of the circular contours on the wafer. Further investigation revealed that the contamination originated from silicon grease that was accidently deposited on the heatsink during preventive maintenance of the implanter end station.

Charging
Build-up of surface charge on an implanted wafer can lead to fluctuations in the size of the incident ion beam and create serious dose variations over the wafer surface[31-34]. The effect of "flooding" the implanted wafer with low energy electrons and thereby reducing the surface potential is shown in Fig. 19 for a poly-on-oxide film that has been implanted (a) without and (b) with electron flooding.

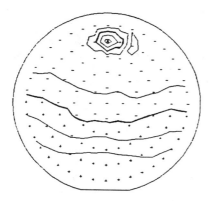

Fig 17: Circular contours at the top of the wafer are the result of wafer handling with contaminated tweezers.

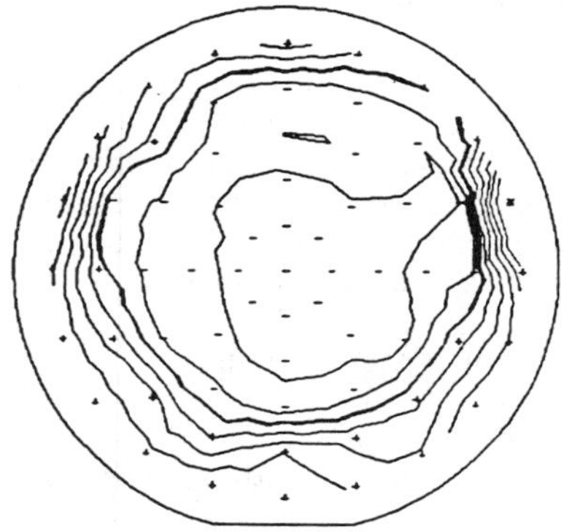

Fig 18: Circular contours around the circumference of the wafer are due to silicon grease. Visual examination shows white, cloudy rings in the same position.

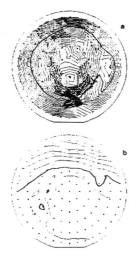

Fig 19: Poly-on-oxide wafer map (a) without electron shower and (b) with electron shower [31].

Fig 20: Contour map indicating axial channeling along Si <100> is the result of a 80 keV boron implant.

Channeling

Axial and planar channeling can have dramatic effects on implanted profiles and transistor characteristics [35, 36]. The effect of direct axial channeling, with an ion beam impinging along the <100> axis, is shown in Fig. 20 for a sheet resistance map of a 50 keV B^+ implant. The region of lower sheet resistance in the center of the wafer results from deeper penetration by the channeled ion beam. Figure 21 shows a subtle example of planar channeling along the <220.> direction from a mechanical-scan implanter with the wafer mounted on a flat heatsink. Figure 22 shows the more extensive planar channeling variation resulting from an implant with an X-Y scan implanter with a convex-domed heatsink.

Particles

One of the most important issues in modern implantation is reduction of particle contamination. This area is being addressed by concerted efforts in understanding the density, nature, and origin of particles generated within implantation systems.   Particle maps, such as shown in Fig. 8, are often capable of indicating the origination of particle sources. In this map, the distribution of particles indicates problems with a dirty vent valve.

Low Dose Implant Monitors

The measurement of low dose implants has been enhanced in recent years with refined four-point mapping techniques, the introduction of the double implant technique [15],   and pulsed C-V techniques [38]. A C-V measurement, shown in Fig. 23, indicates a standing wave pattern resulting from a scan lock-up condition with an electrostatic-scan system. The average doping variation over the wafer was 13%.

Implanter

Failures with the implantation equipment are usually identifiable by the implanter's type of beam scanning (x-y, hybrid, or mechanical), and sometimes by the distinctive traits of the manufacturer's equipment. Each of the following areas that are discussed have the potential of having representative examples from all three types of beam scanning. However, examples of all scanning types are not presented.

Scan Failures

In a high current, mechanical scanning implanter, an unstable ion beam can produce non-uniform doping or "stipping". This effect is especially seroious for low-dose implants where the number of passes of the wafer through the ion beam is low. Fig. 24 illlustrates the effects of a beam dropout during a 2-scan

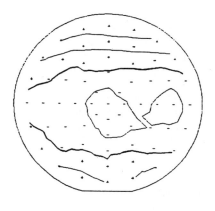

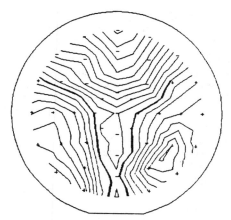

Fig 21: Typical pattern on this wafer is the result of planar channeling in a high current, mechanically-scanned implanter.

Fig. 22: Typical planar channeling pattern on this wafer map resulted from an implant of a wafer that was clamped to a cooled dome in an electrostatically-scanned implanter.

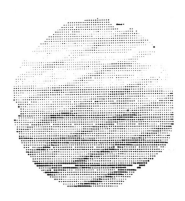

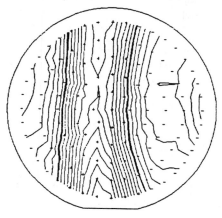

Fig 23: Doping variations from C-V measurements show a standing wave pattern resulting from scan lock-up in an X-Y scanning implanter.

Fig 24: Vertical striping pattern is the result of a beam dropout in a high current, mechanically scanned implanter.

implant. Figure 25 shows the more serious results of a complete stoppage of the slow scan, where the wafer continued to spin in front of the ion beam. The companion diameter scan (Fig. 26), shows low sheet resistance strip along the overdosed region in the middle of the wafer. The effect of repeated beam dropouts is shown in Fig. 27, where four vertical striping paths are evident.

A different type of scanning failure is related to the slow scan drive function of the implanter. One example of this type of failure is illustrated in Fig. 28 with vertical striping patterns that are similar to the previous example. However, this failure was traced to a mechanical problem in the slow scan drive system of the implanter resulting in a non-linear drive sequence. Another example, Fig. 29 shows a slow scan failure, in which the outer edge of the wafer remained in the beam as a result of inadequate scan amplitude. This is confirmed by Fig. 30, a diameter scan of this wafer, which shows an area of very low sheet resistance near the wafer edge.

Figure 31 shows an implant with a hybrid-scanning implanter, which has an electrostatic scan in one direction and a rotational scan in the other direction. The stripes which are parallel to the flat are a result of the fast rotational scan, which gives good uniformity in that direction. The top of the wafer shows higher sheet resistance values due to underscanning by the electrostatic plates. In an electrostatic-scan implanter, a failure in the quadrapole lens power supply may exhibit contour patterns similar to that shown in Fig. 32. The center stripes and large non-uniformity may have been caused by changes in beam shape.

Scan lock-up is another failure mode for the X-Y scanning system. With this type of failure, the X-scan loses frequency calibration and becomes a multiple of the Y-scan frequency. Examples of this effect are shown with an optical dosimetry map (Fig. 33) and a laser-thermal probe map (Fig 6).

<u>Narrow Beam</u>
The size and rate of scanning of the ion beam can effect the dose uniformity, wafer temperature and charging characteristics of the implantation system. Since the beam size can vary with ion type, energy, beam current and alignment of the optical elements of the beamline, a systematic approach involving wafer mapping and direct   measurements of the beam size in the ion implanter tool is required to control this situation. The "striping" effects of a narrow ion beam in a mechanical-scan implanter is shown in Fig. 34, where the striping is shown in the vertical (fast scan) direction across the wafer. A companion diameter scan, Fig. 35, shows the variation of sheet resistance horizontally across the wafer. A more subtle example is shown in Fig 36 where the apparent wafer uniformity is quite good, 0.52%. However, the diameter scan of this wafer (Fig 37) shows small-scale variations due to a narrow beam coupled with an overly large scan speed. The impact of "micro-uniformity" is

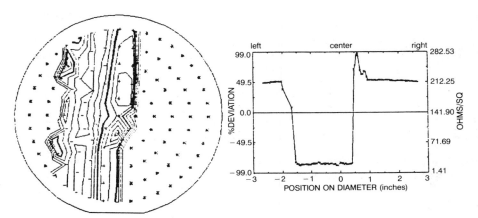

Fig 25: Striping pattern on a wafer resulting from an implanter whose horizontal scan drive completely stopped. The wafer remained in the beam during the remainder of the implant cycle.

Fig. 26: Companion diameter scan for Fig. 25 provides more information about the vertical striping pattern.

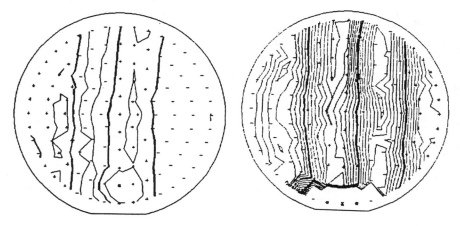

Fig. 27: Four vertical striping patterns resulting from numerous beam dropout during the implant.

Fig: 28: Striping pattern similar to that of Fig. 27, but were caused by a non-linear scan drive sequence.

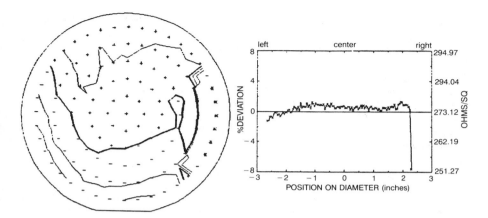

Fig. 29: Pattern on the outer edge of the wafer due to a scan stoppage such that the wafer continued to spin in front of the beam.

Fig. 30: Companion Diameter Scan to Fig. 29 indicating the low area of sheet resistance on the outer edge of the wafer due to the overdose caused by scan stoppage.

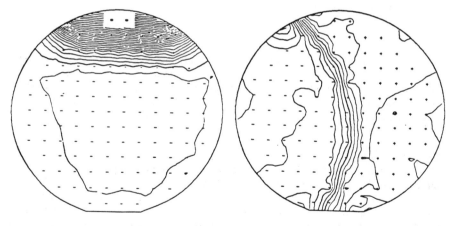

Fig. 31: Sheet resistance contour map showing the effects of underscanning in a hybrid-scanning implanter.

Fig. 32: Contour map shows the effects of failure in the quadrupole lens power supply of an electrostatic- scanning implanter.

coupled with an overly large scan speed. The impact of "micro-uniformity" is accentuated with the larger chip sizes which are characteristic of VLSI/ULSI devices being manufactured on large-area wafers.

Beam Neutralization

An electrostatic-scanning system in which the neutral trap deflection plates are placed before the X and Y scanning plates may produce a pattern such as that shown in Fig. 38 [16,35]. This behavior is due to the formation of neutral atoms within the deflection and Y-scaning plates under poor vacuum conditions.

Wafer Cooling

The use of high power ion beams, presently ranging up to ≈ 4 kW for high dose implants, is driven by the requirements for efficient implantation of large-area wafers. Conventional implanter cooling mechanisms are sufficient to hold the wafer temperature in the range of 70 to 100° C during high power implantation. Figure 39 illustrates swirling patterns, resulting from anomalous dopant activation effects, or "in-situ annealing", on a wafer which was implanted on a heat sink which had previously "burned" by direct exposure to the ion beam.   The increased thermal resistance of the implanted heatsink material resulted in a wafer temperatureduring implant in excess of 150° C.

Anneal

The anneal cycle is an often overlooked factor in the monitoring of the dose uniformity of an implanted wafer. Thermal factors which affect final sheet resistance and uniformity of an implanted wafer include anneal time and temperature, the type of temperature cycle, ambient gas flow rate, as well as wafer size and spacing in the furnace tube or rapid thermal annealer.

Time /Temperature

Insufficient anneal time does not allow for complete dopant activation or diffusion to the proper junction depth. Fig. 40 shows numerous horizontal contours and a higher than expected sheet resistance variation which characterizes a wafer which has been annealed for too short a time. Insufficient temperature of anneal will also affect the level of dopant activation (sheet resistance), as well as dopant distribution (uniformity). The circular contours in Fig. 41 are indications of insufficient annealing in the center portion of the wafer due to thermal gradients within the annealer.

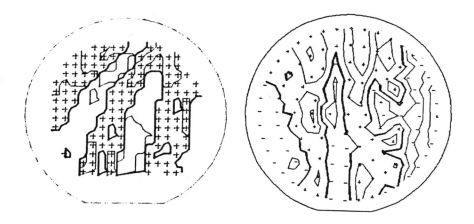

Fig. 33: An optical dosimetry map showing pattern due to a scan lockup in an X-Y scanning implanter.

Fig. 34: Striping pattern resulting from a narrow ion beam.

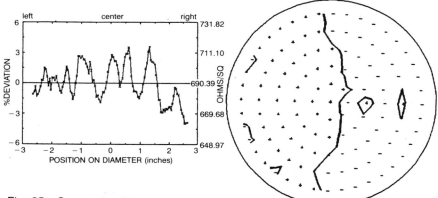

Fig. 35: Companion Diameter Scan to Fig. 34 confirms the diagnosis of ion beam striping.

Fig. 36: Good implant with σ = 0.52%.

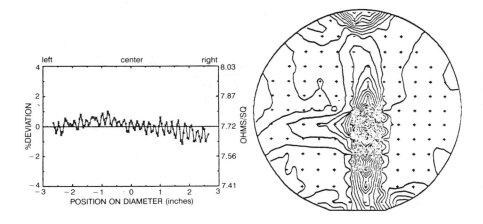

Fig. 37: Companion Diameter Scan to Fig. 36 shows micro-uniformity resulting from a narrow ion beam.

Fig. 38: Contour Map showing effects of neutrals in an electro-static scanning implanter.

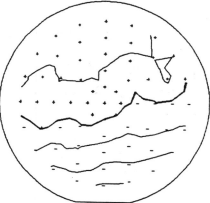

Fig. 39: Results of a wafer implanted when loaded upon a damaged heatsink. The implant parameters: As, 3E14, 100 keV at 23 mA.

Fig. 40: Contour map showing incomplete dopant activation due to insufficient anneal time. Rs = 40.9 Ω/sq. σ = 1.37%.

Ambient Gas Flow

Nitrogen or argon are commonly used to keep atmospheric oxygen and water vapor from entering the chamber during the anneal cycle. Turbulent gas flows in a local area of a furnace tube caused by incorrect flow rates or wafer spacings in a furnace tube can lead to both oxidation from back-flow of atmospheric oxygen and local temperature gradients. Figure 42 shows the locally higher sheet resistance at the top of the wafer associated with a high-flow rate furnace anneal. Figure 43 shows the unusual sheet resistance characteristics that were observed on a wafer annealed in a cracked furnace tube which allowed the ambient atmosphere to mix with the process gas flow.

An alternative anneal technique is that of high pressure oxidation. While offering unique and important advantages over the standard diffusion tube, it is capable of introducing measurable process variations to the annealed implanted wafers [35]. Figure 44 shows a sheet resistance map of a B^+ implanted wafer which was annealed at 1000° C in a high-pressure (10 atmosphere) furnace tube. The top-to-bottom gradient in sheet resistance was caused by a vertical temperature gradient which also produced comparable variations in oxide thickness over the wafer.

Four-Point Probe

The requirements for a four-point probe in a full-wafer mapping system are much more stringent than a probe used for 5 or 9 site samples. In order to measure ion implant dose uniformity of ≈0.75% or less, they must have a repeatability of better than 0.1% in order to properly measure the implant system performance. The impact of a 0.5% noise factor, giving the map a busy or noisy appearance, is shown in Fig. 45 [29].

Multiple Problems

Singular problems with the implant process are easily identifiable and relatively straight-forward to rectify. However, there are many cases when the contour map or other process output is complicated by the effects of more than one problem. As an example, Fig. 46 illustrates a contour map where the variation of horizontal contours from the bottom to the top of the map are typical of a problem with the furnace anneal. The vertical contours on the outer edge of the wafer indicate that the wafer was subjected to an underdosing caused by a scan failure. The asterisks at the bottom of the wafer indicate that the wafer was probed too close to the edge of the wafer by the four point probe. Finally, the companion diameter scan in Fig 47 indicates that the implant cycle was complicated by the use of a narrow ion beam and a high-velocity slow-scan .

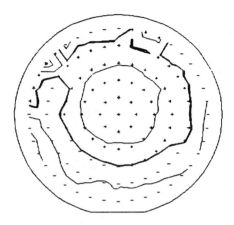

Fig. 41: Circular contours caused by insufficient anneal termperature. Rs = 9.8 Ω/sq, σ = 1.05%.

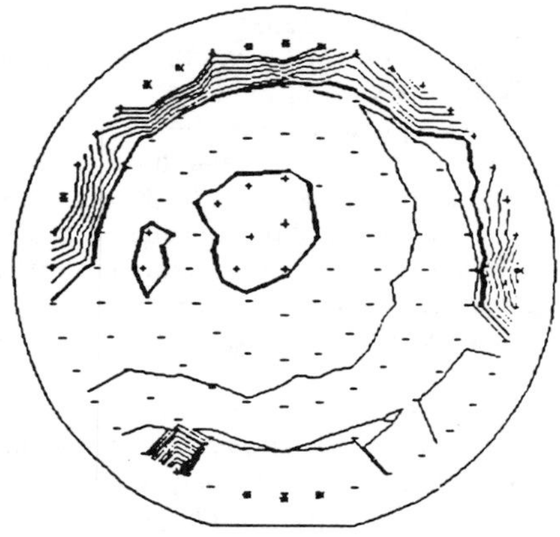

Fig. 42: Crowded contours at the top of the wafer map are due to excessive turbulance of the ambient air flow. Rs = 9.2Ω/sq, σ = 2.6%.

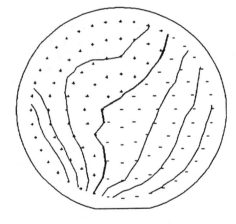

Fig. 43: Unusual contours due to a cracked quartz anneal tube. Rs = 22 Ω/sq; σ = 1.8%.

Fig 44: Contour map showing the variation of sheet resistance due to a high pressure oxidation anneal cycle.

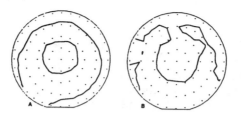

Fig. 45: Two sheet resistance contour maps of the same wafer using identical test condition and two probes: (a) probe with good repeatability resulted 15.88 Ω/sq and σ = 0.35%. (b) probe with poor repeatability on qualification wafer resulted in Rs = 15.72 Ω/sq and σ = 0.48%.

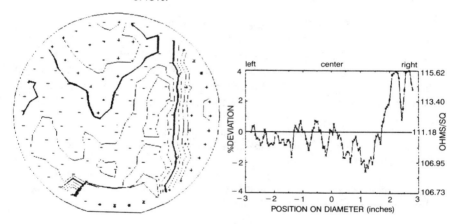

Fig. 46: Wafer map illustrates the complications of process trouble-shooting when three implant process problem occur at the same time.

Fig. 47: Companion Diameter Scan shows that the wafer had a fourth additional problem: slow-scan drive malfunction.

F. Dose Accuracy Surveys

Despite the impressive variety of implant system malfunctions which can lead to dose errors and other process failures, most implant systems respond with diligent attention to detail and can be operated with a certain level of confidence under routine conditions. One important measure of the performance of implantation systems is comparison of performance to other systems. While uniformity and dose repeatability performance of a single implant tool can often be demonstrated with variations of 1% or less, relative and absolute dose accuracy is more difficult to achieve. Comparison of results from various implantation systems under the same nominal implant conditions, known as a "round-robin" experiment, gives a measure of the *relative* dose accuracy of all the systems tested. These kinds of tests serve to highlight anomalous performance of individual implant tools and provide a means of cross-calibration between tools which share common processing functions. When a relatively large sample is taken from numerous locations as well as implantation system types, a kind of industry "report card" can be examined.

Gan and Perloff conducted one of the largest surveys, involving 45 implant locations and 85 systems, during the summer of 1980 [41]. All implants were carried out for a single condition (150 keV B^+ through an oxide film of 1050 Å into Si[100] at a dose of 5×10^{14} ions/cm^2) with a common anneal (30 minutes in dry N_2 at 950° C). The measured relative dose accuracy was 6%. The median uniformity for the 3 inch wafers used in the survey was 0.72%.

Later surveys covering numerous implant dose and energy conditions, incorporated a larger variety of machine types and larger wafer diameters. The early results from these surveys give an optimistic estimate of the usual state of affairs [41-47]. A summary of a number of recent round robin surveys of ion implantation systems is shown in Fig. 48. The sample size, target material and references are listed in Table 3.

Table 3: Dose Accuracy Round Robin Conditions

Survey	Number of implanters	Ion/ Substate	Reference
Gan/Perloff	85	B / Si	41
Current/Markert	14	As,B / Si	42
Current/Keenan	17	As / Si	43
Steeples	≈70	As,B / Si	44
Larson	47	As, B / Si	45
Larson/Kennedy	≈12	As, B /Si	46
Lane	5	Si / GaAs	47

Although individual implant systems and groups of cross-calibrated production tools can be maintained within tight ranges for dose uniformity, repeatabiltiy and relative dose accuracy, these round-robin surveys indicate that the typical relative dose accuracy of standard implant tools is no better than ≈10%.   The absolute dose accuracy performance can be expected to be somewhat worse that this level. The use of ion implantation to provide "standards" for measurement techniques such as SIMS and AES, where high levels of absolute dose accuracy are desired, should therefore be treated with some caution.

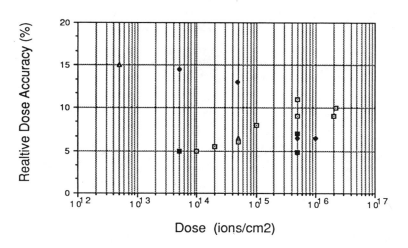

Figure 48. Relative dose accuracy measurements for a number of round robin experiments.

G. Summary

Ion implantation is one of the most versitile and pervasive technologies in modern IC fabrication. The wide range of implantation conditions, such as ion type, energy, beam current and dose, and the stringent requirements for accuracy and repeatability imposed by advanced IC device manufacturing place special demands on any effort towards a comprehensive program for implant process control. The complexity of ion implantation equipment, driven by ever-increasing extensions of the capabilities of these tools, results in a wide variety of failure modes which must be looked for, recognized and corrected. The high volume of wafers which are processed through modern implantation systems also demands that these process control mechanisms be able to quickly raise the alarm for corrective action, before a significant number of wafers are mis-processed. The occasional example of million-dollar losses of IC wafers as a result of undetected implant malfunctions serve as a reminder of the importance of deligent monitoring of process and equipment conditions.

One of the major design goals for later generations of implantation systems is the extensive use of *internal* sensors, coupled to adaptive control mechanisms, so that routine use of *external* control loops will no longer be needed. Examples of internal monitors under consideration for new implant systems include direct dose measurements and sensors for particles, wafer temperature and surface charge levels. Until these advanced tools are developed, the use of the diagnostic and process control measurements, such as those discussed in this chapter, is urgently recommended.

References:

1. I.H. Wilson and K.M. Barfoot, " Measurement and Control of Accelerator Parameters", in **Ion Implantation Science and Technology**, ed. J.F. Ziegler, Academic Press (1984), pp.537-602.
2. **Ion Implantation Techniques**, eds. H. Ryssel and H. Glawischnig, Springer (1982).
3. **Ion Implantation Science and Technology**, ed. J.F. Ziegler, Academic Press(1984).
4. **Ion Beams**, R.G. Wilson and G.R. Brewer, Wiley (1973).
5. **Ion Implantation**, H. Ryssel and I. Ruge, Wiley (1986).
6. **Ion Implantation Technology**, eds. M.I. Current, N.W. Cheung, W. Weisenberger and B. Kirby, North-Holland (1987); also published as Nuc. Inst. and Meth. in Phys. Res., **B21**, Nos. 2-4 (1987). See also previous sessions of this conference series under the title of "Ion Implantation Equipment and Techniques".
7. A.J. Rogers, "In Process Control Systems For Forming Processes", presented at "IEEEProcess Control For Manufacturing", a videoconference produced by the Institute of Electrical and Electronic Engineers (IEEE), 24 Sep 1986.
8. K.H. Zaininger and F.P. Heiman, "The C-V Technique as an Analytical Tool", Solid State Technology, **13**, No 5, (1970) pp 49-56.
9. R.K. Berglund, "Practical Approach to Monitoring of Implanted Layers", Semiconductor International, **27**, No 5, (1984) pp 155-159.
10. C.B. Yarling, "Leakage Currents in Ion Implantation Systems", Semiconductor International, **6**, No 8, pg 110-114, Aug 1983.
11. J.L. Forneris, G.B. Forney, R.A. Cavanagh, G. Hrebin and J.L. Blouse,"Implant Process for Bipolar Product Manuafacturing and Their Effects on Device Yield", Ion Implantation: Equipment and Techniques, eds. H. Ryssel and H. Glawishnig, Springer (1983)407-425.
12. L.A. Larson and M.I. Current, "Metallic Impurities and Dopant Cross-Contamination effects in Ion Implanted Surfaces", Materials Research Soc. Proc. Vol **45** (1985)381-388.
13. L.A. Larson, M.I. Current and C. Healy, "Enhanced Diffusion effects of Dopant Cross-Contamination in Ion Implanted Surfaces", Semiconductor Silicon 1986, Electrochem. Soc. Proc. Vol. **86-4** (1986) 667-677.
14. A.S. Oberai, "Lithography-Challenges of the Future", Solid State Tech. **30**, No. 9 (1987) 123-128.

15. A.K. Smith, W.H. Johnson, W.A. Keenan, M. Rigik, R. Kleppinger, " Sheet Resistance Monitoring of Low Dose Implants Using the Double Implant Technique",Nuc. Inst. Meth. **B21** (1987) 529-536.
16. M.I. Current, D.S. Perloff, and L.S. Gutai, "Wafer Mapping Techniques for Characterization of Ion Implantation Processing", in **Ion Implantation Techniques**, eds. H. Ryssel and H. Glawischnig, Springer (1982) pp.235-254.
17. P.L. Swart, H. Aharoni and B.M. Laquet, "Optical Reflectometry of Radiation Damage in Ion-Implanted Silicon", Nuc. Inst. Meth. **B6** (1985) 365-371.
18. W.L. Smith and M.W. Taylor, "Ultra-high Resolution Dose Uniformity Monitoring with Thermal Waves", SPIE Proc. Vol **530** (1985) 201-214.
19. J.R. Golin, N. Schell, J. Glaze, R. Ozarski, "Latest Advances in Ion Implant Optical Dosimetry", Nuc. Inst. Meth. **B21** (1987) 542-549.
20. J. C. Chen, " Monitoring Low-dose Single implanted Layers with Four-Point Probe Technology", Nuc. Inst. Meth. **B21** (1987), 526-528.
21. J. Schuur, C. Waters, J. Maneval, N. Tripsis, A. Rosencwaig, M. Taylor, W.L. Smith, L. Golding and J. Opsal, " Relaxation of Ion Implant Damage in Silcon Wafers at Room temperature Measured by Thermal Waves and Double Implant Sheet Resistance", Nuc. Inst. Meth. **B21** (1987), 544-558.
22. B. Kirby, L.A. Larson and R-Y. Liang, "Thermal-Wave Measurements of Ion Implanted Silicon", Nuc. Inst. Meth. **B21** (1987), 550-553.
23. R.O. Deming and W.A. Keenan, "C-V Uniformity Measurements",Nuc. Inst. Meth. **B6** (1985), 349-356.
24. J. Pollack, N. Turner, R. Milgate, R. Resnek and R. Hertel, " Particulate Performance of a Wafer Handler for Serial Process Ion Implantation", Nuc. Inst. Meth. **B6** (1985) 202-209.
25. J.C. Cheng, "The E-Prope 200 Wafer Probing System", Silicon Valley Implant Users Group notes, Dec. 1986.
26. "Measuring Radial Resistivity Variations on Silicon Slices", ASTM F81, **10.05**, ASTM Standards.
27. W.A. Keenan, W.H. Johnson, A.K. Smith, "Advances in Sheet Resistance Measurements for Ion Implant Monitoring", Solid State Tech. **28**, No. 6 (1985).
28. W.A. Keenan and A.K. Smith, "Resistivity Monitoring for Ion Beam Processes", SPIE Proc. **530** (1985) 174-181.
29. W.A. Keenan, W.H. Johnson, "Production Monitoring of 200 mm Wafer Processing", **Emerging Semiconductor Technology**, ASTM STD960 (1986) 598-614.

30. W.L. Smith, A. Rosencwaig, D.L. Willenborg, J. Opsal and "M.W. Taylor," Ion Implant Montitoring with Thermal Wave Technology", Solid State Tech. **29**, No.1 (1986) 85-92.
31. M.E. Mack,G. Ryding, D.H. Douglas-Hamilton, K. Steeples, M. Farley, V. Gillis, N. White, A. Wittkower, R. Lambracht, "Wafer Charging and Beam interaction in ion Implantation", Nuc. Inst. Meth. **B6** (1985) 405-411.
32. P.E. Bakeman and A. F. Putliz, "Ion Implantation Electron Flooding Requirements from a Users's Perspective", Nuc. Inst. Meth. **B6** (1985) 399-404.
33. J.M. Hall, H. Glawischnig and W. Holtschmidt, "Charging Studies in Applied Materials Precision Implant 9000 System" Nuc. Inst. Meth. **B21** (1987)350-353.
34. V.K. Basra, C.M. McKenna and S.B. Felch," A Study of Wafer and Device Cahrging During High Current Ion Implantation", Nuc. Inst. Meth. **B21** (1987) 360-365.
35. M.J. Market and M.I. Current, "Mapping of Ion Implanted Wafers", in **Ion Implantation Science and Technology**, ed. J.F. Ziegler, Academic Press(1984) 487-536.
36. M.I. Current, N.L. Turner, T.C. Smith and D. Crane, "Planar Channeling Effects in Si(100)",Nuc. Inst. Meth. **B6** (1985) 336-348.
37. Hancock and C.B. Yarling, "Particle Measurements In Ion Implanters" Semiconductor International, **7**, No 5, (1984)144-150.
38. R.O Deming and W.A. Keenan,"Low Dose Ion Implant Monitoring", Solid State Tech. **28**,No. 9 (1985) 163-167.
40. J. Pollock, N. Turner, R. Milgate, R. Resnek and R. Hertel, "Particulate Performance of a Wafer Handler for Serial Process Ion Implantation", Nuc. Inst. Meth. **B21** (1987)202-209.
41. J.N. Gan and D.S. Perloff," Post-Implant Methods for Charcterizing the Doping Uniformity and Dose Accuracy of Ion Implantation Equipment", Nuc. Inst. Meth. **189** (1981) 265-274.
42. M.J. Markert and M. I. Current , "Characterization of Ion Implanted Silicon-Applications for IC Process Control", Solid State Tech. **26**, No. 11 (1983) 101-106.
43. M.I. Current and W.A. Keenan, "Ion Implant Round Robin" , Nuc. inst. Meth. **B6** (1985) 418-426 and "A Performance Survey of Production Ion Implanters", Solid State Tech., 28, No. 2 (1985)139-146.
44. K. Steeples, "Sheet Resistivity of Silicon Wafers Implanted with a High Current Machine", Nuc. inst. Meth. **B6** (1985) 412-417.
45. L.A. Larson, "A Round-Robin Experiment on Dosing Accuracy", Electrochem. Soc. Abst. Vol **87-1** (1987) 330.
46. L.A. Larson and G.L. Kennedy, "Dosing Accuracy: A Comparison of Several Experiments", Nuc. Inst. Meth. **B21** (1987) 421-423.
47. A. Lane, Chell Instruments, Norfolk, England, private communication.

SAFETY CONSIDERATIONS FOR ION IMPLANTERS

H. Ryssel

Fraunhofer Arbeitsgruppe für Integrierte Schaltungen, Erlangen

P. Hamers

IBM, Böblingen
Federal Republic of Germany

1. INTRODUCTION

Ion implantation has become the most important doping technique for semiconductor devices in recent years. Ion implanters have changed from complicated experimental equipment to computer-controlled, automated machines. Nonetheless, implanters are still dangerous unless they are designed and operated properly. The main hazards result from high voltages, x rays, and from toxic chemicals necessary to produce the desired ions. In general, with the increasing need for high-throughput implanters, currents have increased, as have too the throughputs of poisonous gases. Modern accelerators are equipped with interlocks to protect the operator from harm. However, old accelerators do not have such equipment, and any interlock can be bypassed. In this paper, the hazards typical for implanters will be discussed.

2. HIGH VOLTAGE

The most obvious danger connected with an accelerator used for ion implantation is due to the high voltages used. These include not only the acceleration voltage, which is in the range of several kV to several hundred kV, but also many auxiliary

ION IMPLANTATION:
SCIENCE AND TECHNOLOGY

415

voltages used for beam extraction, scanning, electron suppression, etc. In Table 1, a list of power supplies together with
their typical values is given.

Besides the dangers of burning or electrolytic decomposition, the
most serious danger of high voltages arises from the blocking of
nerve conductance. Since the electrochemical potential of nerve
membranes is in the order of 60 mV over the distance of one nerve
cell, even low voltages interfere with this potential, thus
disabling the individual from disconnecting the electrical power
or pushing an "emergency off" button.

Table 1   Power supplies used for typical medium-current and
             high-current implanters. The maximum ratings do not
             necessarily occur simultaneously.

Purpose	Max. Voltage (V)	Max. Current (mA)
Acceleration	200,000	50
Extraction	60,000	50
Arc	90	20,000
Accel/Decel	2,000	20
Quadrupole	25,000	5
Suppression	1,000	20
Scanner	50,000	5

It is the designer's responsibility to ensure that components
which are at high potential are properly shielded from human
contact under normal operating conditions. The most effective
protection against high voltage is to enclose the affected parts
completely in a Faraday cage (grounded shielding). Occasionally,
such shieldings must be opened to allow access to the machinery
and circuitry. Electrical interlocks should be used to switch off
the high voltage if any door is opened. Additionally, simple but
effective mechanical interlocks such as the drop bar can ensure
that the terminal is connected to ground when any door is open.
Besides this, key-operated switches and door locks should be

provided to prevent access by unauthorized personnel. However, during maintenance, it may be necessary to override interlocks in order to test proper operation; in such situations, there is no substitute for common sense and caution.

The electrical resistance of the human body is rather low. Moreover, it decreases with increasing voltage, since the skin at the contact points may be destroyed immediately at higher voltages, thus reducing the total resistance. In this case, as a good approximation for the resistance between the two hands of a person, a value of approx. 1200 Ω can be assumed. In Fig. 1, the minimum resistances between several points on a human body and a hand are shown.

For an AC current at frequencies of 50 or 60 Hz, a current as low as 15 mA in the arm may result in a spasm, which usually disables the person from removing his hand from the current source. Even more serious is the influence of electrical current upon the heart. Whenever the electrical field at the heart exceeds 50 mV/cm, the cardial nerve signals are surpassed and ventricular fibrillation results, which may be lethal within 3 to 5 minutes due to circulatory failure. In Fig. 1, the cardiac current factors k_H for several current paths through the human body are given. Cardiac irregularity can take place whenever the body current exceeds 80 mA/k_H for AC or 300 mA/k_H for DC. It should be mentioned that in European countries with 220 V voltage and 50 Hz, almost 3 % of all contacts with a power line, especially with a three-phase line, are lethal.

In ion implanters, almost all high-voltage power supplies are capable of providing currents high enough to cause severe damage. High voltage may be sensed by the feeling of a static discharge before a dangerous contact occurs. However, the spark length for high DC voltage equals about 1 inch per 20 kV, which means that a distance of this order corresponds to a direct contact.

Insulating parts such as, for example, coaxial cables may store some electrical charge, even if disconnected and discharged, and they can cause severe electrical shocks. It is, therefore,

recommended to short circuit or ground these parts carefully
before starting maintenance. Besides these maintenance precautions, all assemblies or enclosures containing high voltages
should have "Danger" or "High Voltage" labels on their outer
surfaces.

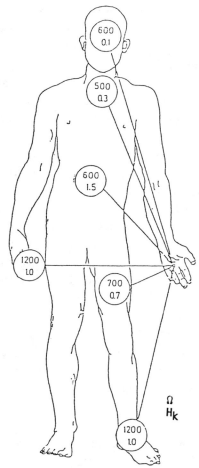

Fig. 1. Heart factor and
electrical resistance of
human body for DC or low
frequency AC between one
hand and several points
on the body

Many techniques are employed by electrical equipment designers to
safeguard the operator as well as the equipment itself. Among the
safety devices used are electronic fuses and circuit breakers for
the power-supply output, as well as current-limiting resistors
for the output and bleeding resistors which discharge the filter
capacitors on the DC power-supply output. As one of the safest
designs, a switched high-voltage power supply can be used to
reduce the filter capacitance as well as the size of the trans-

former. With this design, transient over-current spikes can be monitored, leading to a very rapid switching-off of the equipment when required.

3. RADIATION

Implanters accelerate ions in an electrostatic field. The largest potential difference is usually equal to the desired ion energy. Energetic particles may ionize electrons from the inner shell of the atoms, thus producing characteristic x rays corresponding to the transition energies of the electrons. Moreover, during the retardation of charged particles, bremsstrahlung is emitted. According to the principles of classical electrodynamics, the emitted energy E is given by:

$$\frac{dE}{dt} = \frac{2}{3}\frac{q^2}{c^3} \cdot \left| \frac{d^2x}{dt^2} \right|^2 \tag{1}$$

with q being the charge and c the light velocity. The most abrupt velocity change takes place if charged particles strike a target. Equation (1) is only valid for a statistical average of many charged particles.

An abrupt change of the particle velocity takes place for ions at the target, as well as for electrons which are accelerated in a direction opposite to the ions and strike a positive electrode of the system. Electrons in an accelerator are mainly produced through collisions between ions and residual gas atoms or the walls of the system.

3.1 Electron-Induced X Rays

In order to calculate the intensity of the bremsstrahlung, an electron is considered to be moving from an infinite distance to the K shell of an atom with a nucleus of charge +Zq. By solving the equation of electron motion, under consideration of Eq. (1), one obtains:

$$\frac{dE}{dt} \propto \frac{Z^2}{m^2} \qquad\qquad\qquad (2)$$

indicating that the emitted radiation is proportional to the square of the atomic number of the target material, and is inversely proportional to the mass m of the atoms.

The calculation of the total emitted radiation for thick targets is more complex, since the electron energy decreases, and one has to integrate over all scattering angles and paths of the electrons. For electrons with a kinetic energy of E_{kin}, the following can be derived [2]:

$$E_b = kZE^2_{kin} \qquad\qquad\qquad (3)$$

with $k = 7\times10^{-10}$ eV^{-1}. Using the acceleration voltage V_0 and the electron current I, one obtains for the total bremsstrahlung radiation:

$$P_b = A_0\ ZV_0{}^2\ I \qquad\qquad\qquad (4)$$

with $A_0 = (1.5 \pm 0.3)\ 10^{-9}$ V^{-1}, V the acceleration voltage, and I the electron current.

For the spectral distribution of the emitted radiation as a function of the target material and the emission angle, no general expressions can be given. It is known, however, that the relative number of x rays increases for small angles and high x-ray energies, which means that the most energetic radiation is emitted in the direction of the electron beam.

A semi-empirical expression according to Kramers [3] relates the spectrum integrated over all directions to the energy interval between E and E + dE:

$$E_b = 2A_0Z\ (E_0 - E) \qquad\qquad\qquad (5)$$

where E_0 is the energy of the electrons. The probability that a photon is emitted in this energy interval is:

$$P = 2 A_0 Z \frac{E_0 - E}{E} dE \tag{6}$$

The total energy is obtained by integration:

$$P_b = 2A_0 Z I \int_0^{E_0} (E_0 - E) \, dE = A_0 Z V_0^2 I \tag{7}$$

which is identical to Eq. (4).

The bremsstrahlung spectra are superimposed by characteristic x rays which depend on the atomic number of the target. Because of the impulse conservation law, the most intense x-ray line is emitted from the strongly-bound electrons of the K shell. Characteristic energies of K- and L-shell x-ray radiation are given in Table 2.

Table 2. Characteristic x-ray energies of some elements [4]

Element	Atomic Number	K Energy (keV)	L Energy (keV)
Be	4	0.111	-
C	6	0.284	-
Al	13	1.560	0.118
Si	14	1.834	0.149
Fe	26	7.112	0.846
Mo	42	20.000	2.865
W	74	69.530	12.100

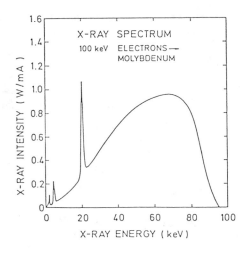

Fig. 2. Typical x-ray intensity spectrum vs electron energy for a medium-Z target

In Fig. 3, the critical points for x-ray production are shown. These are typical of the extraction electrodes in all implanter types and, in addition, the beam-defining aperture in medium-current post-acceleration implanters. A reduction in x-ray production is possible by preventing the electrons from entering the field regions. This is accomplished by using negatively-biased electrodes in front of the field regions, as is shown schematically in Fig. 3. Use of suppression electrodes is also necessary to avoid extraction of electrons from the beam plasma, thus destroying the space-charge neutrality of the beam. This would lead to a beam blow-up (depending on the current density) and additional x-ray production by electrons generated when the ion beam strikes the walls of the system. Therefore, an interlock on this voltage is to be recommended, and is provided in all commercial accelerators.

The electron current across the extraction gap in a well-designed and well-operated implanter may be as high as 10 % of the extraction current.

Since the x-ray production is proportional to the square power of the acceleration voltage, the x-ray emission is especially strong in the case of a post-analysis implanter. The electron current in the acceleration tube of a two-stage implanter is also usually a few percent of the ion current. However, in case of a poor vacuum, misaligned electrodes, or lack of (or improper) suppression, it can be much higher. It is generally limited by the power supply.

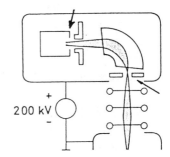

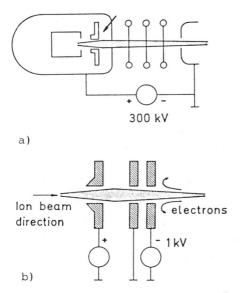

a)

b)

Fig. 3 a) Critical points of x-ray generation of single-gap and two-gap implanters
b) Electron suppression electrode

The material of the electrodes which are struck by the electrons is of great importance in x-ray production. To reduce such production in two-stage accelerators, the beam-defining aperture is made of carbon (Z = 6) or aluminum (Z =13). The latter, however, is only possible for low-current implanters. Sometimes a coating of beryllium (Z = 4) is also used [1]. The extraction electrode is often made of molybdenum (Z = 42) or, preferably, carbon (Z = 6).

3.2 Ion-Induced X Rays

The x-ray production caused by ions is much less than that due to electrons (approx. by 6 orders of magnitude), since ions have a much lower velocity than electrons accelerated with the same voltage. Ions can, however, produce characteristic x rays by knocking electrons from inner shells. These electrons produce bremsstrahlung when they strike a target and have, at maximum, double the speed of the ions. The maximum is, therefore, [5]:

$$E_e = \frac{E_{ion}}{250 \cdot M} \tag{8}$$

where M is the mass of the ion. Characteristic x rays are only produced with a small cross-section. The K_α energy of silicon is 1.83 keV.

3.3 Radiation Units

Corpuscular radiation and electromagnetic radiation, as well as all types of charged or neutral particles, are detected and measured by their electrical interaction with matter. There are two classes of units. One of them counts the photons or particles, such as the flux (e.g., photons/sec) and the Curie (in radioactive decays/sec). The other group of units, the Roentgen, Gray or Sievert, measures the deposited energy. Since in commercially-used implanters, only x-ray radiation is released, the second type of unit is more convenient.

The Roentgen (R) is the most commonly used unit by far to measure the radiation dose. As defined by the International Commission on Radiological Units and Measurements, one Roentgen is the quantity of x radiation which creates in one cubic centimeter of dry air under standard conditions (0°C, 760 torr) ions carrying 1 esu of electricity of either sign. Thus, 1 R creates $1.61 \cdot 10^{12}$ ion pairs in 1 g dry air. The SI unit for the absorbed dose (D) is the Gray measured in J per kg. An older but still used unit is the rad. The rad (rd) defines the amount of radiation which deposits 100 ergs per gram in an arbitrary material.

Different types of radiation show quite different path lengths in the absorbing medium, due to the different stopping cross sections, which results in different relative biological effectiveness (q). For this reason, each type of radiation was assigned values of q, and another radiation unit, the dose equivalent (D_q) measured in Sievert (Sv) or the rem (Roentgen Equivalent Man) was devised. It holds

$$D_q = q \cdot D$$

The q factors for all types of radiation are given in Table 3. For x rays, the value of q is unity, by definition. Thus, a rad of x rays equals one rem or one Gray equals one Sievert. Moreover, the quantity of radiation defined by the Roentgen in air or in tissue, and by the rad, are all approximately the same to within ten or fifteen percent.

Table 3. q factors for different types of radiation

Radiation	q
x ray, γ	1
β	1
α	20
protons	10
neutrons	5..15

Table 4 summarizes the physiological effects upon humans of whole-body exposure received within a few hours. In Table 5, the radiation-safety standards of several countries are given. These safety standards are calculated on the assumption that the risk of lethal radiation damage is equal to the risk of all other types of accidents through one's life.

Table 4. Effects of whole-body exposure received within a few
hours

1	rem	No detectable change
10	rem	Blood changes detectable
100	rem	Some injury; no disability
250	rem	Injury and disability
500	rem	50 % deaths occur within 30 days

Table 5. Radiation-safety standards in several countries

USA	0.25 mr/hr
Germany	0.25 mr/hr
Japan	0.06 mr/hr
England	0.75 mr/hr

3.4 Shielding

X-ray radiation from a point source can be attenuated by absorption or scattering. In scattering, the radiation is spread
over a wide volume, whereas in absorption, the radiation is
transformed into heat. There are four interaction mechanisms of
electromagnetic radiation such as x rays, each showing characteristic dependence on the density and binding strengths of the
electrons of the shielding material. In most of these interactions, the energy spectrum of the radiation is shifted towards
a lower energy, which is finally absorbed as heat.

In Rayleigh scattering, i.e. elastic scattering of photon by
electrons which dominates at low energies, the radiation is
distributed isotropically, without absorption.

In the photo effect, a photon transfers its energy to an
electron, exciting it to a higher energy level or acually
ejecting the electron from its bound state in the target atom.
Consequently, absorption edges are observed exactly at those
energies at which, in the inverse process of x-ray production,
characteristic x rays are generated. The photoeffect is the
dominant process for low-energy photon absorption. As the photon
energy increases beyong the K shell for a given absorber,
photoelectric absorption loses significance, because the binding
energy of the electrons is small relative to the incident photon
energy. In this energy range, Compton scattering becomes
dominant.

In the Compton process, a photon collides with an electron; part of the energy is transferred to the electron, and a scattered photon is emitted with a lower energy. Compton scattering shows relatively little variation in the energy range between 10 and 500 keV.

At very high energies, i.e. above 1.02 MeV, the photon energy is high enough for pair formation, which means the production of an electron-positron pair. All these mechanisms are controlled by the laws of conservation of energy and momentum, which leads to the different cross sections as a function of the incoming photon energy and the target atomic number. The E and Z dependencies are listed in Table 6.

Table 6. E and Z dependence in various attenuation processes

Process	Absorption	E dependence	Z dependence
Elastic scattering	no	$\approx E^{-2}$	Z^2
Photo effect	yes	$\approx E^{-3-..3.5}$	$Z^4 .. Z^5$
Compton effect	yes	$\approx E^{-1/2}$	Z
Pair formation	yes	$\log E$	$Z .. Z^2$

Figures 4 and 5 show the relative importance of the attenuation mechanisms and the total mass attenuation coefficient μ/p, as well as the mass attenuation for a light element, Al, and a heavy element, Pb. Materials chosen for shielding are generally high-Z elements with high specific density.

As shown in Figs. 4 and 5, the attenuation coefficients for the four attenuation mechanisms may be added to the total coefficient μ_0:

$$\mu_0 = \mu_R + \mu_P + \mu_C \ (+\mu_{pair}) \tag{9}$$

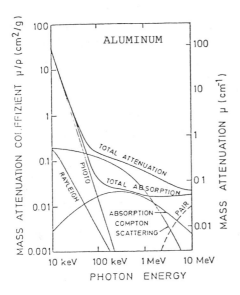

Fig. 4. Mass attenuation
and absorption coefficient
for Al, showing the relative
contributions of the different scattering mechanisms

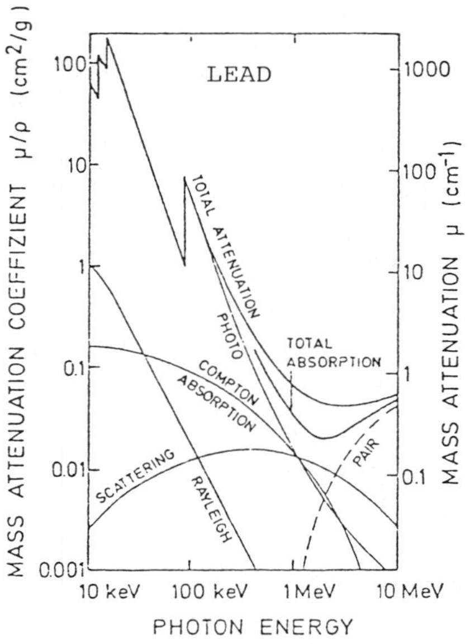

Fig. 5. Mass attenuation
and absorption coefficient
for Pb, showing the relative
contributions of the different scattering mechanisms

For a monochromatic x-ray beam, the intensity of the beam after
passing through an absorber of thickness x is given by:

$$I = I_0 \, \exp(-\mu_0 x) \tag{10}$$

with I_0 the incident intensity and μ_0 the total attenuation
coefficient.

The bremsstrahlung radiation from an implanter will, however,
have a continuous energy spectrum extending up to the
acceleration energy. Since the total absorption coefficient μ_0
decreases steadily with photon energy in the low energy region,
the selective absorption of the softer x rays renders the
spectrum towards the high-energy part of the spectrum. This
approximation is almost exact in shielding situations where the
incident radiation is reduced by many orders of magnitude.

3.5 Biological Effects

The absorption of ionizing radiation by tissue produces excited atoms, molecules, ions and radicals, resulting in chemical processes which influence the biological reactions inside the cells. The most probable process is the oxidation of enzymes which contain a SH (sulfhydryl) group by a diffusing OH^- radical. The chemically-active SH groups are destroyed and, as a consequence, the enzyme is blocked. The biological consequences of high-dose radiation exposure are, therefore, similar to the damage caused by certain poisons, e.g. nitrogen mustard gas $((ClCH_2-CH_2)_2-NCH_2CH_2)$.

New international recommendations assume strict proportionality between dose and biological effects, i.e., no threshold dose is assumed. Therefore, small doses are to be avoided; the likelihood of any effect, however, is correspondingly small.

If N persons are exposed to a dose equivalent D_q (in Sv) the number n of persons who become ill is:

$$n = N \cdot f \cdot D \qquad (11)$$

where f is the risk factor given for some diseases in Table 7.

Table 7. Risk factors for malignant tumors

Disease	Risk factor f (Sv^{-1})
leukemia	$2.3 - 6.3 \cdot 10^{-4}$
bone cancer	$5 \cdot 10^{-4}$
thyroid cancer	$5 \cdot 10^{-4}$
lung cancer	$2 \cdot 10^{-3}$
breast cancer	$2.5 \cdot 10^{-3}$
all other tissue	$5 \cdot 10^{-3}$
total	$\leq 2 \cdot 10^{-2}$

Especially important for monitoring the effects of radiation damage are blood tests, since blood is continuously produced within the bones. Through a radiation dose of the order of 10 rem, a dose-dependent blood change (increase in the number of leukocytes, followed by an increase in erythocytes) is induced.

Commercial implanters use lead to shield the operator from the radiation. In experimental equipment or during troubleshooting, extreme care and a continuous monitory of the radation is necessary.

3.6 Examples of Shielding Calculations

In the following section, three examples of shielding cal- culations are given for a medium-current implanter, a high- current implanter, and a high-voltage accelerator.

The total number of x rays transmitted through a shielding of thickness x is given by:

$$\Phi_t = \Phi_0 \cdot \exp\ ((-\mu(E)x) \tag{12}$$

which can be written according to Eq.(6) as:

$$\Phi_t = 2A\ Z\ Io \int_0^{E_0} \frac{E_0 - E}{E} \cdot \exp.\ (-\mu(E)x)\ dE \tag{13}$$

Assuming these photons are emitted isotropicly, the x-ray flux Φ_r in a distance from the source is:

$$\Phi_r = \frac{\Phi_t}{4\pi r^2} \tag{14}$$

According to Jaeger [7], the conversion from x-ray flux to dose per hour in the energy range from 70 keV to 2 MeV is given by:

$$D\ (R/hr) = 1.8 \cdot 10^{-9}\ \Phi_r E \tag{15}$$

where Φ_r is the x-ray flux (photons cm^{-2} sec^{-1}), E is the photon energy in keV. Since, for x rays, R $\approx$ rad (or Gray $\approx$ Sievert), the tolerable dose D can be taken immediately from Table 5. The total number of x-ray photons which can be tolerated is then given by:

$$\Phi = 6.5 \cdot 10^9 \ r^2 \ D/E \tag{16}$$

To calculate the shielding thickness x, Eq.(13) has to be evaluated. For this purpose, the energy range is divided into small intervals for which $\mu(E)$ is taken from Fig. 5. The shielding thickness x is then determined by summing up all intervals iteratively.

For a medium-current implanter with an ion current up to 1 mA, extraction currents up to 5 mA are used. In this case, the reverse electron current is approx. 2 mA, which is stopped in the stainless steel wall of the vacuum system. At an accelerator voltage of 120 kV, the lead shielding should be 6.5 mm thick to reduce the x-ray dose below the recommended limit of 0.25 mrem/hr.

In the case of a high-current implanter running with an ion-beam current of several mA at 80 keV, the reverse electron current is assumed to reach 10 mA. With a molybdenum aperture (Z = 42), the resulting x-ray dose at the distance of 1 m from the source behind 6.3 mm of lead will be 0.009 mrem/hr. This clearly shows the potential x-ray hazard at higher acceleration voltages. Although shielding is partly done by the vacuum tube, as a rule of thumb, the lead shielding for energies higher than 100 keV should be at least 5 mm thick.

With van de Graaff accelerators having energies higher than 2 MeV, the lead shielding should be in the order of several cm in thickness, although the electron currents are low. With these accelerators, special shielding precautions may be taken due to nuclear reactions with some type of ion beams.

It is important to note that when interlocks are jumpered during maintenance, extremely high x-ray radiation can result if the shielding doors are open.

4. TOXIC SUBSTANCES

The most common dopants in ion implantation are arsenic, boron, and phosphorus, which account for roughly 95 % of all implanted ions in silicon. For other semiconductors, however, other dopant ions are also used. Of importance are III-V compound semi-conductors with the main dopants beryllium, sulphur, selenium, and silicon.

The most convenient way to produce ions is to introduce a gas or a gaseous compound of the desired species into the ion source. Alternatively, evaporating a solid (e.g. arsenic or phosphorus) in an oven attached to the source may be done, thus eliminating gas bottles and producing fewer dissociation products, which results in a higher target current of the derived species for a given extraction current. Disadvantageous, however, is the limited operating time of the source due to discontinuous refilling and additional warm-up and cool-down time.

In production implanters, arsenic, boron and phosphorus are usually produced from AsH_3, BF_3, and PH_3, but other compounds such as AsF_3, BCl_3 or even solid arsenic, antimony and phosphorus are also used. In most implanters, gases are stored in high-pressure lecture bottles. Often, mixtures with non-toxic carrier gases are used, e.g., 10% PH_3 or AsH_3 in hydrogen. Together with appropriate regulators and manifolds, the bottles are mounted in a gas cabinet close to the ion source. A toxic exhaust vent is used to reduce any danger to personnel from small leaks in the gas system. When the door of the gas cabinet is open, the air velocity should be in excess of 30m/sec [1]. In the event of malfunction of the exhaust system, the implanter must automatically close the gas valves installed in the high pressure side and stop operation. Employees must simultaneously be made aware of the defect by acoustic and optical alarms. To assure timely detection of leaks in the gas system, the installation of a gas monitor is recommended. Such monitors are offered for sale by various manufacturers and are sensitive enough to detect very small leaks.

During correct operation of the implanter, all toxic substances are confined to the vacuum system. Since only a fraction of the gas in the source is actually ionized and extracted as an ion beam, the major part of the source feed material is deposited inside the system or chemically bonded. Large amounts, however, are also absorbed by pump oil and given off to the environment through the pump exhaust. The exhaust gases from the pumps should, therefore, also pass a chemical filter (for instance, copper sulfate for arsine).

The contaminated materials resulting from maintenance and cleaning, as well as used vacuum oil, must be handled as toxic waste and be disposed of in accordance with the laws of the respective country.

If cryopumps are used, attention must be paid that a proper venting of toxic substance is guaranteed during regeneration or in the event that the pump is defective or the compressor fails.

In every country laws exist which regulate production, transportation, handling, and disposal of toxic materials. Since October 1986, ordinances for toxic substances are in effect in Europe, and in the USA the General Industry Standards, both of which regulate the use of toxic and carcinogenic substances. Always included in these regulations are the industrial threshold limit values. For instance, in the USA these values are defined by OSHA [8], and the Federal Republic of Germany by The Deutsche Forschungsgemeinschaft (DFG) [9].

In Table 8 [8 - 11] the most commonly used ion source materials are presented.

Table 8. Toxic substances used with ion implanters

a.) Silicon dopants

Element	Feed	Form	PEL/TLV (ppm or mg/m^3)	Comment
Arsenic	As	s	0.2	dust potentially carcinogen
	AsH$_3$	g	0.05	pyrophoric, garlic like odor
	AsF$_3$	g	3	pungent odor
	GaAs	s		N/A
Boron	BCl$_3$	l/g	5	pungent irritating odor
	BF$_3$	l/g	3	pungent irritating odor
	B$_2$H$_6$	g	0.1	sicky sweet, pyrophoric
Germanium	GeH$_4$	g	0.2	pungent odor
Phosphorus	P	s		N/A
	PH$_3$	g	0.3	odor of decaying fish
	PCl$_3$	g	1.5	pungent odor
	PF$_3$	g	1	pungent odor
	PF$_5$	l	3	pungent odor
Antimony	Sb	s	0.5	carcinogen, no lower limit
	Sb$_2$O$_3$	s		N/A
	SbH$_3$	g	0.1	garlic like odor
Silicon	SiH$_4$	g	0.5	repulsive, pyrophoric

b.) Toxic implanter components

Substance	Form	PEL/TLV (ppm or mg/m^3)	Comment
Beryllium	s	0.002	dust, percutaneous incorporation, strongly carcinogen (IDLH 300 mg/m^3)
Pump oil	l	10	contaminated with toxic substances
Freon	l	1000	accumulated in liver
SF$_6$	g	1000	
SF$_4$	g	0.1	

c.) III - V - Dopants

Element	Feed	Form	PEL/TLV (ppm or ms/m^3)	Comment
Beryllium	Be	s	0.002	strongly carcinogen
	BeCl$_2$	s	0.002	strongly carcinogen
Selenium	Se	s	0.1 (0.2)	
	CdSe	s	0.1	
	SeO$_2$	s	0.1	
	SeH$_2$	l/g	0.05	rotten radish odor
Silicon	Si	s	---	
	SiH$_4$	g	5	pyrophoric, repulsive odor
Sulfur	S	s	---	
	H$_2$S	g	20 (10)	rotten eggs odor
Tellurium	Te	s	0.1	
Zinc	Zn	s	---	
	ZnCl$_2$	s	1	

Most of the doping elements can be classified in the groups

1) pure elements and oxides
2) hydrides
3) halides

The first group consists, for a large part, of solid, water-soluble substances. These substances can, if taken up by the body over a long period of time, accumulate, and thus increase the risk of cancer. Arsenic can, for example, lead to the development of hyperkeratosis and melanoma, and in the event of accumulation of larger amounts, may lead to carcinomas of the skin, respiratory tract and liver.

The hydrides are, as a rule, gaseous, and thus can reach the blood stream through the lungs, where they can block the nervous system by chemical reaction. Arsine possesses a strong haemolytic effect and, in concentrations of 1-10 ppm [13], inhaled over several hours, can be dangerous to life (lethal). The seldom used compound hydrogen selenide belongs to the most toxic compounds used in implantation. Concentrations of even less than 0.2 ppm [14], if repeatedly or chronically inhaled, can lead to selenosis with the symptoms: vomiting, metallic aftertaste and fatigue. Concentrations of 0.3 ppm [14], inhaled over several hours, are life-threatening.

The third group, the halides, are less toxic than the substances in the previous groups. They should, however, also be handled with the necessary precautions. Halides can, by way of hydrolysis, cause acid burns of the skin and, if inhaled, also cause acid burns of the lungs and mucous membranes. After longer exposure, the development of pulmonary edema is frequent [14].

As a general principle, a safety mask should be worn when changing gas bottles. An effective protection, however, can only be offered by an external-air-independent compressed-air breathing apparatus (SCUBA-Type respirator). Not suitable are activated carbon filters, as these become exhausted rapidly and gas may then pass unobstructed.

The changing of gas bottles, and also all other maintenance and repair activities on parts which come in contact with gas flow or ion radiation, should be conducted only by persons who have received special training. However, not only persons responsible for machine servicing, but also those employed in the area should receive a safety training, in order to guarantee that in a case of safety problems each one really knows how he should react.

In addition, for every machine and for every area, safety directives should be developed, in accordance with pertinent regulations.

This directive should include the following items:

a) General information and remarks regarding the chemicals used,
b) Procedures for the cleaning and maintenance of the apparatus,

   c) A description of the bottle changing operation and
      precautionary measures,
   d) Appropriate conduct in the event of a disturbance or defect
      (for instance, a leakage)
   e) An emergency list (including telephone numbers, persons to
      contact)

During all cleaning operations, including oil changes on the
vacuum pumps, rubber gloves or similar protective covering should
be worn. If it is to be expected that dust arises during these
operations it is recommended that a dust-mask be worn. Furthermore, during grinding and sanding operations it is advisable to
use a special exhaust area, for instance, a walk-in exhaust hood,
as these activities can cause a concentration of contaminants in
the immediate surroundings which exceeds the maximum allowed
level.

The disposal of contaminated material should not be part of the
normal rubbish. All matter, such as cleaning rags or pump oil,
must be disposed of as toxic waste, according to the respective
laws.

Table 9. Scrubber methods for toxic gas

Element	Method	Reagent	Carrier	Product
Arsine	dry bed	$CuSO_4$	silicagel	Cu_3As_2, As_2O_3
Diborane	dry bed	$KMnO_4$	Al_2O_3	B_2O_3, MnO_2
Chlorine	scrubber	NaOH	H_2O	NaOCl
HCl	scrubber	NaOH	H_2O	NaCl
Fluorine	dry bed	Al_2O_3	-	AlF_3
Phosphine	dry bed	$CuSO_4$	silicagel	P_2O_5
Silane	dry bed	$CuSO_4$	silicagel	SiO_2

The disposal of one-way lecture gas bottles poses a large
problem. In order to avoid getting impurities, which may be
present in the bottle residues the gas bottles are never com-

pletely emptied. This residue gas, however little it may be, should not be released into the environment. Rather, it should be properly destroyed, according to the given regulations. There are several possibilities to achieve this goal - one of these is the use of a gas scrubber, in which the gas is chemically bound, whereby the exhausted filter (acceptor material) must also be disposed of as toxic waste. According to which sort of gas is involved, a scrubber with a dry bed or with a wash solution (scrubber) may be implemented. The dry bed washer presents a less problematical solution. In Table 9 various examples of residue gas disposal are presented [15].

The authors would like to thank many colleges, especially K. Haberger, who contributed to this paper.

REFERENCES

1. R. Bustin and P.H. Rose, in: Ion Implantation Equipment ed.: H. Ryssel and H. Glawischnig (Springer, Berlin, 1983)

2. W. Riezler and W. Walcher, Kerntechnik (Teubner, Stuttgart, 1958)

3. H. A. Kramers, Philos. Mag. 46, 836 (1923)

4. Handbook of Chemistry and Physiks, pp. E 184 - 194 (CRC Press, 1977)

5. T. A. Cahill, in: New Use of Ion Accelerators, ed.: J. F. Ziegler (Plenum Press, New York, 1975)

6. E. Sauter, Grundlagen des Strahlenschutzes (Tiemig, München, 1983)

7. T. Jäger, Principles of Radiation Protection Engineering (McGraw Hill, New York, 1965)

8. Occupational Safety and Health Administration: General Industry Standards, 29 CFR 1920, U.S. Government Printing Office, Washington DC

9. Maximale Arbeitsplatzkonzentration und Biologische Arbeitsstofftoleranzwerte, DFG (Verlag Chemie GmbH, Wein heim, 1983)

10. W. Riezler and W. Walcher, Kerntechnik (Teuber, Stuttgart 1958)

11. G.K. Herb, R.E. Caffrey, E.T. Eckroth, Q.T. Jarrett, C.L. Fraust, and J.A. Fulton, Solid State Technology 26, 185 (1983)

12. Registry of Toxic Effects of Chemical Substances, U.S. Government Printing Office, Washington DC

13. Vogl, Heigel, Schaefer, Handbuch des Umweltschutzes (ecomed Verlagsgesellschaft mbH, Landsberg, 1987)

14. R. Kuehn, K. Birett, Merkblätter Gefährliche Arbeitsstoffe (ecomed Verlagsgesellschaft mbH, Landberg, 1987)

15. B. Reimann, Internal Newletter 1987 (Messer Griesheim GmbH, Düsseldorf)

EMISSION OF IONIZING RADIATION
FROM ION IMPLANTERS

Constantine J. Maletskos,* Ph.D., CHP
and
William R. Ghen

Varian/Extrion Division
Gloucester, Massachusetts 01930

United States of America

ABSTRACT

Implanters emit unintended ionizing radiation in the form of x rays because accelerated stray electrons strike atoms of gases and implanter structure. The physics of this process and the sources of x rays in the implanter structures are presented. The dependence of x-ray production on operating conditions are described as well as the methods of radiation reduction by proper design and shielding. Quantities and units of radiation exposure are presented in both the new International System of Units (SI) and the previous special units. A short but general discussion of the biological effects of ionizing radiation describes stochastic (random) and non-stochastic effects and provides a perspective for setting exposure-rate limits for implanters. Finally, some general principles for performing radiation surveys are given and the responsibilities of users and owners of implanters are provided.

TOPICS

1. INTRODUCTION
2. PRODUCTION OF IONIZING RADIATION
3. IMPLANTER LAYOUT AND OPERATION
4. DEPENDENCE OF RADIATION ON OPERATING CONDITIONS
5. METHODS OF RADIATION REDUCTION
6. SHIELDING CONSIDERATIONS
7. ENGINEERING FOR RADIATION CONTROL
8. QUANTITIES AND UNITS OF RADIATION EXPOSURE
9. BIOLOGICAL EFFECTS OF IONIZING RADIATION
10. EXPOSURE LIMITS FOR IMPLANTERS
11. PRINCIPLES FOR PERFORMING RADIATION SURVEYS
12. RESPONSIBILITIES OF IMPLANTER OWNERS AND USERS

* Consultant, Gloucester, Massachusetts

1. INTRODUCTION

Ion implanters play a major role in the production of semi-conductor devices. The function of implanters is to introduce atoms onto or below the surface of materials in order to transform the electrical or mechanical properties of these materials. The accelerated positive ions impinge on specific target materials. The interaction of the positive ions with atoms in the residual gas of the evacuated system or with the inner surfaces of the vacuum-containing system or with other internal components produces electrons that can be accelerated up to the energy of the implanter. The stopping of the electrons results in the emission of ionizing radiation from the implanter. This chapter discusses this unintended radiation, its production, and its control. As implanter beam currents are reaching 30 to 50 mA and energies are in the few MeV domain, attention to the control of radiation emission becomes more important. The discussion in this chapter is confined mainly to implanter energies $\leq$500 keV.

Emission of ionizing radiation from implanters has been discussed from various standpoints by Bustin and Rose (1982), Maletskos and Hanley (1983), Ryssel and Haberger (1984) and Baldwin et al. (1988). For this chapter, theory is minimized and emphasis is placed on descriptive aspects in order to improve understanding, on practical aspects of reducing the emission of radiation, on the details of radiation surveys, and on a more thorough perspective of the biological effects of radiation, the recommended dose limits for the exposure of people, and the setting and attainment of exposure limits for the implanters.

2. PRODUCTION OF IONIZING RADIATION

Acceleration of charged particles (positive or negative) results in the production of electromagnetic radiation (photons). The acceleration can be negative, i.e., deceleration, in the case of an electron being stopped in a target material, to result in photons called x rays (originally called bremsstrahlung, i.e., braking radiation), as emitted in a medical x-ray machine. The acceleration can be positive, in the case of an electron traveling in a circle and constantly changing direction, again to result in x rays, as emitted from a synchrotron.

The relation of the x-ray (bremsstrahlung) intensity per target atom, I, to the particles and target atoms involved is given by the following expression:

$$I \sim Z^2 z^2 / m^2 \qquad (1)$$

where Z is the atomic number of the target atom and where z is the atomic number and m is the mass of the incident particle. Because the intensity is inversely proportional to the square of the mass of the incident particle, protons and other heavier charged particles will produce more than a million times less radiation than electrons. Thus, the positive-ion beams of implanters will produce negligible x rays on striking the target wafers and are of no consideration in the x-ray emission of implanters when compared to the emission resulting from electrons as described in Section 3.

Electrons generally strike thick materials in implanters, thick with respect to the range of the electrons in the materials, and hence, thick-target theory applies. The total x-ray intensity, $I_{e,T}$, per incident electron is then given by

$$I_{e,T} = kZE_e^2 \qquad (2)$$

here k is a constant and E_e is the initial electron energy. Note that the dependence on atomic number is now Z and not Z^2 as in Equation 1. The fraction of the initial electron energy is equal to kZE_e, which translates into a few percent for low-Z materials and electron energies <500 keV. Even so, this amount can represent a substantial emission of x rays. The intensity per unit time (i.e., the power, P_T)) from which the exposure rate can be determined is given by

$$I_T/t = P_T = k_1 Z E_e^2 i \qquad (3)$$

where t is the time, k_1 is another constant and i is the electron current. Thus, the radiation emission is proportional to the current and to the square of the electron energy. As the current and voltage of accelerators continue to increase, the radiation emission also increases and will require various methods to reduce it, including shielding. The emission is also proportional to the atomic number of the target material and, hence, the use of low-Z materials in those areas where accelerated electrons strike will reduce the emission. A practical material with the lowest atomic number is beryllium (Z = 4) and can be used in specific areas.

At low energies (<<500 keV), the angular distribution of the
emitted x rays resulting from electrons striking a thin target are
mainly in a lobe perpendicular to the electron direction as shown in
curve A of Figure 1*. As the electron energy increases, the angular
distribution moves forward as indicated by the lobe marked curve B,
and at higher energies (greater than a few MeV) the distribution is
predominantly in a narrow lobe in the forward direction (curve C).
Although the interaction is more complicated, similar angular-
distribution characteristics are observed for thick targets that
apply to the implanter situation.

The x rays emitted by the deceleration of charged particles, in
this case electrons, can have energies between zero and a maximum
that is equal to the energy of the charged particle. A distribution
of x-ray energies results in a spectrum where the most abundant
energies are the intermediate energies as shown in Figure 2.
Electrons striking the target material can eject electrons from the
atoms and the filling of the vacancies results in the emission of
characteristic x rays of fixed energies as shown in the figure.
Also, characteristic x rays can be emitted from certain types of
radiation detectors used to determine x-ray spectra, and these are
shown in the figure as well. Depending on the electron energies and
the implanter materials involved, only the general shape of the
spectrum may be applicable. Even the low-energy portion will be
missing as indicated by the sharp cutoff at energy E_c (see, also,
Section 4). The actual transition will be smoother, but as the
x rays traverse more implanter components, less of the lower-energy
x rays will be emitted from the implanter, i.e., the spectrum is

---

* Most of the figures in the text are schematics without numerical
values and are intended to demonstrate a principle rather than actual
data. For implanters, numerical data would be different for each
implanter type or model and should be obtained specifically for the
implanter of interest. The same concept applies to the figures
concerned with the discussion on the biological effects of ionizing
radiation.

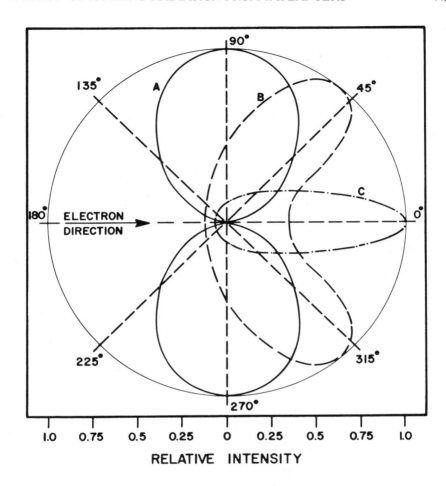

Figure 1   Schematic of the angular distribution of the relative x-ray intensity. The figure is a cross-sectional plane that contains the electron. The lobes shown are three-dimensional lobes of revolution about the electron axis. Curve A is for electrons with energy <<500 keV; curve C is for electron energies >1 MeV; and curve B is for an intermediate electron energy. These lobes are representative of an interaction with a single electron or with a thin target. For thick targets, the angular distribution is a complicated function of electron energy and absorber atomic number but results in similar lobe characteristics. For implanters, lobes of type B would be more characteristic, although the lobes observed outside the implanter surfaces would be highly distorted because of the variable absorption of x rays by the implanter structures.

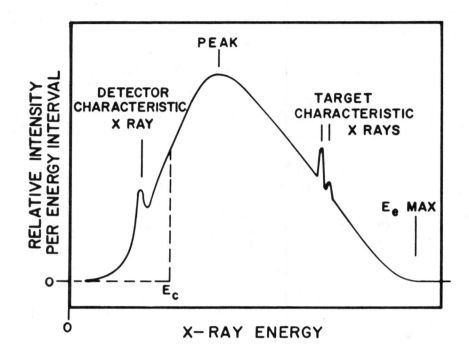

Figure 2   Schematic of a spectrum of x rays emitted from the stopping of electrons in a thick target. The energies of the x rays range from zero to a maximum energy equal to the electron energy. Electrons ejected from atoms are later replaced; this process results on characteristic x rays that superimpose on the spectrum. These characteristic x rays can arise from the atoms of the target material or from the atoms of an x-ray detector that has a linear response with energy, such as the iodine in a NaI(Tl) scintillation detector. In an implanter, the structures can absorb the lower-energy x rays and these x rays would not penetrate to the outside of the implanters. This condition is illustrated by the dotted line that shows a sharp cutoff energy although, actually, the transition would be smoother. If the atomic number of the target is low, the characteristic x rays would occur in the low-energy region and would not be observed. As the lower x-ray energies are absorbed, the more energetic x rays remain; this conditioning is called hardening of the spectrum.

hardened or transformed into a higher-energy spectrum. This hardening has a bearing on the shielding considerations described in Section 6.

3. IMPLANTER LAYOUT AND OPERATION

Production ion implanters using direct current (DC) voltage-gradient acceleration are available for beam energies up to 500 keV. The type of acceleration in conventional implanters can be either single stage or dual stage. Each type has advantages and disadvantages, and each has different radiation characteristics. In Figure 3, the beamline is shown schematically for both types with the principle regions of x-ray generation indicated. Positive-ion-induced x-ray generation in conventional implanters is insignificant, but electron generated x rays can occur any place in the beamline where fast electrons collide with atoms, either residual gas or beamline structures. Other possible mechanisms of radiation generation such as electron or ion acceleration or deflection are not significant.

3.1 Two-Stage Implanters

The majority of production implanters now in operation use two-stage acceleration. In two-stage implanters, beam analysis takes place between the two stages, thus reducing the beam current that must be handled in the second or final acceleration stage.

The principal regions of x-ray generation are the extraction region and the acceleration region, and the associated electron "target" areas. The extraction region is a narrow gap between the ion source and the extraction electrode. This is a region of comparatively poor vacuum and many ion-gas-molecule collisions occur, producing electrons that are accelerated toward the ion-source exit aperture by the extraction field, typically 25 to 60 kV.

Electrons accelerated toward the ion source in hot-filament sources will strike the arc chamber, usually molybdenum (Z=42) or the tungsten filament (Z=74). Thus, a large quantity of x rays can be formed. Fortunately, the low accelerating voltage results in a low-energy radiation spectrum, and structural materials in the ion-source chamber may provide enough shielding so that no additional shielding is needed. Shielding considerations are discussed in Section 6.

MEDIUM CURRENT, 2 STAGE ACCELERATION

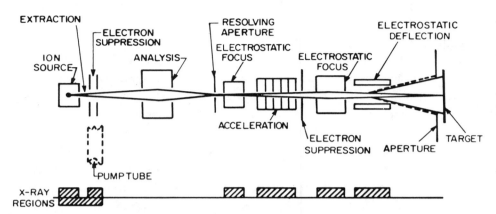

HIGH CURRENT, SINGLE STAGE ACCELERATION

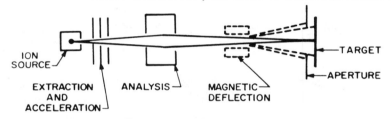

Figure 3   Regions of x-ray generation in ion implanters. Upper: Schematic representation of medium-current ion implanter using two-stage acceleration and electrostatic focus and beam deflection. The pump tube, if used, is a source of x-ray generation only under fault conditions. Lower: High-current ion implanter using single-stage acceleration and mechanical scanning, or mechanical scanning and magnetic-beam deflection.

Electrons formed in the ion beam beyond the extraction region are prevented from entering the extraction gap by means of a negative-suppression or acceleration-deceleration ("accel-decel") voltage applied to the extraction electrode or to an electrode immediately behind the extraction electrode. The negative voltage isolates electrons downstream of the extraction region from the positive ion source bias voltage, thus, preventing this bias voltage from stripping the electrons from the beam, preserving the beam space-charge neutrality for optimum beam transport. Suppression voltages in two-stage medium current implanters are usually 2 to 3 kV. High current implanters require higher suppression voltage to overcome the beam space charge, usually 10 to 20 kV. Secondary electrons produced at the extraction electrode cannot be suppressed.

The acceleration region is where the most troublesome radiation is generated even though pre-acceleration analysis significantly reduces the accelerated ion-beam current. Two-stage medium-current implanters use acceleration voltages up to 475 kV (the full voltage of the implanter less the extraction voltage). The acceleration structure consists of a series of electrodes separated by insulating spacers. Each electrode consists of a plate containing a circular aperture through which the ion beam passes. In normal operation, the positive side of the acceleration power supply is connected to the beam-entrance end of the acceleration tube, with the beam-exit end at electrical ground. Field-gradient uniformity of the acceleration-tube gaps is maintained by a resistor string connecting the electrodes throughout the entire length of the acceleration tube. However, high-current implanters use active power-supply units on each of a fewer number of gaps. Electrons formed by ion-gas and ion-structure collisions within the acceleration tube are accelerated toward the analyzer region, and those electrons that are accelerated through the entire length of the acceleration tube will attain the highest energies.

Electrons that pass through the resolving aperture are deflected to the side of the beamline as soon as they enter the field of the analyzer magnet preventing them from reaching the extraction region.

Thus, the maximum electron energy in the implanter is equal to the acceleration voltage or the extraction voltage, whichever is greater, but the ion-beam energy is equal to the sum of the extraction and acceleration voltages. For operation in the decel mode, see Section 3.3.

Many two-stage implanters use a resolving aperture or slit at the focal point of the analyzing magnet. The back, or ion-beam-exit side of this resolving aperture becomes the impact surface for most of the high-energy electrons from the acceleration region. Graphite (Z=6) or beryllium (Z=4) are usually used to minimize the generation of x-ray radiation from this area. Beryllium is the lowest-Z material that has structural properties and vacuum characteristics compatible with this application. Beryllium has toxicity concerns that can limit its use in beamline applications to areas where it can be totally protected from ion-beam impact and sputtering.

Some two-stage implanters use a focusing lens at the ion-beam-entrance end of the acceleration tube to improve beam transmission through the acceleration tube and beamline. The lens voltages range from 20 to 50 kV and may be sufficient to produce, by themselves, significant low-energy radiation. Also, the electron beam from the acceleration tube can strike parts of the lens or lens-housing structure, generating significant x-ray radiation within the lens housing. External shielding around the resolving aperture and lens housing usually is required to keep radiation from this area within acceptable limits.

X rays generated within the acceleration tube present a difficult shielding requirement. The radiation originates from a cylindrical-like source along the inside of the acceleration tube, and high voltage requirements prohibit the use of metallic shielding around the acceleration tube except in the form of non-contacting rings or bands, where possible. Non-conductive shielding material, usually lead oxide, has been incorporated into the acceleration tube insulators in some applications. Additional general shielding is usually required in the implanter walls surrounding the acceleration tube region. Electrons in the beam beyond the acceleration tube are prevented from entering the beam-exit end of the acceleration tube by a suppression electrode with a negative voltage (1 to 3 kV). The last electrode in the acceleration tube is often used as the suppression electrode.

Two-stage medium current implanters normally implant one wafer at time. The ion beam is uniformly dispersed over the surface of a stationary wafer by deflecting the beam in two perpendicular directions to produce a "raster" scan. Electrostatic scanning is used, with triangular voltage waveforms applied to deflection plates in the scanner system. As higher beam energies and larger diameter wafers have evolved, higher scan voltages have been required.

Low-energy electrons in the scan region will be accelerated toward the positive scan electrodes at up to the maximum peak-to-peak scan voltage (positive on one electrode and negative on the opposite electrode, both voltages tracking with the acceleration voltage). Scan voltages (up to 40 kV peak-to-peak) are high enough in some implanters to generate significant radiation within the scan housing and sometimes external shielding is required. Also, significant radiation arises from the lens voltages that provide post-acceleration electrostatic focusing. Finally, control of vacuum pressure is critical in this region in order to minimize the production of the free electrons that directly affect x-ray production, beam purity, and scan-waveform-induced, wafer-implant nonuniformity.

Implanters require electron bias voltages in the Faraday system for dose-accuracy purposes, and some implanters use electron-flood systems for wafer charge neutralization. The voltages (a few hundred volts) are so low that any x rays generated will be attenuated by the beamline housing and will be of no concern.

Two-stage implanters require the ion-source vacuum system be in the electrically isolated high-voltage terminal structure. In some systems, due to space constraints, the vacuum pump is located outside of the high-voltage terminal, and is connected to the ion-source chamber by an insulating pump tube. The pump tube is not a source of radiation under normal operating conditions, but under some fault conditions, it is possible to develop a glow discharge over a portion of the pump tube. The pump tube is divided into segments with metallic separators that are connected by a resistor network for voltage-gradient uniformity. Failure of one of the resistors can result in severe gradient non-uniformity which can start a glow discharge inside the pump tube. This glow discharge can supply enough electrons, that are accelerated by the acceleration voltage, to generate some x-ray radiation near the high-voltage-terminal end of the pump tube, even when there is no beam of positive ions.

3.2 Single-Stage Implanters

Single-stage implanters evolved as low-energy, high-current machines. The semiconductor-device industry has required increasingly high-beam-current, high energy implanters. Today, single-stage implanters are capable of 10-mA beam currents at up to 160 keV.

The major advantage of single-stage acceleration is beam-line power distribution design simplicity (Figure 3). The entire beamline except the ion source and extraction/acceleration electrodes are at electrical ground. Ion source power supplies (filament, arc) are the only ones requiring isolated electrical power. Separate extraction and acceleration power supplies are not required.

Pre-analysis acceleration requires the acceleration of the entire positive-ion beam. As a consequence, an acceleration power supply with a much higher-current capability is necessary. The single-stage configuration keeps the major radiation region confined to the ion-source area. The acceleration of all the positive ions, represented by the full extraction current, to full beam energy results in a high electron current accelerated toward the ion source. The ion-source region then becomes the major area of x-ray generation. Because of the high pre-analysis beam current, electron suppression becomes very important in the control of this radiation. High current, single-stage implanters use suppression voltages up to 12 kV.

Single-stage high-current implanters do not use electrostatic beam scanning because of the inherent space-charge problems associated with electrostatic deflection of high-current-density ion beams. Therefore, focused positive-ion beams in combination with mechanical, or with magnetic and mechanical, scanning systems are used to implant batches of wafers uniformly and to meet the thermal limits of the wafer photo-resist coating. No significant radiation is produced in the scan regions of these machines.

3.3 Implanter Decel Operation

Some implanters are equipped with a capability of decel operation for low-energy implants. In the decel mode, the final acceleration stage is operated with reversed (negative) polarity. This stage, then, will isolate electrons in the region downstream of the decel electrode from the ion source positive bias, thus, also acting as an electron suppressor during decel operation. Medium-current implanters usually use decel voltages of only a few kV, so that photons produced by collisions involving electrons accelerated by the decel voltage will have insufficient energy to penetrate the implanter structure. However, some highcurrent implanters can use considerably higher decel voltages (up to 60 kV). The higher electron current associated with high ion-beam current may produce significant radiation at the areas of electron impact during operation at high decel voltage. Electron impact in decel operation will be downstream of the decel region.

3.4 Implanters in the MeV Domain

Implanters that use DC acceleration and air insulation between the high-voltage enclosure and terminal are limited to about 500 keV because of the physical machine size required to produce the necessary air gaps.

Higher effective energies can be obtained with some ions by using a doubly or triply charged ion beam, at considerably reduced beam current, or with other methods of acceleration.

A two-stage, air-insulated implanter, for example, may implant a 1-mA ion beam at 500 keV, using a singly charged beam, and about 100 μA at an effective 1 MeV using a doubly charged beam. Because no actual accelerating fields greater than 500 keV exist, there is no increase in radiation when implanting with multiply charged beams, and radiation usually decreases due to the lower ion beam and electron currents.

For implanting at energies much above 1 MeV, combination DC and radiofrequency (RF) acceleration is sometimes used. For this method, a DC beam is further accelerated by passing through a series of acceleration stages in which the accelerating fields are produced by synchronizing the frequency of the RF voltage to the ion-beam velocity. Beam transmission in RF acceleration sections is usually inefficient, resulting in many secondary electrons produced by beam collisions along the structure. Because the electron beam velocity will be much higher than the ion beam velocity, electron synchronization will not occur, and electron energies above the RF voltage of one stage are unlikely. Radiation in the RF section will be limited to that produced by electrons of one stage of RF acceleration. Tandem DC Van de Graff types are also used. These can be high radiation producers and require exceptionally good internal shielding.

At energies above 1 MeV, accelerated protons (H^+) and dopant ions can produce nuclear reactions, in their interaction with implanter structural materials and with wafers, to result in induced radioactivity and in photons and neutrons. Implanters in this energy region should be evaluated for radiation emission on a case-by-case basis. Shielding and other radiation-protection requirements could be considerably different from those for implanters with lower energies.

4. DEPENDENCE OF RADIATION PRODUCTION ON OPERATING CONDITIONS

From the equations of Section 2, the main parameters that control the production of radiation are the electron current and energy. In practice, neither of these two parameters is known. The electron current is not metered, although it can be estimated to some extent as the difference between net acceleration power supply current (the difference between power supply current with beam on and beam off) and total beam Faraday current. This does not account for beam lost between the acceleration system and the Faraday system.

The bulk of the electrons may attain energy in the acceleration tube corresponding to the difference between the total voltage and the extraction voltage. Electrons can attain energies in other parts of the implanter corresponding to other voltages, always less than the acceleration voltage. Actually, there is a range of voltages for each operating voltage that contributes to the electron energies because the electrons may interact with gases and structures before the full distance over which the total voltage gradient is applied, can be traversed. In spite of the above restrictions, it is, nevertheless convenient, practical and relatively realistic to relate the production of x rays to the operating current and voltage.

Figure 4 shows the relation between the x-ray intensity and the positive-ion-beam current at a particular acceleration voltage. Curve A is the curve predicted by theory, while curve B is a typical curve that would be observed in an operating implanter with sufficient shielding removed to make the measurements. The intensity does increase with increasing current, but the break in the slopes occurs because of changing beam optics in order to attain the higher currents.

Figure 5 shows the relation between x-ray intensity and the acceleration voltage at a particular beam current. Curve A is a parabola predicted by theory that the exposure rate is proportional to the square of the electron energy, E_e^2. Curve B is a typical curve that would be observed in an operating implanter. At the lower voltages (lower-electron energies), this curve does not reflect the true intensity because the radiation is highly attenuated by the intrinsic components of the implanter. As the voltage increases, the intensity increases more rapidly than the relation to E_e^2 would imply because less attenuation by the structures occurs for the higher-energy x rays. Depending on how the implanter is constructed, the latter part of curve B can approach the relation to E_e^2.

The suppression system is designed to remove or prevent electrons from entering regions where they can be accelerated. The effectiveness of the system depends on the design and location of the suppression electrodes, but, once installed, only the voltage on the system can control the efficiency of suppression. Figure 6 shows the relation between the x-ray intensity and the negative suppression voltage at a particular beam current and acceleration voltage. Curve A indicates the general relation of decreasing intensity with increasing negative voltage. The portion of the curve marked B indicates the relation when the operating current cannot be maintained at lower suppression voltages and, hence, the intensity is reduced by the practical necessity to maintain stable output. The

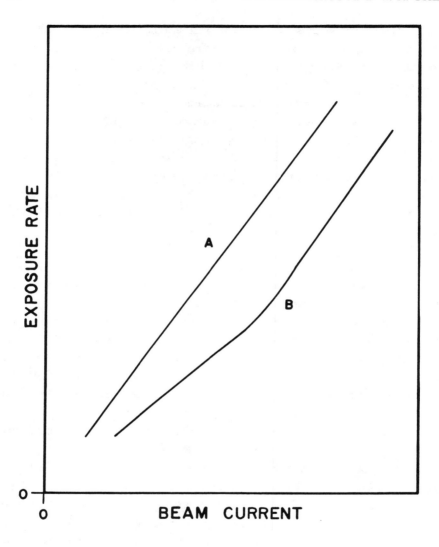

Figure 4   Schematic representation of the exposure rate from an implanter as a function of the positive-ion beam current, which acts as a surrogate for the electron current that is difficult to determine, at a particular acceleration voltage.   Curve A is the linear curve that is expected from theory and that is observed over ranges of beam current. When the beam current covers a wide range, the optics of the beam can change in order to attain the beam current and curve B is a more typical relation between exposure rate and beam current.

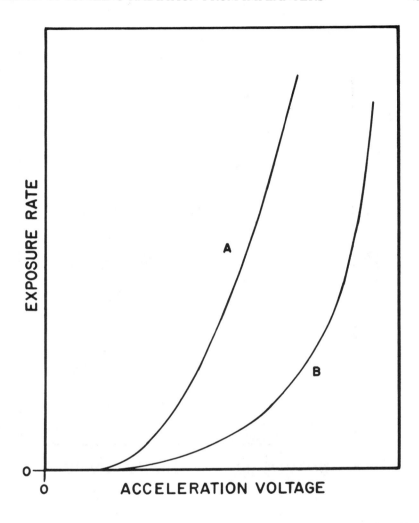

Figure 5   Schematic representation of the exposure rate from an implanter as a function of the positive-ion acceleration voltage, which acts as surrogate for the electron acceleration voltage, at a particular beam current.   Curve A represents the theoretical situation where the exposure rate is proportional to the square of the acceleration voltage.   In an implanter, the structures attenuate the lower-energy x rays more readily so that the exposure rate depends on the combination of attenuation and on the square of the acceleration voltage as represented by curve B.

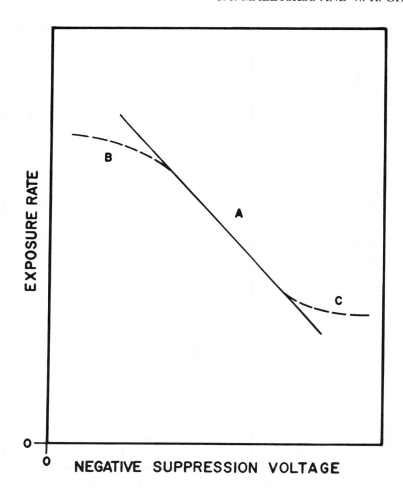

Figure 6   Schematic representation of the exposure rate from an implanter as a function of negative suppression voltage at a particular beam current and acceleration voltage. Negative suppression voltage repels electrons and prevents them from entering regions where they can be accelerated. As the negative voltage is increased, the repulsion is improved and the exposure rate decreases as indicated by curve A. As the negative voltage is increased, the efficiency of repulsion will eventually diminish and the exposure rate will tend to level off as indicated by curve C. At lower negative voltages, it becomes more difficult to maintain the beam current and the exposure increases less rapidly as indicated by curve B.

portion marked C indicates the condition when the efficiency of the suppression cannot be improved appreciably by further increasing the suppression voltage. Implanter suppression systems are normally designed for operation in this condition.

Depending on the implanter design, beam focusing may or may not play a significant role in the x-ray intensity of the implanters. Generally, the intensity is not affected significantly. Fortunately, implantation efficiency and uniformity are so important that an implanter is not likely to be operated for a significant time in a poorly focused condition.

Finally, the x-ray intensity is strongly dependent on the type of gas that is used, although there is no theory or rule of thumb that can predict a relation between intensity and the implanting gas. In specific implanters, experience has shown that, under the same operating conditions, x-ray intensity tends to increase in order with the following gases: boron (BF_3), phosphorus (PH_3), nitrogen (N_2), arsenic (AsH_3), argon (Ar) and krypton (Kr). Silicon (SiF_4) and oxygen (O_2) produce even higher intensities. Oxygen results in the highest intensities of the gases enumerated, but its high corrosiveness on implanter parts limits its use. Molecular and dimer beams can result in more radiation than atomic beams for the same current and voltage. For oxygen and nitrogen, radiation from the dimer beam can exceed that from the atomic beam by several times; thus, e.g., O_2^+ can result in a much higher level of radiation than O^+. Other type implanters and different operating conditions can result in significant changes in the radiation characterisitcs of the gases mentioned here. Because specific information on the radiation output cannot be given for all types of ion beams, only those gases should be used that have been specified by the manufacturer. If other gases are to be used, a radiation survey should be made to determine the acceptability of the new species under which it is to be used.

The angular distribution of the x rays, i.e., the lobe geometry is not necessarily the same for all gases. Account of this phenomenon must be taken in implanter shielding (Section 5) and in the specifications of the implanter.

5. METHODS OF RADIATION REDUCTION

Ion implanters have become a standard piece of manufacturing equipment in the semiconductor industry. As such, it is desirable to control the external radiation from implanters such that no special restrictions or precautions are needed for implanter operators, and nearby and passing workers. For this reason, the radiation from all external surfaces of implanters is kept at or below the recommended radiation exposure rates, under all expected operating conditions designated by the implanter manufacturer's specifications. As discussed in Section 2 and from Equation 3, reduction of electron current and use of low-Z targets for electrons can significantly reduce the total radiation generated. Additional radiation reduction must be accomplished by suitable shielding.

5.1 Electron Current Reduction

Electrons in the beamline are produced by ion collisions with beamline structures and residual gas in the beamline. Beamline-structure collisions are minimized by good beam alignment and proper beam focusing and positioning.

The number of ion-gas collisions are reduced by improved vacuum in the critical areas, and by operating the implanter ion-source gas-supply system with the lowest gas flow commensurate with stable ion-beam operation. Beamline cleanliness is also important, as beamline deposits retain gasses, especially water vapor, whenever they are exposed to air during wafer exchange, ion source replacement, beamline maintenance, etc. Low-energy electrons in the beamline are prevented from entering critical electron-acceleration regions by means of suitable negative suppression voltage.

5.2 Shielding for Radiation Control

Semiconductor-device technology is forcing implanter manufacturers to produce higher-energy implanters, and manufacturing requirements have demanded ever increasing beam currents for increased wafer throughput. Production ion implanters are now capable of on-target beam power of several kilowatts. One method of reducing the exposure rate is by increasing the distance between the source of radiation and the point of interest. The relation with distance is of the form $1/r^n$, where r is the distance and n is a number varying between 1 and 2. When the dimensions of the radiation source and detector are small compared to the distance between them, n=2, and the relation is the well-known inverse-square law. When the dimensions are large compared to the distance, the value of n decreases progressively towards 1. Because the housings of implanters (high-voltage enclosures and other parts) are already quite large for other reasons, the small decrease in exposure rate to be gained by increasing the dimensions of the housings is not advantageous in terms of space and cost. For this reason, reduction of radiation to acceptable levels depends on the use of appropriate shielding material rather than distance.

TABLE 1 Implanter Materials in Relation to Control of Radiation

ELEMENT	Z (ATOMIC NUMBER)	
Beryllium (Be)	4	
Carbon (C)	6	Preferred
Aluminum (Al)	13	Electron Targets
Chromium (Cr)	24	
Iron (Fe)	26	
Nickel (Ni)	28	
Copper (Cu)	29	
Molybdenum (Mo)	42	Preferred
Tungsten (W)	74	Shielding
Lead (Pb)	82	

Table 1 lists the commonly used materials in implanter ion sources, ion-source chambers and beamline structures, with their atomic numbers. Lead is included as it is the principal shielding material used.

From the standpoint of shielding reliability, shielding should be provided by the implanter structure itself wherever possible. Most ion-source and beamline vacuum-chamber walls are aluminum or stainless steel. Where only low-energy radiation is generated, stainless-steel walls frequently provide sufficient shielding.

Where additional shielding is required, the preferred method is to attach it permanently to the appropriate implanter structures. For the most efficient utilization of shielding material, it should be located as close as possible to the origin of the radiation.

When shielding cannot be incorporated feasibly into implanter structures, it should be securely attached to interlocked covers, panels, doors, etc. When shielding cannot be interlocked and is removable or is incorporated in removable parts, it is necessary to label them, identifying them as radiation shields that must be placed on the appropriate structures and to preclude future replacement with unsuitable materials. In turn, "shield required" labels should be placed on structures under the removable shielding where the labels will serve as reminders for the reinstallation of the shields.

5.3 Implanter Operation in Excess of Specifications

Manufacturers provide implanter operating specifications to which the users are expected to adhere.

Excessive ion-beam energy is usually prevented by design limitations on acceleration power supplies, but modification of voltage-control circuits or replacement with power supplies of higher-voltage capability can result in implanter operation beyond original design requirements. Ion-beam current is often limited by beam transmission efficiency, rather than by the current capability of the acceleration power supply. Wafer-throughput considerations often encourage implanter users to try to operate at maximum obtainable beam current. It is possible to operate some implanters at 50% above published beam-current specifications. Implanters usually do not have beam overvoltage and overcurrent interlock protection (except power-supply overcurrent protection). Also, source materials not included in the specifications, may be used that will result in greater radiation output. Some protection from excessive radiation during implanter out-of-specifications operation is usually provided by conservative shielding design based on low working exposure-rate limits (Section 10).

Out-of-specifications operation, although impossible to prevent, is not recommended unless appropriate precautions are taken as discussed in Sections 10 and 12.

6. SHIELDING CONSIDERATIONS

An important method of reducing the radiation emission from implanters is by the use of shielding. The interaction of photons (x rays) with matter takes place by three main processes that result in the absorption and scattering of the photons, and, in one of these processes, the eventual production of more photons.

The first of these processes is the photoelectric process. In this case, the photon interacts with the atom as a whole, the photon is completely absorbed and an electron is ejected from the atom, most often the electron in the K shell if the photon energy is larger than the K-shell binding energy. The ejected electron receives all the energy of the photon less the binding energy and the photon no longer exists. The photoelectric process is predominant at lower photon energies and at higher atomic numbers as shown in Figure 7 and depends strongly on the atomic number $(z^4$ to $z^5)$. Thus, lead can be ~ 100 times more effective than an equal thickness of aluminum.

In the second process, the Compton process, because the photon has sufficient energy, the photon interacts with an electron in an atom as if the electron were free. The result of this interaction, where both photon momentum and energy play a role, is that the incident photon results in another photon of lower energy and in a different direction, the balance of the energy being transmitted to the electron that moves in a third direction. Thus, two effects contribute to the attenuation of photons by the Compton process: the absorption of photon energy, analogous to the photoelectric effect, and the scattering of photons with lower energies at all angles. The Compton process is proportional to the atomic number (Z). Photons can be scattered toward a point of interest when the photons were not directed originally toward this point. This condition is called broad-beam geometry and the photon attenuation is less than the total that might be anticipated. The extreme situation of scattering is in the backward direction resulting in what is called skyshine. X rays emitted in a vertical direction, as from the top of an implanter,

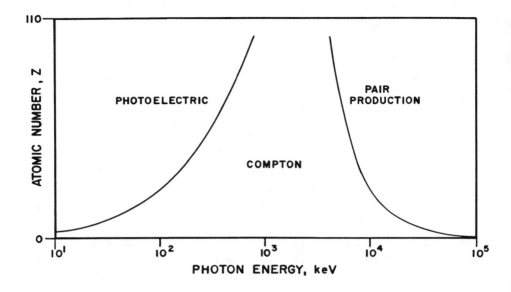

Figure 7   Relation between the atomic number of the absorber to the energy of the photons interacting with the absorber for the three main processes in the interaction of photons with matter. The lines separating each of the two sets of processes occur where their corresponding attenuation coefficients are equal to each other. Pair production can occur only above 1.02 MeV. For implanters with energies ≤500 keV, the photoelectric process is applicable and materials with higher atomic number, such as lead (Z = 82), are good candidates for shielding.

can be scattered by the atoms in air to contribute to the exposure rate around the vicinity of the implanter and may be important if the implanter top is not shielded sufficiently.

The third process, pair production, is important at energies above 1.02 MeV. In this process, the photon interacts with the field of a charged particle, especially that of the field of the charged nucleus of an atom, and is completely absorbed. In place of the photon, two electrons appear, a positron–electron pair, that share all the energy of the photon less the rest mass, 0.51 MeV, of each. The positron eventually interacts with an electron to annihilate each other, and their rest masses appear as two photons of 0.51 MeV each moving in opposite directions. The pair-production process is proportional to the square of the atomic number (Z^2). This process may become important for implanters that are operated well above 1 MeV. However, it is not important for implanters that operate at lower energies and that attain high energies with multiply charged ions because the energy of the electron stream is still at the lower energies.

The relationship of these three processes with energy are shown for lead in Figure 8. The photoelectric attenuation coefficient decreases monotonically except for the discontinuities at the K and L edges. For low-Z elements, a K edge would not be observed down to 0.01 MeV. The Compton absorption coefficient rises to a peak at 0.51 MeV and then declines, while the pair-production coefficient starts at 1.02 MeV and continues to increase. The sum of these three effects forms the total absorption coefficient by which photons no longer exist. Finally, the Compton scattering coefficient also decreases monotonically with energy and, if all the scattered photons do not get scattered back to the point of interest, this coefficient adds to the total absorption coefficient to yield the total attenuation coefficient. When x rays penetrate beyond the faces of implanters, some photons are scattered to the point of interest and, hence, the applicable attenuation coefficient lies somewhere between the total absorption and the total attenuation coefficients.

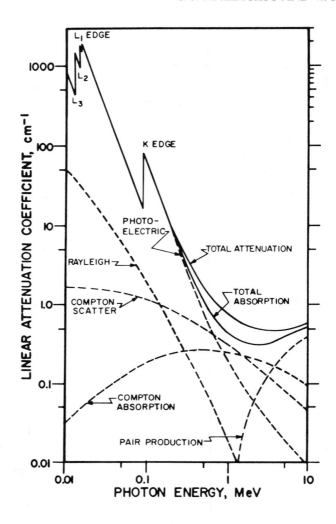

Figure 8   Linear attenuation coefficients for lead as a function
of photon energy. The three main processes, photoelectric, Compton
and   pair   production,   are   described   in   the   text.   Rayleigh
scattering, not described in the text, is coherent scattering at
small   forward   angles without change in the photon energy. Rayleigh
scattering   is   not   very   important   at   low energies and low atomic
numbers.   At   higher   atomic   numbers,   Rayleigh scattering can be
comparable   or   greater   than   Compton   scattering at low energies as
shown   in   the   figure,   but   Rayleigh scattering is still one to two
orders   of   magnitude   less than the total absorption and attenuation
coefficients.   Note   the   discontinuities at the K and L edges when
the photon energies are equal to the binding energies of these inner
electrons.

In passing through an absorber, photons interact in a very small thickness of the absorber in a random fashion such that a fraction of the photons are removed based on the coefficients described above. In the next small thickness the same fraction is removed from the remaining incident photons, and so on. Mathematically this interaction translates into the equation

$$N = N_{o}e^{-\mu x} \tag{4}$$

where N_{o} is the number of initial photons, N is the number remaining after passing through a thickness x of absorber, and μ is the attenuation coefficient appropriate to the conditions. N/N_{o} is the fraction remaining or the transmission. When x is the distance (inches or cm), μ is the linear attenuation coefficient (inches^{-1} or cm^{-1}, i.e., inverse inches or cm). The transmission curves in Figures 9 and 10 are based on Equation 4 with $\mu = \mu_{a}$, the total absorption coefficient, not accounting for any removed by scattering and, hence, providing conservative thicknesses of the materials shown.

In actual practice, x rays are emitted by the implanters with a spectrum of photon energies. Each energy in this continuum would result in a different transmission so that the composite transmission curve would be concave upwards in Figures 9 and 10. The transmission of low-energy x rays would be much less than the transmission of the higher-energy photons in the spectrum. However, because inner structures in implanters will attenuate the lower-energy x rays, the spectrum that is observed beyond the implanter surfaces contains proportionately more higher-energy photons, i.e., the spectrum has been hardened. Because of this condition, the transmission curves in Figure 9 and 10 are not only conservative but apply more closely to the real conditions. Thus, these curves may be useful for planning and design purposes but should not be used for construction. More realistic curves can be obtained by a calculational method that divides up the x-ray spectrum into energy intervals, determines the attenuation of each bin and then sums of each of these as described by Bustin and Rose (1982). Alternatively, experimental data from thick-target sources may be used as described by Malestskos and Hanley (1983). However, the reliable method for determining the adequacy of the shielding in an implanter is by actual exposure-rate measurements with properly calibrated instruments.

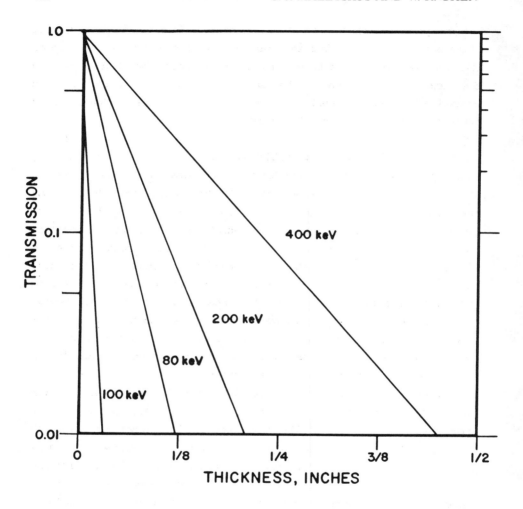

Figure 9   Transmission of photons as a function of lead thickness for the energies indicated. The transmission was determined with the total absorption coefficients shown in Figure 8 at the respective energies. As discussed in the text, these transmission curves are conservative for planning and design purposes in that these curves represent attenuation for broad-beam conditions where the Compton-scattered photons are assumed to reach the point of interest. Note the reversal of the order for the transmission at 80 and 100 keV because of the K edge shown in Figure 8.

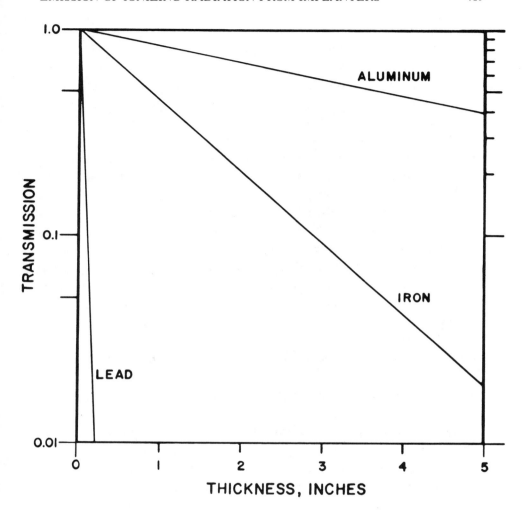

Figure 10 Transmission of photons as a function of thickness for aluminum, iron and lead for 200-keV photons. As described in Figure 9, the transmission is based on the total absorption coefficients. Even massive amounts of aluminum are not very useful. Iron in the form of stainless-steel structures can help and in the form of magnets can be effective. Lead surpasses both materials by a large margin.

7. ENGINEERING FOR RADIATION CONTROL

Modern ion implanters, if improperly operated, or inadequately shielded, are capable of emitting x-ray radiation which greatly exceeds the recommended limits. Several engineering controls are incorporated into implanters which are designed to prevent unexpected increases in radiation. The commonly used ones are as follows:

Shield interlocks. All covers, panels, doors, etc. that are part of the radiation-control system and that permit normal implanter operation when open or removed, must be interlocked to disable the acceleration-voltage power supply. In some cases, these parts also require interlocks for electrical safety.

Vacuum interlocks. Vacuum interlocks are usually incorporated in ion implanters for product quality considerations. The same interlock system can be used as a radiation interlock provided the set point is below the threshold of significant radiation and provided that the interlock action disables the ion beam, rather than diverting it to a beam dump.

Suppression voltage interlocks. Adequate suppression voltage is extremely important in preventing excessive radiation. Unfortunately, loss of suppression voltage has little effect on other implanter functions and is likely to be unnoticed. Therefore, a dependable, fail-safe interlock system is essential. The acceleration power supply is normally interlocked through the suppression interlock system.

Radiation exposure-rate monitors. Self-contained radiation monitors are available that can be supplied as part of an ion implanter. Radiation monitors have adjustable set points, and are provided with a set of electrical contacts that can be used for alarm actuation or ion beam interlocking, as well as direct reading exposure-rate displays. Though such monitors may not be required, but if used, care must be taken in positioning them to assure that radiation produced by abnormal implanter operation will be detected. A radiation monitor should be used as a redundant radiation-safety system, rather than a substitute for good implanter design and other engineering controls for radiation.

8. QUANTITIES AND UNITS OF RADIATION EXPOSURE

The radiation output of implanters in terms of intensity and power (intensity per unit time), discussed in Sections 2 and 4, are not sufficient parameters for purposes of radiation protection. What is required, in addition to these, is an accounting of the interaction of the x rays with air or tissues in order to determine the energy deposited in these media. This energy arises from the ionization of the atoms in the media to produce energetic electrons and from the further ionization by these electrons until the energy of all the electrons is totally expended. Collection of electrons produced in a given mass of air provides an indication of the radiation field to which a person may be exposed. The biological effects of radiation, in turn, are related to the energy deposited in organs and tissues as discussed in Section 9.

The radiation field in air, into which a person may enter, is characterized by the quantities exposure (X) and its derivative with time, exposure rate ($\dot{X}$). Exposure is the total charge of all the electrons produced by photons ionizing a unit mass of air when those electrons are all stopped in air. In the International System of Units (SI), the unit of exposure is coulombs per kilogram (Table 2). The special unit for exposure (now being phased out) is the Roentgen.

The absorbed dose (D) and its time derivative, absorbed-dose rate ($\dot{D}$), is the mean energy imparted per unit mass of matter. The matter can be air, inanimate material and, for purposes of biological effects, tissues, organs, and partial and whole bodies of people. The SI unit of absorbed dose is the Gray and the special unit is the rad (Table 2).

The radiations emitted by implanters are x rays (i.e., photons). However, there are other ionizing radiations, such as electrons, protons, alpha particles, other charged particles and neutrons, the biological effects of which may not be the same as those of photons for the same absorbed dose. The quantities dose equivalent (H) and its time derivative, dose-equivalent rate ($\dot{H}$), provide a means of placing all ionizing radiations on a common basis insofar as biological effects are concerned. The dose equivalent is given by H = DQN, where Q is the quality factor (the average value of the

Table 2   Summary of Radiation Quantities and Units

Quantity		S.I. Unit		Special Unit	
Name	Symbol	Name	Symbol	Name	Symbol
Exposure	X	Coulombs per kilogram	C/kg	Roentgen	R
		$1 R = 2.58 \times 10^{-4}$ C/kg (exactly)			
Absorbed Dose	D	Gray	Gy (= 1 J/kg)	rad	rad
		100 rad = 1 Gy			
Dose Equivalent	H	Sievert	Sv (= 1 J/kg)	rem	rem
		100 rem = 1 Sv			

relative biological effectiveness) of the radiation under consideration and N is the product of other modifying factors, if any (N = 1 at this time). The reference radiation is 250-keV photons. Hence, for the low-energy x rays emitted by implanters Q = 1 and the absorbed dose and dose equivalent are numerically equal, both in SI and special units. The SI unit of dose equivalent is the Sievert and the special unit is the rem (Table 2).

In terms of the special units and for photons with energies less than a few MeV, 1 R in air is approximately equal to 1 rad in air which, in turn, is approximately equal to 1 rad in most tissues and organs and, finally, equal to 1 rem. Because of this special situation, these units are used interchangeably. This is not good practice and the distinction between the units should be observed. This practice is less possible with the new SI system.

Exact definitions of these quantities and units and the qualifications that apply to them are described in ICRU Report 33 of the International Commission on Radiation Units and Measurements (ICRU, 1980). A summary of the units described above and in ICRU Report 33 are shown in Table 2. For purposes of making implanter radiation surveys and meeting regulatory limits, the quantity used is the exposure rate as discussed in Section 10.

9. BIOLOGICAL EFFECTS OF IONIZING RADIATION

The biological effects of ionizing radiation result from the ionization of atoms during the interaction of the radiation with the atoms in the cells of tissues and organs. Ionization events are followed by a series of physical and chemical events over short and long time periods that may result in no effects on the cells, in effects from which the cells may recover and return to normal, or in effects that may be potentially harmful to the cells.

There are two classes of radiation-induced biological effects in exposed persons. These are stochastic and nonstochastic effects. Stochastic effects are those in which the probability of occurrence of these effects increases with increasing absorbed dose. The severity of the effect, however, does not depend on the dose. Injury to cells or parts of cells, such as the genes in the somatic cells (i.e., all the body cells other than the germ cells) leads to the induction of cancer (solid tumors and leukemia). Injury to the germ cells leads to changes in the genetic material that can be transmitted to progeny. Cancer and genetic effects are observed many years after exposure to radiation, and it is assumed that there is no threshold dose below which no effect takes place, even though there is no experimental proof of this. Stochastic effects can result from acute exposure (short exposure to higher dose equivalents) or from chronic exposure (long continuous exposure to lower dose equivalents) and may be observed above 0.1 to 0.2 Sv (10 to 20 rem).

Nonstochastic effects are those in which the severity of the occurrence increases with increasing dose, generally from acute exposure. In this case, the response time is much shorter and the severity increases as more and more cells are damaged or killed, i.e., a degeneration or degradation of the tissues or organs takes place to the point where they are clinically significant. Non-stochastic effects do have a threshold below which the effects are not observed and some effects may be reversible, the tissues and organs returning to normal, if the dose is not too large. Examples of nonstochastic effects include lens opacification (cataracts), changes in the red and white cells of blood, loss of hair and lowered sperm production in the male. A familiar effect is erythema (reddening) and blistering of the skin, much as they occur on exposure to ultraviolet and infrared rays. People vary in their sensitivity to these effects such that higher dose equivalents may be required to produce the same effect that occurs in others at lower dose equivalents as indicated in Figure 11. Nonstochastic effects occur mainly under acute exposures. Blood changes can be observed with sensitive measurements at dose equivalents of about 0.1 to 0.2 Sv (10 to 20 rem). A dose equivalent of 5 to 7 Sv (500 to 700) (received under acute conditions, i.e., less than 24 h) to an area of skin is the threshold for erythema, while 1 to 1.5 Sv (100 to 150 rem) (received under acute conditions) to the whole body will result in nausea. In both cases, if the dose equivalents are not very much greater than indicated, recovery will take place. In the limit, whole-body exposure to about 3 Sv (300 rem) will result in the deaths of 50 percent of the exposed persons in about one month without medical care, while all will die at significantly larger dose equivalents.

The sensitivity of organs and tissues is variable on exposure to radiation. Thus, for example, for cancer induction, the breast is more sensitive than red bone marrow which, in turn, is more sensitive than the thyroid. Bone is one of the least sensitive tissues. The sensitivity of people to radiation also is variable, with persons in bad health more sensitive than those in good health. Rapidly proliferating cells are more sensitive than those that proliferate slowly; thus, the young are more sensitive than the old. Recent evidence indicates that the developing embryo and fetus, more sensitive than the young, may be even more sensitive than previously thought. Irradiation of the embryo or fetus may result in the nonstochastic effects of anatomical or functional teratogenesis. The

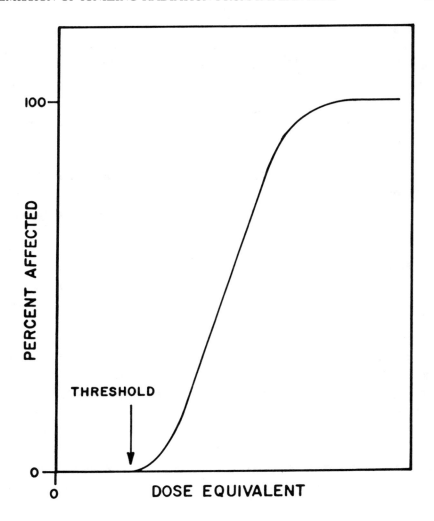

Figure 11 Schematic relation between the number of persons affected in a population as a function of the amount of radiation received by each of the persons in that population expressed as the dose equivalent for nonstochastic biological effects as described in the text. Such effects have a real threshold below which the effect is not observed. The varying sensitivity of people for this effect results in an S-shaped curve as indicated. At a high-enough dose equivalent, all the persons will show the effect. For any one person, a nonstochastic effect increases with increasing dose equivalent and, at lower dose equivalents, some nonstochastic effects are reversible, much as sunburn is reversible in response to photons with ultraviolet energies.

parts of the body that are irradiated play a role in the effects to be observed relative to irradiation of the whole body (i.e., whole-body and partial-body irradiation). Finally, there are many other variables that play a role such as the characteristics of the radiation, the measurements for and calculation of dose equivalents, the exposure period, the latent period (time between exposure and detection of the cancer, many years), the condition of the person and the natural cancer incidence.

The biological basis for establishing radiation exposure limits is based on the study of the biological response to the radiation dose. The biological variable might be the incidence of cancer in a population for stochastic effects or the degree of severity for nonstochastic effects. The physical variable, the agent producing the effect, would be the radiation expressed as the dose equivalent. Such dose-response studies have been conducted with various species of animals and have also involved epidemiological investigations of people accidently exposed to radiation or specifically exposed for therapeutic purposes. These studies result in dose-response curves, and a typical one for cancer in some organ or tissue is shown in Figure 12. Curve L shows a linear or direct proportionality to the dose equivalent, while curve LQ shows a linear response at lower dose equivalents to be followed at higher dose equivalents with a quadratic response (proportional to the square of the dose equivalent). Data from the study of irradiated people always comes from high dose equivalents and these data have to be extrapolated to the lower dose equivalents for purposes of setting exposure limits. These curves always pass through zero, indicating no threshold dose for the effect under study. From such results, conservative estimates are made for setting the risks of cancer in a population, i.e., estimates that are high enough to account for the lack of detailed knowledge and, hence, to provide maximum protection from exposure to radiation. The exposure limits are then set so that the risks from radiation are comparable to the risks faced and accepted by people in the safer industries and in their daily activities.

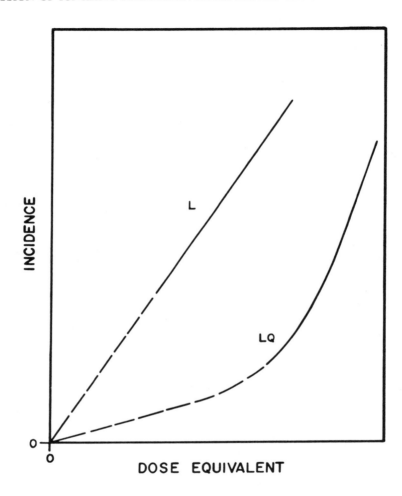

Figure 12 Schematic relation between the incidence of a stochastic biological effect, expressed as a fraction of a population, as a function of the amount of radiation received by a particular organ or tissue in each of the persons in that population expressed as the dose equivalent. Data for such results are obtained at high dose equivalents (solid lines) and must be extrapolated to lower dose equivalents (dotted lines), from which dose limits can be recommended. Curve L represents a linear response applicable to the incidence of some cancers and genetic effects. Curve LQ represents a combination of linear (at lower dose equivalents) and quadratic (at higher dose equivalents) responses for other cancers. Whenever there is doubt in the determination of the response curve, the linear response is chosen to develop conservative dose limits, i.e., limits with the higher margin of safety.

Because the sensitivity for some cancers is so low, exposure limits for them could become large enough to result in nonstochastic effects. Thus, limits for nonstochastic effects also are set and are set sufficiently below the threshold so that the more sensitive persons will not show such effects.

Finally, from all the studies and analyses and by consensus of experts, recommendations for exposure are prepared by national and international organizations. These recommendations are then translated into regulations that form the guidance to be followed. Exposure limits are set for workers, the public, the young and are also set for special situations such as emergencies. Some examples of such limits are shown in Table 3 taken from NCRP Report No. 91 (NCRP, 1987). Similar limits can be found in ICRP Publication 26 (ICRP, 1977).

A comprehensive review of the biological effects of ionizing radiation and the development of the risk values can be found in the report of the National Academy of Sciences (NAS/NCR, 1980) and in the report of the United Nations (UNSCEAR, 1982).

10. EXPOSURE LIMITS FOR IMPLANTERS

Historically, implanters have been designed to be placed wherever users chose to locate them in their facilities. Implanters are, therefore, constructed to produce exposure rates that are low enough so that personnel working with them do not have to be classified as radiation workers.

Implanters can be included in that category of devices that emit radiation called electronic products. Examples of such devices are television receivers and computer display terminals, airport x-ray baggage inspection systems, and gas and aerosol (smoke) detectors. All such devices come under a main category called consumer products that either emit radiation or contain radioactive materials. The contribution of consumer products to the total annual dose equivalent (3.6 mSv or 360 mrem) to the population of the United States from various sources of radiation is about 3 percent, when natural background contributes 82 percent and medical exposures (diagnostic and therapeutic x rays and nuclear medicine) contribute 17 percent (NCRP, 1987b).

TABLE 3 Some Examples of Recommended Dose Limits [a,b]

Type of Exposure	Annual Dose Equivalent[c]	
	mSv	rem
Occupational		
stochastic	50	5
nonstochastic		
lens of eye	150	15
all others	500	50
guidance: cumulative	10 x age	1 x age
Public		
stochastic		
continuous or frequent	1	0.1
infrequent	5	0.5
nonstochastic	50	5
Embryo, fetus		
total	5	0.5
per month	0.5	0.05

[a] From NCRP (1987a)
[b] Excluding medical exposures
[c] Even though the dose limits are acceptable limits, the intent is to keep exposures always as low as reasonably achievable (ALARA).

In the U.S., the exposure limits for electronic products are 36 pC/kg s (0.5 mR/h) at 5 cm from all the accessible surfaces of the products as set by the Center for Devices and Radiological Health of the Food and Drug Administration of the Department of Health and Human Services. Other countries and jurisdictions may set the same value or different values, usually lower. The lowest exposure limit is that in Japan set at 4.3 pC/kg s (0.06 mR/h).

The distance of 5 cm relates to the method of measurement. For reference purposes, exposure is best measured with an ionization chamber, properly designed and fabricated and with a calibration traceable to a national standard, which measures the ionization in air directly and, hence, the exposure rate. At these low exposure rates, a portable chamber with a volume of 300 to 500 cm^3 (20 to 30 in^3) is necessary to develop sufficient current to measure (10^{-14} to 10^{-13} A). Thus, while the face of the chamber can be placed at the surfaces of the implanter, the effective geometric center of the chamber is about 5 cm (2 in.) from the surfaces.

The accessible surfaces need to be defined with some common sense. If the implanter is totally enclosed with metallic or plastic panels, there is no problem in the definition of accessible. The bottom of an implanter that cannot be reached, may or may not be an accessible surface. If the implanter is placed on a floor that rests on the ground, the bottom surface is not accessible. However, if the implanter is placed on a floor level above ground, then the bottom surface becomes accessible for the area underneath this floor, and measurements at the ceiling of the floor below would be required if the bottom face was not available at the manufacturer's plant. Some implanters may have components that are movable, such as electronic and control consoles on wheels to allow access to the structures behind them. The surfaces on the implanter exposed behind such movable components then become accessible under the definition.

The exposure limit is defined such that the exposure rate from an electronic product does not exceed this limit. In order to meet this requirement, the working exposure limit must be set below the recommended limit by an amount such that the chance of exceeding the recommended limit is small. Most electronic products to which the public might be exposed are designed with fixed operating characteristics so that the radiation emitted is always the same. The operating conditions of implanters, however, can be varied so that increases in exposure rates can result. Deliberately operating above implanter specifications can result in increased radiation. Increased radiation also can result when operating with surfaces that are not cleaned well, with poorer vacuum pressure, or with poorer focusing conditions that may still allow the job at hand to be performed adequately although not optimally. To account for these conditions, a further reduction in the working limit would be required.

If the recommended limit is taken as 36 pC/kg s (0.5 mR/h), for example, a working limit can be derived as follows. The x rays are generated at random and arrive at the survey meter at random. Thus, the reading of the meter will fluctuate about an average value. In addition, the electron current may not be steady but can fluctuate for various transient reasons. If it is assumed that these fluctuations are nearly Gaussian in nature, then the fluctuation can be expressed as a standard deviation. For the exposure rate under consideration, the standard deviation is about 15 percent and the chance of being beyond two standard deviations, 30 percent, is 1 in 20, an acceptable chance. The working limit becomes, then, 70 percent of the recommended value or 25 pC/kg s (0.35 mR/h). To account for increases in radiation because of deviations from specifications, a factor of 2 may be used reducing the working limit to 12.5 pC/kg s (0.18 mR/h) or in round numbers to 15 pC/kg s (0.2 mR/h). When the recommended limit is as low as 4.3 pC/kg s (0.06 mR/h), the working limit becomes disproportionately lower than the example, because the standard deviation is greater at the lower exposure rates, to a value of about 1.4 pC/kg s (0.02 mR/h). This value is approximately equal to the exposure rate of natural background. To reach this goal, considerably more shielding, and hence weight, and much more time to make a thorough radiation survey are required.

Exposure limits are set on the basis of the potential exposure period of an individual. In the example of 36 pC/kg s (0.5 mR/h), the dose equivalent rate to a person standing flush against the face of the high-voltage enclosure of an implanter would correspond to about 2.5 μSv/h (0.25 mrem/h) because of the photon attenuation in the body. If the person literally stands there for one hour per week for 52 weeks, then the annual dose equivalent would be 130 μSv (13 mR). These values are low compared to the dose limits for the public (Table 3). Thus, in the normal operation of a well designed and constructed implanter, the dose equivalent to the operator at the end station, where the exposure rate is equal to background, or to the passerby near any part of the implanter, is small compared to that from natural sources and will not produce detectable stochastic and nonstochastic biological effects.

11. PRINCIPLES FOR PERFORMING RADIATION SURVEYS

Modern ion implanters are designed to be used in a manufacturing environment without requiring any special procedures to protect operating, maintenance, and other nearby or passing personnel from radiation exposure above the recommended dose limits. Ion implanters are very large, complex devices, frequently with many available options, and capable of a wide range of operating conditions. These factors, along with normal manufacturing tolerances, as well as some factors that are not known, suggest that implanter-to-implanter radiation non-uniformity within a model or product line is to be expected. Accordingly, individual implanter radiation surveys, rather than model-qualification or statistically sampled surveys, become an important part of implanter radiation safety. The individual implanter radiation survey is intended to verify conformance to appropriate radiation exposure-rate specifications, or to identify non-conformance and define corrective measures.

An implanter user does not have to make a survey as long as operation is within specifications and changes have not been made. The user can always choose to perform a survey, and must perform one if operation is to be beyond the specifications or if electrical or structural changes have been made.

11.1 Presurvey Checks and Implanter Operating Conditions

A radiation survey should start with a careful inspection to confirm that all fixed and removable shielding is properly installed and correctly labeled where required. The inspection should confirm that all interlocks involving the production of radiation and requiring shielding are functioning correctly. The choice of the gas appropriate to the radiation test also should be confirmed. This gas should be that gas listed under the implanter specifications that provides the ion species that will produce the highest radiation. For many implanters, this gas will be argon. Argon is usually available in implanters for use as a "clean-up" gas or as a vaporizer carrier gas.

In making the survey, the operating conditions are those specified for the implanter by the manufacturer. These conditions are the specified current and the full acceleration voltage. Maletskos and Hanley (1983) recommended that the implanter should be operated at the maximum sustainable current. Because the sustainable current reached by different persons is variable, the present recommendation of using the specified current along with a reduced working exposure-rate limit (Section 10) results in the same or better protection and in more consistent surveys. The negative suppression voltage should have the specified value and, for those implanters with variable suppression voltage, the correct voltage should be set. The ion beam is usually directed to a beam flag or beam "dump" that can handle the ion-beam thermal load for the time required, up to a few hours, for a complete radiation survey. If source-material, electrical or mechanical changes are made, exposure-rate measurements are necessary in accel and decel modes as well as under de-analyze and defocused conditions.

11.2 Survey Instruments

A variety of portable, battery-powered radiation survey instruments are available that can be used for assessing the exposure rates from implanters. The instruments consist of a metering unit with either an attached detector or a detector on a cable. Detectors useful for implanter surveys include an ionization chamber that can provide absolute values of exposure rate, a Geiger- Muller (GM) tube, more sensitive than the chamber, that can provide an exposure rate relative to the chamber value, and a sodium iodide scintillation crystal (NaI(Tl)) with a photomultiplier tube, which has, at least, 10 times more sensitivity than the GM tube and which can be used for rapid scanning of the implanter surfaces.

These detectors have responses to photons that depend on the photon energy. Representative responses for a GM tube and an ionization chamber are shown in Figure 13. For implanters with energies <500 keV, the relative response of a GM tube, depending on the type, can be 6 or 7 times the response at higher energies at which the GM tube is normally calibrated. Because the x rays are emitted in a spectrum of energies, it is not easy to determine the overresponse applicable to the measurement of the exposure rate of a particular implanter, and an energy-correction factor must be obtained by comparing the exposure rate from a GM survey meter with that from an ionization chamber in the same position. These factors can range up to 5. Their determination must be made for each implanter type and/or energy, with shielding installed, because the x-ray spectra are different for each type or energy. Because ionization chambers are not as sensitive as GM tubes, this cross calibration is made at exposure rates that are high enough for accurate readings in both instruments. If an energy-calibration factor is not applied, either too much shielding will be installed or an implanter will be presumed to exceed the exposure-rate limits when, in fact, it does not.

11.3 Survey Procedures

The accessible surfaces cover large areas. The lobe characteristics are not necessarily exactly the same among implanters of the same type, and thus, all the accessible areas must be surveyed.

Areas with potentially high exposure rates can be missed easily when a grid pattern is used to simplify the survey. Instead, a tight zig-zag pattern should be used to cover essentially the whole area and to probe for streaming of radiation from areas such as edges of doors, door handles and locks, screw holes, hinges, corners of enclosures or joints between panels, etc., where gaps or openings in shielding material may allow localized radiation leaks or streaming.

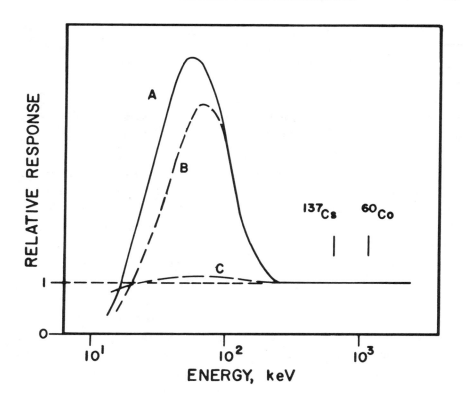

Figure 13   Schematic of the relative response of photon detectors as a function of photon energy. Photon detectors such as GM tubes and ionization chambers are calibrated in the radiation fields produced from radioactive sources such as [137]Cs (662 keV) and [60]Co (1110 and 1130 keV) as shown in the figure.

The response at lower energies increases because of the increase in the efficiency in the interaction of the photon with the detector. The change in response is expressed as a response relative to that at the calibration energy. Curve A is typical of a bare cylindrical GM tube, while curve B is typical of this GM tube covered with a shield thick enough to exclude charged particles, namely electrons and beta rays. The relative response can be as large as 6 or 7 or more. Curve C represents a well designed and constructed air-equivalent chamber compensated to be nearly energy independent, with a response that may be as high as 1.2. The relative responses return to unity and become less than unity at very low energies because the detector wall materials start to attenuate the photons before the sensitive regions of the detectors are reached.

The bottom of implanters may not be accessible at the time of manufacture. If the implanter is located in an area where personnel access under the implanter is possible, that area must be surveyed as well.

The large areas to be surveyed can be done in a reasonable time with the use of a very sensitive detector such as the sodium iodide scintillation detector. This detector, uncalibrated, serves as a spotter of areas above a minimum but acceptable exposure rate. The particular areas can be marked, and, then, the exposure rates at these areas can be determined with the cross-calibrated GM tube and verified, when possible, with the calibrated ionization chamber. The cross-check between GM tube and ionization chamber should be made whenever possible as a continuing check on the energy-correction factor because changes, unknown to the surveyor, may be made in the implanter, electrical or mechanical, that can change the correction factor.

As with all surveys, survey forms provide a convenient means of recording the shielding inspection, the operating conditions, the data of the survey, and the calculations to determine the final net exposure rates. If surveys are made, records of surveys should be retained in accord with user established procedures.

If additional shielding is required, the thickness of the shielding material, e.g., lead, should be determined by actual measurement with a sheet of lead large enough to minimize photon scattering into the detector from regions near the point of interest. The additional shielding should be installed in a permanent form, with interlocks or with labels as the situation requires.

11.4 Precautions

It is not advisable to remove covers or other shielding material to get sufficient exposure rates for ionization chamber readings. High-voltage shock hazards can exist with covers off or doors open (and interlocks bypassed) and the removal of shielding material can shift the x-ray energy spectrum and change the energy-correction factor.

Ionization chambers and the phototubes of sodium iodide detectors have components that can be permanently damaged by strong magnetic fields. These detectors must be kept clear of implanter ion-source and analyzer magnets.

The electronic circuitry used in all types of survey meters can be affected by RF fields. Only instruments that are not sensitive to these fields should be used for implanter surveys.

Ionization chamber calibration is based on a uniform exposure rate throughout the volume of the chamber. Inaccurate readings will result when measuring radiation from small "leaks", where the x-ray stream impinges only on a portion of the chamber volume. When the ionization chamber is used to establish the energy-correction factor, locations should be chosen where the exposure rate is reasonably uniform throughout an area which is larger than the cross section of the ionization chamber.

Radiation measuring instruments, like other precision instrumentation, require periodic calibration. This is especially important where the measurement involves personnel safety and regulatory compliance. Calibration should be done by an accredited facility using radiation sources that are traceable to a national standards laboratory.

12. RESPONSIBILITIES of IMPLANTER OWNERS AND USERS

In the operation of implanters, there are five major potential hazards to personnel: high voltage, hazardous gases and their residues, moving mechanical components, fire, and ionizing radiation. Thus, appropriate attention to safety is required in order to protect the operators and other personnel from potential harm.

One of the more important aspects of safety is the training of personnel. Training is as important for radiation exposure as it is for the other hazards even though the exposure rates are low. Personnel should be informed that implanters do emit ionizing radiation, that the exposure rates are at or below the recommended limits, and that implanters are similar to other electronic devices that emit radiation with which they are familiar or use as noted in Section 10. Operation of the implanters within the manufacturer's specifications is the best way of minimizing radiation exposure.

There are occasions where changes might be made in implanters used for production or research. The types of changes that might be considered include, for example, running beyond specifications, use of nonspecified gases, changes in operation, and changes in mechanical and electrical structures. Such changes may result in increased exposure rates. These changes can be made if in-house or outside expertise is available to perform radiation surveys. The changes may be acceptable or additional shielding may be added to accommodate the changes. Some of the additional shielding may be removable because of other constraints. Such shielding is either interlocked or labelled with signs that state the implanter can be operated only after the shield is replaced. Operating with such shields removed or with bypassed interlocks, for whatever reasons, is not acceptable unless expertise in radiation protection is available to monitor the situation.

Owners of implanters should be knowledgeable in the information discussed above. There are two other items, however, of which owners should also be aware. In the United States, some states require that implanters be registered with a department of health, of labor or of environment. Registration fees may be required as well as periodic re-registration. Other countries may have similar registration requirements. Sometimes an organization may sell implanters to a third party. If the third party is not knowledgeable about implanters, the party should be informed that implanters emit radiation in addition to the existence of the other potential hazards, that complete manuals and specifications are included and that their contents should be adhered to, and that there may be registration requirements. Implanter owners who transfer ownership or the location of implanters should inform the implanter manufacturer so that current customer lists can be maintained for possible future safety notifications. A radiation survey after reassembly and before routine operation would be most advisable.

REFERENCES

Baldwin, D.G., King, B.W. and Scarpace, L.P. (1988). "Ion Implanters: Chemical and Radiation Safety", Solid State Tech. 31, 99-105.

Bustin, R. and Rose, P.H. (1982). "Safety and Ion Implanters", pages 105-120 in Ion Implantation Techniques, Ryssel H. and Glawischnig, H., Eds. (Springer-Verlag, New York).

Evans, R.D. (1955). The Atomic Nucleus (McGraw-Hill, New York).

ICRP (1977). International Commission on Radiological Protection, Recommendations of the International Commission on Radiological Protection, ICRP Publication 26 (Pergamon Press, New York).

ICRU (1980). International Commission on Radiation Units and Measurements, Radiation Quantities and Units, ICRU Report 33 (International Commission on Radiation Units and Measurements, Washington, D.C.).

Maletskos, C.J. and Hanley, P.R. (1983). "Radiation Protection Considerations of Ion Implanter Systems," IEEE Trans. Nucl. Sci. NS-30, 1592-1596.

NAS/NRC (1980). National Academy of Sciences/National Research Council, The Effects on Populations of Exposure to Low Levels of Ionizing Radiations: 1980, Committee on the Biological Effects of Ionizing Radiations of the National Academy of Sciences/National Research Council (National Academy Press, Washington, D.C.).

NCRP (1987a). National Council on Radiation Protection and Measurements, Recommendations on Limits for Exposure to Ionizing Radiation, NCRP Report No. 91 (National Council on Radiation Protection and Measurements, Bethesda, Maryland).

NCRP (1987b). National Council on Radiation Protection and Measurements, Ionizing Radiation Exposure of the Population of the United States, NCRP Report No. 93 (National Council on Radiation Protection and Measurements, Bethesda, Maryland).

Ryssel, H. and Haberger, K. (1984). "Ion Implantation: Safety and Radiation Considerations", pages 603-627 in Ion Implantation Science and Technology, Ziegler, J.F., Ed. (Academic Press, New York).

UNSCEAR (1982). United Nations Scientific Committee on the Effects of Atomic Radiation, Ionizing Radiation: Sources and Biological Effects, 1982 Report to the General Assembly, with annexes, Publication E.82.IX.8 (United Nations, New York).

INDEX